Oscar Castillo, Witold Pedrycz, and Janusz Kacprzyk (Eds.)

Evolutionary Design of Intelligent Systems in Modeling, Simulation and Control

Studies in Computational Intelligence, Volume 257

Editor-in-Chief
Prof. Janusz Kacprzyk
Systems Research Institute
Polish Academy of Sciences
ul. Newelska 6
01-447 Warsaw
Poland
E-mail: kacprzyk@ibspan.waw.pl

Further volumes of this series can be found on our homepage: springer.com

Vol. 236. Natalio Krasnogor, Belén Melián-Batista, José A. Moreno-Pérez, J. Marcos Moreno-Vega, and David Pelta (Eds.)
Nature Inspired Cooperative Strategies for Optimization (NICSO 2008), 2009
ISBN 978-3-642-03210-3

Vol. 237. George A. Papadopoulos and Costin Badica (Eds.)
Intelligent Distributed Computing III, 2009
ISBN 978-3-642-03213-4

Vol. 238. Li Niu, Jie Lu, and Guangquan Zhang
Cognition-Driven Decision Support for Business Intelligence, 2009
ISBN 978-3-642-03207-3

Vol. 239. Zong Woo Geem (Ed.)
Harmony Search Algorithms for Structural Design Optimization, 2009
ISBN 978-3-642-03449-7

Vol. 240. Dimitri Plemenos and Georgios Miaoulis (Eds.)
Intelligent Computer Graphics 2009, 2009
ISBN 978-3-642-03451-0

Vol. 241. János Fodor and Janusz Kacprzyk (Eds.)
Aspects of Soft Computing, Intelligent Robotics and Control, 2009
ISBN 978-3-642-03632-3

Vol. 242. Carlos Artemio Coello Coello, Satchidananda Dehuri, and Susmita Ghosh (Eds.)
Swarm Intelligence for Multi-objective Problems in Data Mining, 2009
ISBN 978-3-642-03624-8

Vol. 243. Imre J. Rudas, János Fodor, and Janusz Kacprzyk (Eds.)
Towards Intelligent Engineering and Information Technology, 2009
ISBN 978-3-642-03736-8

Vol. 244. Ngoc Thanh Nguyen, Rados law Piotr Katarzyniak, and Adam Janiak (Eds.)
New Challenges in Computational Collective Intelligence, 2009
ISBN 978-3-642-03957-7

Vol. 245. Oleg Okun and Giorgio Valentini (Eds.)
Applications of Supervised and Unsupervised Ensemble Methods, 2009
ISBN 978-3-642-03998-0

Vol. 246. Thanasis Daradoumis, Santi Caballé, Joan Manuel Marquès, and Fatos Xhafa (Eds.)
Intelligent Collaborative e-Learning Systems and Applications, 2009
ISBN 978-3-642-04000-9

Vol. 247. Monica Bianchini, Marco Maggini, Franco Scarselli, and Lakhmi C. Jain (Eds.)
Innovations in Neural Information Paradigms and Applications, 2009
ISBN 978-3-642-04002-3

Vol. 248. Chee Peng Lim, Lakhmi C. Jain, and Satchidananda Dehuri (Eds.)
Innovations in Swarm Intelligence, 2009
ISBN 978-3-642-04224-9

Vol. 249. Wesam Ashour Barbakh, Ying Wu, and Colin Fyfe
Non-Standard Parameter Adaptation for Exploratory Data Analysis, 2009
ISBN 978-3-642-04004-7

Vol. 250. Raymond Chiong and Sandeep Dhakal (Eds.)
Natural Intelligence for Scheduling, Planning and Packing Problems, 2009
ISBN 978-3-642-04038-2

Vol. 251. Zbigniew W. Ras and William Ribarsky (Eds.)
Advances in Information and Intelligent Systems, 2009
ISBN 978-3-642-04140-2

Vol. 252. Ngoc Thanh Nguyen and Edward Szczerbicki (Eds.)
Intelligent Systems for Knowledge Management, 2009
ISBN 978-3-642-04169-3

Vol. 253. Akitoshi Hanazawa, Tsutom Miki, and Keiichi Horio (Eds.)
Brain-Inspired Information Technology, 2009
ISBN 978-3-642-04024-5

Vol. 254. Kyandoghere Kyamakya, Wolfgang A. Halang, Herwig Unger, Jean Chamberlain Chedjou, Nikolai F. Rulkov, and Zhong Li (Eds.)
Recent Advances in Nonlinear Dynamics and Synchronization, 2009
ISBN 978-3-642-04226-3

Vol. 255. Catarina Silva and Bernardete Ribeiro
Inductive Inference for Large Scale Text Classification, 2009
ISBN 978-3-642-04532-5

Vol. 256. Patricia Melin, Janusz Kacprzyk, and Witold Pedrycz (Eds.)
Bio-inspired Hybrid Intelligent Systems for Image Analysis and Pattern Recognition, 2009
ISBN 978-3-642-04515-8

Vol. 257. Oscar Castillo, Witold Pedrycz, and Janusz Kacprzyk (Eds.)
Evolutionary Design of Intelligent Systems in Modeling, Simulation and Control, 2009
ISBN 978-3-642-04513-4

Oscar Castillo, Witold Pedrycz,
and Janusz Kacprzyk (Eds.)

Evolutionary Design of Intelligent Systems in Modeling, Simulation and Control

Springer

Prof. Oscar Castillo
Tijuana Institute of Technology
Department of Computer Science
P.O. Box 4207
Chula Vista CA 91909
USA
E-mail: ocastillo@hafsamx.org

Prof. Janusz Kacprzyk
Polish Academy of Sciences
Systems Research Institute
Newelska 601-447 Warszawa
Poland
E-mail: kacprzyk@ibspan.waw.pl

Prof. Witold Pedrycz
University of Alberta
Dept. Electrical and Computer Engineering
Edmonton, Alberta T6J 2V4
Canada
E-mail: pedrycz@ee.ualberta.ca

ISBN 978-3-642-26083-4 e-ISBN 978-3-642-04514-1

DOI 10.1007/978-3-642-04514-1

Studies in Computational Intelligence ISSN 1860-949X

Typeset & Cover Design: Scientific Publishing Services Pvt. Ltd., Chennai, India.

Printed in acid-free paper

9 8 7 6 5 4 3 2 1

springer.com

Preface

We describe in this book, new methods for evolutionary design of intelligent systems using soft computing and their applications in modeling, simulation and control. Soft Computing (SC) consists of several intelligent computing paradigms, including fuzzy logic, neural networks, and evolutionary algorithms, which can be used to produce powerful hybrid intelligent systems. The book is organized in four main parts, which contain a group of papers around a similar subject. The first part consists of papers with the main theme of evolutionary design of fuzzy systems in intelligent control, which consists of papers that propose new methods for designing and optimizing intelligent controllers for different applications. The second part contains papers with the main theme of evolutionary design of intelligent systems for pattern recognition applications, which are basically papers using evolutionary algorithms for optimizing modular neural networks with fuzzy systems for response integration, for achieving pattern recognition in different applications. The third part contains papers with the themes of models for learning and social simulation, which are papers that apply intelligent systems to the problems of designing learning objects and social agents. The fourth part contains papers that deal with intelligent systems in robotics applications and hardware implementations.

In the part of Intelligent Control there are 5 papers that describe different contributions on evolutionary optimization of fuzzy systems in intelligent control. The first paper, by Ricardo Martinez-Marroquin et al., deals with the design of membership functions of a fuzzy logic controller for an autonomous mobile robot using ant colony optimization. The second paper, by Ricardo Martinez et al., deals with the evolutionary optimization of type-2 fuzzy logic systems applied to the control of linear plants. The third paper, by Cynthia Solano-Aragon and Arnulfo Alanis, studies a multi-agent system with fuzzy logic control for autonomous mobile robots in known environments. The fourth paper, by Gerardo Mendez and Angeles Hernandez, proposes hybrid interval type-1 non-singleton type-2 fuzzy logic systems as equivalent to type-2 adaptive neuro-fuzzy inference systems. The fifth paper, by Ieroham Baruch and Rosalba Galvan-Guerra, proposes a centralized direct and indirect neural control of distributed parameter systems.

In the part of Pattern Recognition there are 5 papers that describe different contributions on achieving pattern recognition using hybrid intelligent systems. The first paper, by Martha Pulido et al., describes ensemble neural networks with fuzzy logic integration for complex time series prediction. The second paper, by Ricardo Munoz et al., describes modular neural networks with fuzzy logic integration for face, fingerprint and voice recognition and the optimization of the network architecture with hierarchical genetic algorithms. The third paper, by Erika Ayala et al., describes the optimization of modular neural networks with fuzzy integration using a genetic algorithm with application to face recognition. The fourth paper, by Magdalena Serrano et al., proposes an intelligent hybrid system for person identification using biometric measures and modular neural networks with fuzzy integration of responses. The fifth paper, by Monica Beltran et al., deals with modular neural networks with fuzzy response integration for human signature recognition.

In the part of Learning and Social Simulation there are 3 papers that describe different contributions for creating learning objects and social intelligent agents. The first paper by Mario Garcia and Brunett Parra, describes a hybrid recommender system architecture for obtaining learning objects. The second paper, by Dora Luz Flores et al., deals with the application of fuzzy semantic networks for interaction representation in social simulation. The third paper, by Carelia Gaxiola et al., describes a fuzzy personality model based on transactional analysis for socially intelligent agents and robots.

In the part of Robotics and Hardware Implementations several contributions are described on the application evolutionary methods for achieving optimization of fuzzy systems in robotics applications and also hardware implementations of fuzzy systems. The first paper, by Nohe Cazarez et al., describes a new method for controlling unstable non-minimum phase systems with fuzzy logic. The second paper, by Selene Cardenas, describes a new genetic approach for the optimization of walking patterns of a biped robot. The third paper, by Oscar Montiel et al., deals with the design and simulation of the type-2 fuzzification stage with active membership functions and its hardware implementation on FPGAs. The fourth paper, by Roberto Sepulveda et al., deals with a methodology to test and validate a VHDL inference engine of a type-2 fuzzy system with the Xilinx system generator. The fifth paper, by Roberto Sepulveda et al., deal with the modeling and simulation of the defuzzification stage of a type-2 fuzzy controller using the Xilinx system generator and Simulink.

In conclusion, the edited book comprises papers on diverse aspects of evolutionary methods, fuzzy models, soft computing techniques and hybrid intelligent systems. The book addresses theoretical aspects of the models and methods as well as application papers, ranging from intelligent control, pattern recognition, robotics and hardware implementations.

June 30, 2009

Oscar Castillo
Tijuana Institute of Technology, Mexico

Janusz Kacprzyk
Polish Academy of Sciences, Poland

Witold Pedrycz
University of Alberta, Canada

Contents

Part I
Intelligent Control

Optimization of Membership Functions of a Fuzzy Logic Controller for an Autonomous Wheeled Mobile Robot Using Ant Colony Optimization

Ricardo Martínez-Marroquín, Oscar Castillo, and José Soria

Tijuana Institute of Technology, Tijuana México
ocastillo@hafsamx.org

Abstract. In this paper we describe the application of a Simple ACO (S-ACO) as a method of optimization for membership functions' parameters of a fuzzy logic controller (FLC) in order to find the optimal intelligent controller for an Autonomous Wheeled Mobile Robot. Simulation results show that ACO outperforms a GA in the optimization of FLCs for an autonomous mobile robot.

1 Introduction

Nowadays, fuzzy logic is one of the most used methods of computational intelligence and with the best future; this is possible thanks to the efficiency and simplicity of Fuzzy Systems since they use linguistic terms similar to those that human beings use.

The complexity for developing fuzzy systems can be found at the time of deciding which are the best parameters of the membership functions, the number of rules or even the best granularity that could give us the best solution for the problem that we want to solve.

A solution for the above mentioned problem is the application of evolutionary algorithms for the optimization of fuzzy systems. Evolutionary algorithms can be a useful tool since its capabilities of solving nonlinear problems, well-constrained or even NP-hard problems. Among the most used methods of evolutionary algorithms we can find: Genetic Algorithms, Ant Colony Optimization, Particle Swarm Optimization, etc.

This paper describes the application of evolutionary algorithms, such as the Ant Colony Optimization as a method of optimization of the parameters of the membership functions of the FLC in order to find the best intelligent controller for an Autonomous Wheeled Mobile Robot.

This paper is organized as follows: Section 2 shows the concept of Ant Colony Optimization and a description of S-ACO which is the technique that was applied

O. Castillo et al. (Eds.): Evolutionary Design of Intelligent Systems, SCI 257, pp. 3–16.
springerlink.com

for optimization. Section 3 presents the problem statement and the dynamic and kinematic model of the unicycle mobile robot. Section 4 shows the fuzzy logic controller proposed and in Section 5 it's described the development of the evolutionary method. In the Section 6 the simulation results are shown. Finally, Section 7 shows the Conclusions.

2 S-ACO Algorithm

Ant Colony Optimization (ACO) is a probabilistic technique that can be used for solving problems that can be reduced to finding good path along graphs. This method is inspired on the behavior presented by ants in finding paths from the nest or colony to the food source.

The S-ACO is an algorithmic implementation that adapts the behavior of real ants to solutions of minimum cost path problems on graphs [11]. A number of artificial ants build solutions for a certain optimization problem and exchange information about the quality of these solutions making allusion to the communication systems of the real ants [5].

Let us define the graph $G=(V,E)$, where V is the set of nodes and E is the matrix of the links between nodes. G has $n_G=|V|$ nodes. Let us define L^K as the number of hops in the path built by the ant k from the origin node to the destiny node. Therefore, it is necessary to find:

$$Q=\{q_a,\ldots,q_f \mid q_1\in C\} \tag{1}$$

Where Q is the set of nodes representing a continuous path with no obstacles; $q_a,\ldots,q_f$ are former nodes of the path and C is the set of possible configurations of the free space. If $x^k(t)$ denotes a Q solution in time t, $f(x^k(t))$ expresses the quality of the solution. The general steps of S-ACO are the followings:

- Each link (i,j) is associated with a pheromone concentration denoted as τ_{ij}.
- A number $k=1,2,\ldots,n_k$ are placed in the nest.
- On each iteration all ants build a path to the food source (destiny node). For selecting the next node a probabilistic equation is used:

$$p_{ij}^k(t)\begin{cases}\dfrac{\tau_{ij}^k}{\sum_{j\in N_{ij}^k}\tau_{ij}^{\alpha}(t)} & if\;\; j\in N_i^k\\ 0 & if\;\; j\notin N_i^k\end{cases} \tag{2}$$

Where, N_i^k is the set of feasible nodes (in a neighborhood) connected to node i with respect to ant k, τ_{ij} is the total pheromone concentration of link ij, and α is a positive constant used as again for the pheromone influence.

- Remove cycles and compute each route weight $f\left(x^k(t)\right)$. A cycle could be generated when there are no feasible candidates nodes, that is, for any i and any k, $N_i^k = \varnothing$; then the predecessor of that node is included as a former node of the path.
- Pheromone evaporation is calculated with equation (3):

$$\tau_{ij}(t) \leftarrow (1-\rho)\tau_{ij}(t) \tag{3}$$

Where $\rho \in [0,1]$ is the evaporation rate value of the pheromone trail. The evaporation is added to the algorithm in order to force the exploration of the ants, and avoid premature convergence to sub-optimal solutions [11]. For $\rho = 1$ the search becomes completely random [11].

- The update of the pheromone concentration is realized using equation (4):

$$\tau_{ij}(t+1) = \tau_{ij}(t) + \sum_{k=1}^{n_k} \Delta\tau_{ij}^k(t) \tag{4}$$

Where $\Delta\tau_{ij}^k$ is the amount of pheromone that an ant k deposits in a link ij in a time t.

- Finally, the algorithm can be ended in three different ways:
  - When a maximum number of epochs has been reached.
  - When it has been found an acceptable solution, with $f\left(x_k(t)\right) < \varepsilon$.
  - When all ants follow the same path

3 Problem Statement

The model of the robot considered in this paper is a unicycle mobile robot (see Figure 1), that consists of two driving wheels mounted of the same axis and a front free wheel.

A unicycle mobile robot is an autonomous, wheeled vehicle capable of performing missions in fixed or uncertain environments. The robot body is symmetrical around the perpendicular axis and the center of mass is at the geometrical center of the body. It has two driving wheels that are fixed to the axis that passes

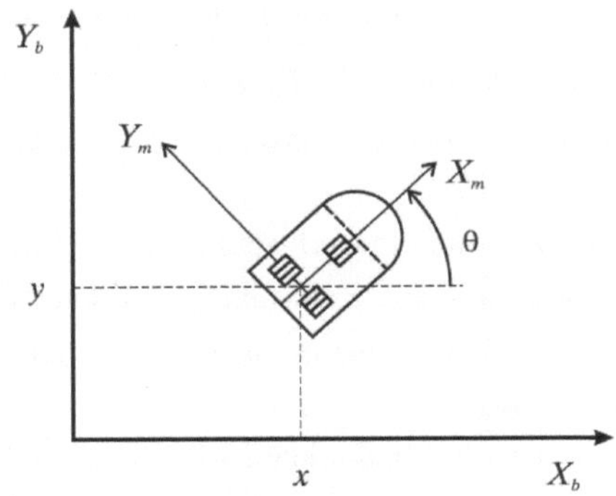

Fig. 1. Wheeled Mobile Robot [10]

through C and one passive wheel prevents the robot from tipping over as it moves on a plane. In what follows, it´s assumed that motion of the passive wheel can be ignored in the dynamics of the mobile robot presented by the following set of equations [8]:

$$M(q)\dot{\vartheta} + C(q,\dot{q})\vartheta + D\vartheta = \tau + F_{ext}(t) \tag{5}$$

$$\dot{q} = \underbrace{\begin{bmatrix} \cos\theta & 0 \\ \sin\theta & 0 \\ 0 & 1 \end{bmatrix}}_{J(q)} \underbrace{\begin{bmatrix} v \\ w \end{bmatrix}}_{\vartheta} \tag{6}$$

Where $q = (x, y, \theta)^T$ is the vector of the configuration coordinates; $\vartheta = (v, w)^T$ is the vector of linear and angular velocities; $\tau = (\tau_1, \tau_2)$ is the vector of torques applied to the wheels of the robot where τ_1 and τ_2 denote the torques of the right and left wheel respectively (Figure 1); $F_{ext} \in \mathbb{R}^2$ uniformly bounded disturbance vector; $M(q) \in \mathbb{R}^{2\times 2}$ is the positive-definite inertia matrix; $C(q,\dot{q})\vartheta$ is the vector of centripetal and Coriolis forces; and $D \in \mathbb{R}^{2\times 2}$ is a diagonal positive-definite damping matrix. Equation (6) represents the kinematics of the system, where (x, y) is the position of the mobile robot in the X-Y (world) reference frame, θ is the angle between heading direction and the x-axis v and w are the angular and angular velocities, respectively.

Furthermore, the system (5)-(6) has the following non-holonomic constraint:

$$\dot{y}\cos\theta - \dot{x}\sin\theta = 0 \tag{7}$$

which corresponds to a no-slip wheel condition preventing the robot from moving sideways[9]. The system (6) fails to meet Brockett's necessary condition for feedback stabilization [2], which implies that anon-continuous static state-feedback controller exists that stabilizes the close-loop system around the equilibrium point.

The control objective is to design a fuzzy logic controller of τ that ensures:

$$\lim_{t\to\infty}\left\|q_d(t)-q(t)\right\|=0 \tag{8}$$

for any continuously, differentiable, bounded desired trajectory $q_d \in \mathbb{R}^3$ while attenuating external disturbances.

A more detailed description can be found on reference [10].

4 Fuzzy Logic Control Design

In order to satisfy the control objective it is necessary to design a fuzzy logic controller for the real velocities of the mobile robot. To do that, a Takagi-Sugeno fuzzy logic controller was designed, using linguistic variables in the input and mathematical functions in the output. The error of the linear and angular velocities (v_d, w_d respectively), were taken as inputs variables, while the right (τ_1) and left (τ_2) torques as outputs. The membership functions used on the input are trapezoidal for the negative (*N*) and positive (*P*), and a triangular was used for the zero (*C*) linguistic terms. The interval used for this fuzzy controller is [-50 50] [10].

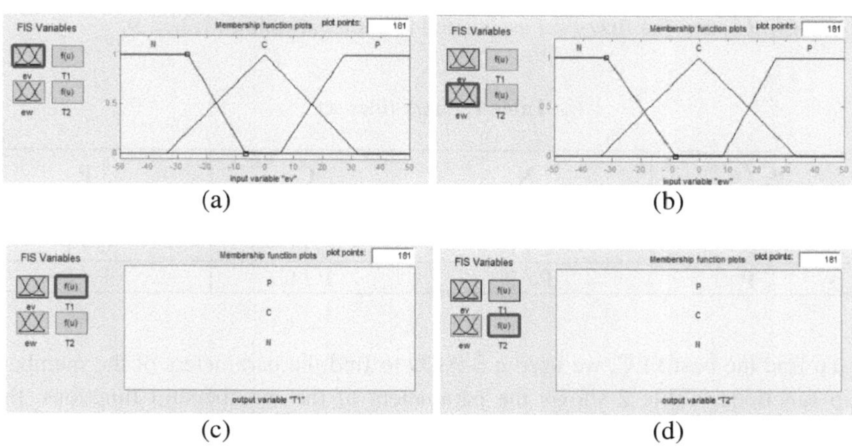

Fig. 2. (a) Linear velocity error. (b) Angular velocity error. (c) Right output (τ_1). (d) Left output (τ_2).

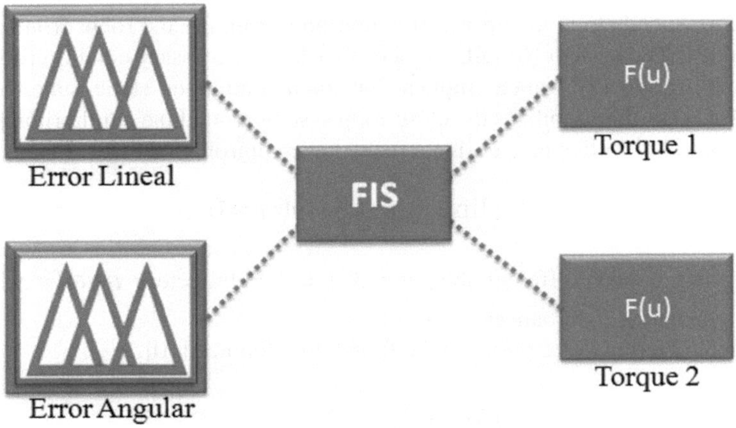

Fig. 3. Fuzzy Logic Controller Architecture.

Figure 2 shows the input and output variables, and figure 3 shows the general FLC architecture.

The rule set of the FLC contains 9 rules, which governs the input-output relationship of the FLC and this adopts the Takagi-Sugeno style inference engine [10], and it is used with a single point in the outputs, this mind that the outputs are constant values, obtained using weighted average defuzzification procedure. In Table 1 we present the rule set whose format is established as follows:

$$\text{Rule } i: \text{if } e_v \text{ is } \mathrm{G}_1 \text{ and } e_w \text{ is } G_2 \text{ then } F \text{ is } G_3 \text{ and } N \text{ is } G_4$$

where $G_1..G_4$ are the fuzzy set associated to each variable i=1,2,…,9.

Table 1. Fuzzy rules set

e_v / e_w	**N**	**C**	**P**
N	N / N	N / C	N / P
C	C / N	C / C	C / P
P	P / N	P / C	P / P

To find the best FLC, we used a S-ACO to find the parameters of the membership functions. Table 2 shows the parameters of the membership functions, the minimal and maximum values in the search range for the S-ACO algorithm to find the best fuzzy logic controller.

It is important to remark that values shown in Table 2 are applied to both inputs and both outputs of the fuzzy logic controller.

Table 2. Parameters of the membership Functions.

MF TYPE	POINT	MINIMAL VALUE	MAXIMAL VALUE
Trapezoidal	a	-50	-50
	b	-50	-50
	c	-15	-5.1
	d	-1.5	-0.5
Triangular	a	-5	-1.8
	b	0	0
	c	1.8	5
Trapezoidal	a	0.5	1.5
	b	5.1	15
	c	50	50
	d	50	50
Constant (N)	a	-50	-50
Constant (C)	a	0	0
Constant (P)	a	50	50

5 ACO Architecture

A S-ACO algorithm was applied for the optimization of the membership functions for the fuzzy logic controller. For developing the architecture of the algorithm it was necessary to follow the next steps:

1. Marking the limits of the problem in order to eliminate unnecessary complexity.
2. Representing the architecture of the FLC as a graph that artificial ants could traverse.
3. Achieving an adequate handling of the pheromone but permitting the algorithm to evolve by itself

5.1 Limiting the Problem and Graph Representation

One of problems found on the development of the S-ACO algorithm was to make a good representation of FLC. First we reduced the number of elements that the method needed to find by deleting the elements whose minimal value and maximal values are the same (see Table 2) and therefore if they were included they will not change any way. Table 3 shows the parameters of the membership functions included in the search.

The next step was to represent those parameters shown in table 3; to that, was necessary to discretize the parameters in a range of possible values in order to represent every possible value as a node in the graph of search. The level of discretization between minimal and maximal value was of 0.1 (by example: -1.5, -1.4, -1.3,..., -0.5).

Table 3. Parameters of the Membership Functions Included in S-ACO Search

MF TYPE	POINT	MINIMAL VALUE	MAXIMAL VALUE
Trapezoidal	c	-15	-5.1
	d	-1.5	-0.5
Triangular	a	-5	-1.8
	c	1.8	5
Trapezoidal	a	0.5	1.5
	b	5.1	15

Table 4 shows the number of possible values that each parameter can take.

Table 4. Number of Possible Values of the Parameters of Membership Functions

MF TYPE	POINT	COMBINATIONS
Trapezoidal	c	100
	d	15
Triangular	a	33
	c	33
Trapezoidal	a	15
	b	100

Figure 4 shows the search graph for the proposed S-ACO algorithm, the graph can be viewed as a tree where the root is the nest and the last node is the food source.

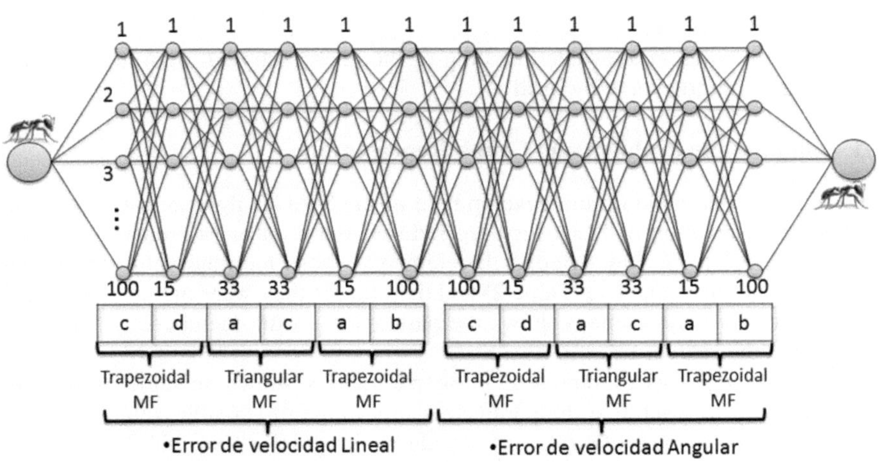

Fig. 4. S-ACO Architecture

5.2 Updating Pheromone Trail

An important issue is that the update of pheromone trail be applied in the best way possible. In this sense we need to handle the evaporation (Equation 3), and increase or deposit of pheromone (Equation 4), where the key parameter in evaporation is denoted by ρ that represents the rate of evaporation and in deposit of pheromone is denoted by $\Delta\tau$ that represents the amount of pheromone that an ant k deposits in a link ij in a time t. For ρ we assign a random value and Equation 9 shows the way how the increase of pheromone is calculated.

$$\Delta\tau = \frac{(e_{\max} - e_k)}{e_{\max}} \tag{9}$$

Where $e_{\max} = 10$ is the maximum error of control permitted and e_k is error of control generated by a complete path of an ant k. We decided to allocate $e_{\max} = 10$ in order to stand $\Delta\tau \in [0,1]$.

6 Simulation Results

In this section we present the results of the proposed controller to stabilize the unicycle mobile robot, defined by Equation (5) and Equation (6), where the matrix values

$$M(q) = \begin{bmatrix} 0.3749 & -0.0202 \\ -0.0202 & 0.3739 \end{bmatrix},$$

$$C(q,\dot{q}) = \begin{bmatrix} 0 & 0.1350\dot{\theta} \\ -0.150\dot{\theta} & 0 \end{bmatrix},$$

$$\text{and } D = \begin{bmatrix} 10 & 0 \\ 0 & 10 \end{bmatrix}$$

were taken from [6]. The evaluation was made through computer simulation performed in MATLAB® and SIMULINK®.

The desired trajectory is the following one:

$$\vartheta_d(t) = \begin{cases} v_d(t) = 0.2(1 - \exp(-t)) \\ w_d(t) = 0.4\sin(0.5t) \end{cases} \tag{10}$$

Table 5. S-ACO Results of Simulations for FLC Optimization

Iterations	Ants	α	ρ	Average Error	Time
20	10	0.2	random	1.5589	00:01:30
20	10	0.2	random	1.451	00:01:34
25	10	0.2	random	1.5566	00:01:46
25	10	0.2	random	1.4767	00:01:51
25	10	0.2	random	1.4739	00:02:05
25	10	0.2	random	1.6137	00:02:08
25	10	0.2	random	1.6642	00:01:54
25	100	0.2	random	1.3484	00:20:30
25	100	0.2	random	1.3413	00:18:44
25	100	0.2	random	1.3360	00:18:31
25	100	0.2	random	1.2954	00:18:32
25	100	0.2	random	1.4877	00:18:41
25	100	0.2	random	1.2391	00:18:31
10	15	0.2	random	1.6916	00:01:14
10	15	0.2	random	1.4256	00:01:09
40	65	0.2	random	1.2783	00:19:17
40	65	0.2	random	1.4011	00:19:45
40	65	0.2	random	1.2216	00:19:33
40	65	0.2	random	1.2487	00:19:49
50	70	0.2	random	1.3782	00:26:09
50	70	0.2	random	1.0875	00:27:35
50	70	0.2	random	1.4218	00:33:45
50	70	0.2	random	1.475	01:08:48
25	80	0.2	random	1.4718	00:14:55
25	80	0.2	random	1.4212	00:15:00
25	80	0.2	random	1.3221	00:14:52
25	80	0.2	random	1.1391	00:15:41
50	80	0.2	random	1.2148	00:28:43
62	**50**	**0.2**	**random**	**1.0322**	**00:24:49**
50	80	0.2	random	1.1887	00:29:55
50	80	0.2	random	1.2158	00:29:56
60	90	0.2	random	1.3493	00:41:56
60	90	0.2	random	1.3060	00:39:48
60	90	0.2	random	1.3161	00:40:00

and was chosen in terms of its corresponding desired linear v_d and angular w_d velocities, subject to the initial conditions

$$q(0) = (0.1, 0.1, 0,)^T \text{ and } \vartheta(0) = 0 \in \mathbb{R}^2$$

The gains γ_i, i=1, 2, 3 of the kinematic model (see [10]) are $\gamma_1 = 5$, $\gamma_2 = 24$ and $\gamma_3 = 3$ were taken from [10].

6.1 S-ACO Algorithm Results for the Optimization of the FLC

Table 5 shows the results of the FLC, obtained varying the values of maximum iterations and number of artificial ants, where the highlighted row shows the best result obtained with the method. Figure 5 shows the evolving of the method.

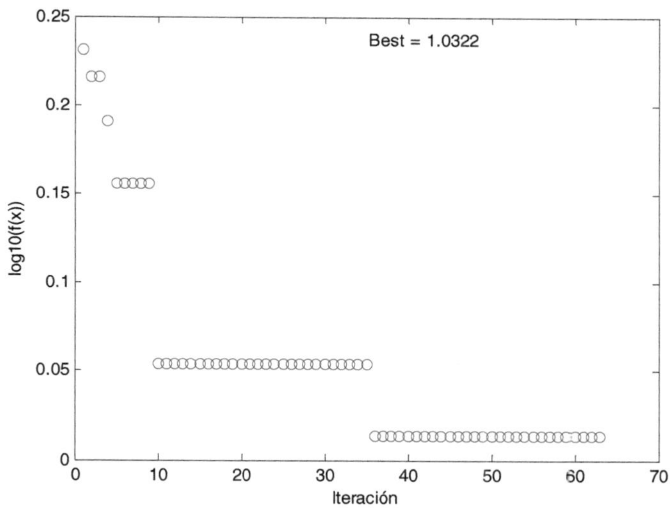

Fig. 5. Evolution of the S-ACO for FLC Optimization.

Figure 6 shows the membership functions of the FLC obtained by S-ACO algorithm.

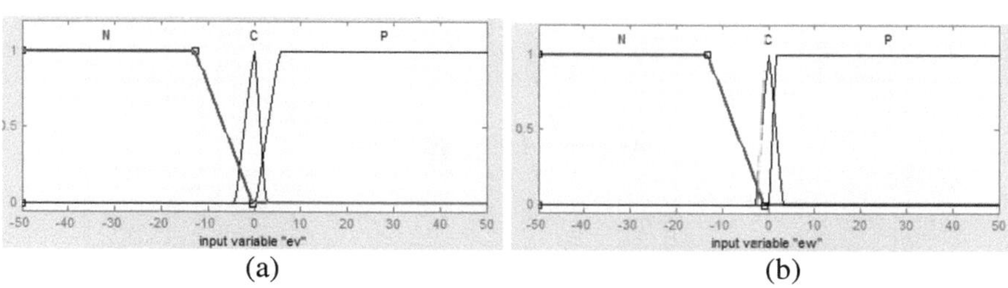

Fig. 6. (a) Linear velocity error, and (b) angular velocity error optimized by S-ACO algorithm.

Figure 7 shows the block diagram used for the FLC that obtained the best results. Figure 8 shows the results of linear and angular errors, and Figure 9 shows the output results of the fuzzy controller that represents the torque applied to the wheels of the autonomous mobile robot.

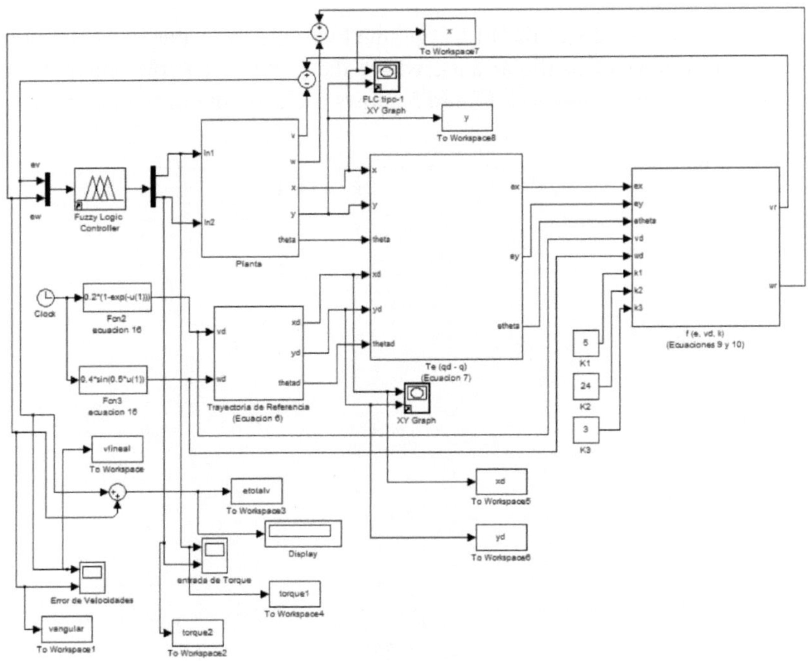

Fig. 7. Block diagram for simulation of the FLC

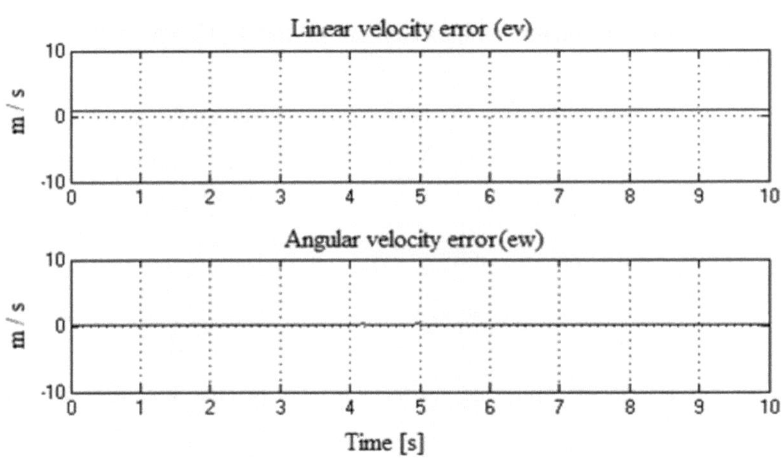

Fig. 8. Linear and angular velocity errors

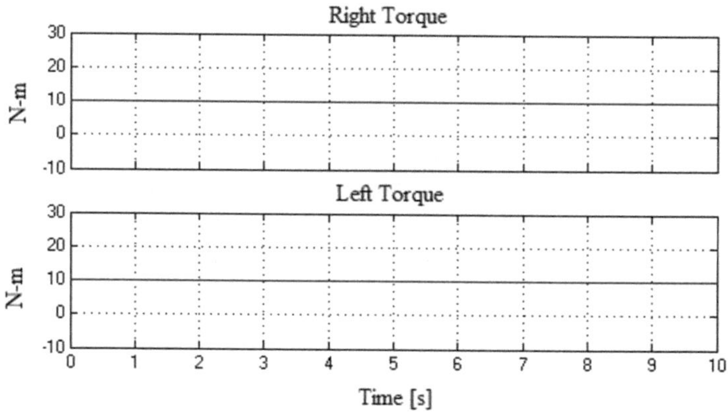

Fig. 9. Right and left torques

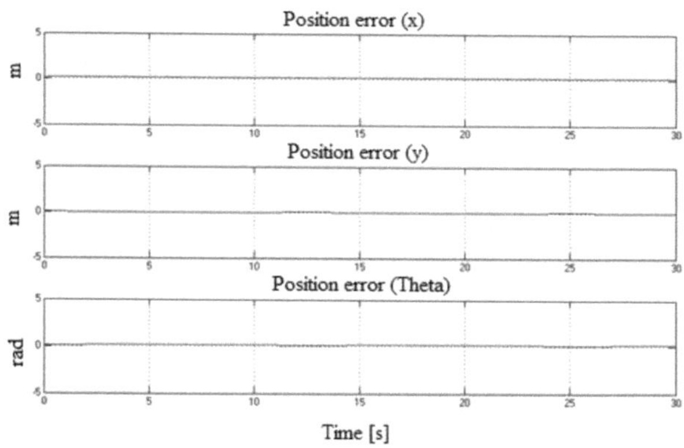

Fig. 10. Position errors in x, y, θ .

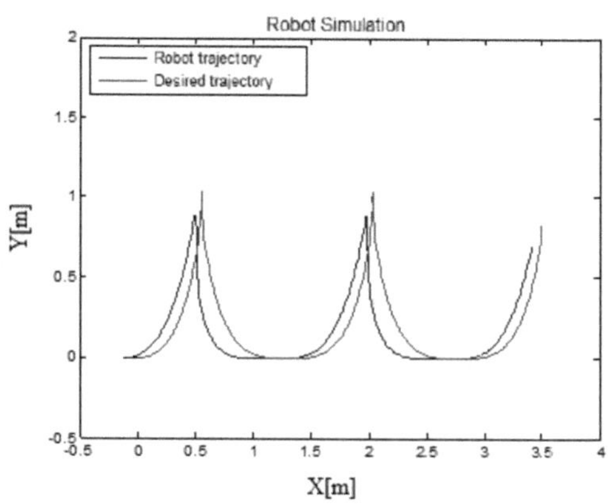

Fig. 11. Obtained trajectory

The positions errors of the autonomous mobile robot can be observed in Figure 10. Figure 11 shows the desired trajectory and obtained trajectory.

7 Conclusions

A trajectory tracking controller has been designed based on the dynamics and kinematics of the autonomous mobile robot through the application of ACO for the optimization of membership functions for the fuzzy logic controller with good results obtained after simulations.

Acknowledgment. We would like to express our gratitude to the CONACYT and Tijuana Institute of Technology for the facilities and resources granted for the development of this research.

References

[1] Bloch, A.M.: Nonholonomic mechanics and control. Springer, New York (2003)
[2] Brockett, R.W.: Asymptotic stability and feedback stabilization. In: Millman, R.S., Susman, H.J. (eds.) Diferential Geometric Control Theory, p. 181. Birkhauser, Boston (1983)
[3] Castillo, O., Melin, P.: Soft Computing for Control of Non-Linear Dynamical Systems. Springer, Helderberg (2001)
[4] Dorigo, M., Birattari, M., Stützle, T.: Ant Colony Optimization. IEEE Computational Intelligence Magazine, 28–39 (November 2006)
[5] Dorigo, M., Stützle, T.: Ant Colony Optimization, Bradford, Cambridge, Massachusetts (2004)
[6] Duc Do, K., Zhong-Ping, Y., Pan, J.: A global output-feedback controller for simultaneous tracking and stabilizations of unicycle-type mobile robots. IEEE Trans. Automat. Contr. 20(3), 589–594 (2004)
[7] Engelbrecht, A.P.: Fundamentals of Computational Swarm Intelligence. Wiley, England (2005)
[8] Lee, T.-C., Song, K.-T., Lee, C.-H., Teng, C.-C.: Tracking control of unicycle-modeled mobile robot using a saturation feedback controller. IEEE trans. Contr. Syst. Technol. 9(2), 305–318 (2001)
[9] Liberzon, D.: Switching in Systems and control. Bikhauser (2003)
[10] Martínez, R., Castillo, O., Aguilar, L.: Intelligent control for a perturbed autonomous wheeled mobile robot using type-2 fuzzy logic and genetic algorithms. JAMRIS 2(1), 1–11 (2008)
[11] Porta-García, M., Montiel, O., Sepulveda, R.: An ACO path planner using a FIS for a Path Selection Adjusted with a Simple Tuning Algorithm. JAMRIS 2(1), 1–11 (2008)
[12] Sepúlveda, R., Castillo, O., Melin, P., Montiel, O.: An Efficient Computational Method to Implement Type-2 Fuzzy Logic In control Applications, Analysis and Design of Intelligent Systems Using Soft Computing Techniques. Advances in Soft Computing 41, 45–52 (2007)

Evolutionary Optimization of Type-2 Fuzzy Logic Systems Applied to Linear Plants

Ricardo Martinez[1], Oscar Castillo[2], Luis T. Aguilar[3], and Antonio Rodriguez[4]

[1] PhD Student of Computer Science in the Universidad Autónoma de Baja California Tijuana, México
e-mail: mc.ricardo.martinez@hotmail.com
[2] Tijuana Institute of Technology, Tijuana México
e-mail: ocastillo@hafsamx.org
[3] Instituto Politécnico Nacional, Centro de Investigación y Desarrollo de Tecnología Digital, PMB 88, P.O. Box 439016 San Ysidro CA., 92143-9016; Fax: +52(664)6231388
e-mail: luis.aguilar@ieee.org
[4] Universidad Autónoma de Baja California, Tijuana México
e-mail: ardiaz@uabc.mx

Abstract. We describe a different kind of evolutionary methods to optimize a type-2 fuzzy logic controller (FLC) applied to linear plants. The evolutionary method used is a genetic algorithm to find the optimal FLC for the plant control. The plant receives a linear signal of input controlled by an optimized FLC, obtaining as result the control and the stability of the plant. Simulations results were made in Simulink showing the effectiveness of the proposal.

1 Introduction

The evolutionary methods are used for different purposes such as optimization, solution search's processes and other kind of applications. This study is about several evolutionary methods applied to the autonomous linear plants. To this purpose, we use a type-2 fuzzy logic system to develop the optimal controller. Previously we optimize a type-2 fuzzy logic controller for an autonomous mobile robot for trajectory tracking, where the genetic algorithms were used to find the optimal controller obtaining good results applied some kind of perturbation. For the study of the evolutionary methods, we use a transfer function to test the optimal type-2 fuzzy logic controller. One of the evolutionary methods is genetic algorithms that we used to find the parameters of the membership functions, using the genetic operators, mutation and crossover obtaining an Optimal FLC for the plant control.

This paper is organized as follows: Section 2 presents the theoretical basis and problem statement, 2.1 presents an introductory explanation of Type-2 Fuzzy Logic, subsection 2.2 presents the basics of Evolutionary Methods. Section 3 introduces the controller design where a genetic algorithm is used to select the parameters. Robustness properties of the closed-loop system are achieved with a type-1 fuzzy logic control system using a Takagi-Sugeno model where the error

O. Castillo et al. (Eds.): Evolutionary Design of Intelligent Systems, SCI 257, pp. 17–31.
springerlink.com

and the change of error, are considered the linguistic variables. Section 4 provides a simulation study of the plant using the controller described in Section 3. Finally, Section 5 presents the conclusions.

2 Theoretical Basis and Problem Statement

This section describes the theoretical basis of the paper as well as the problem definition. Some basics about type-2 fuzzy systems and genetic-fuzzy systems are first presented.

2.1 Type-2 Fuzzy Logic Systems

If we have a type-1 membership function, as in Figure 1 (a), and we are blurring it to the left and to the right as illustrated in Figure 1 (b), then, for a specific value x', the membership function (u'), takes on different values, which are not all weighted the same, so we can assign an amplitude distribution to all of those points. Doing this for all $x \in X$, we create a three-dimensional membership function –a type-2 membership function– that characterizes a type-2 fuzzy set [1, 14]. A type-2 fuzzy set $\tilde{A}$, is characterized by the membership function:

$$\tilde{A} = \{((x,u), \mu_{\tilde{A}}(x,u)) \mid \forall x \in X, \forall u \in J_x \subseteq [0,1]\} \tag{1}$$

in which $0 \le \mu_{\tilde{A}}(x,u) \le 1$. Another expression for $\tilde{A}$ is,

$$\tilde{A} = \int_{x \in X} \int_{u \in J_x} \mu_{\tilde{A}}(x,u)/(x,u) \qquad J_x \subseteq [0,1] \tag{2}$$

where $\int\int$ denote union over all admissible input variables x and u. For discrete universes of discourse $\int$ is replaced by $\sum$ [14]. In fact $J_x \subseteq [0,1]$ represents the primary membership of x, and $\mu_{\tilde{A}}(x,u)$ is a type-1 fuzzy set known as

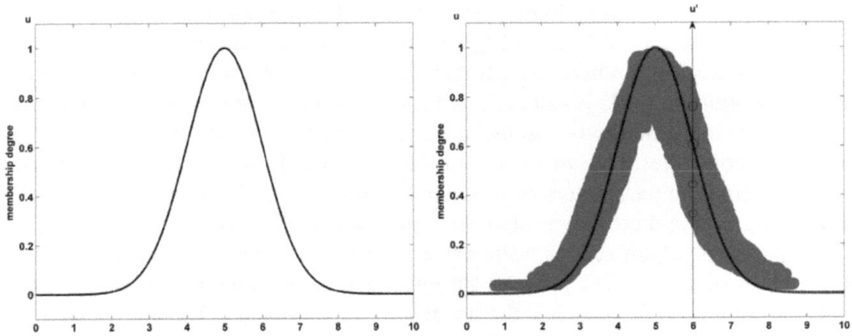

Fig. 1. a) Type-1 membership function and b) Blurred type-1 membership function.

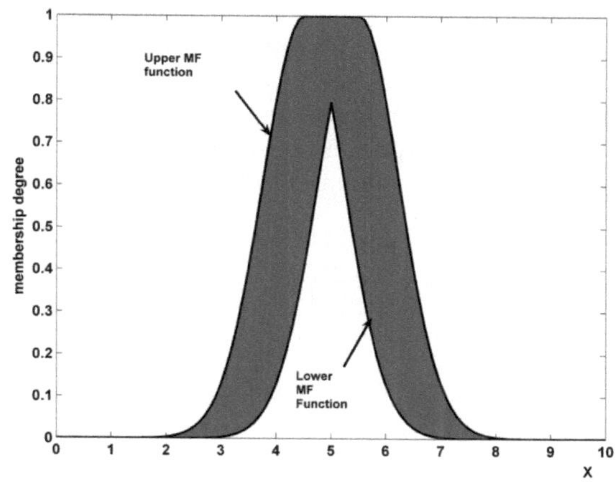

Fig. 2. Interval type-2 membership function.

the secondary set. Hence, a type-2 membership grade can be any subset in [0,1], the primary membership, and corresponding to each primary membership, there is a secondary membership (which can also be in [0,1]) that defines the possibilities for the primary membership [20].

This uncertainty is represented by a region called footprint of uncertainty (FOU). When $\mu_{\tilde{A}}(x,u)=1, \forall u \in J_x \subseteq [0,1]$ we have an interval type-2 membership function, as shown in Figure 2. The uniform shading for the FOU represents the entire interval type-2 fuzzy set and it can be described in terms of an upper membership function $\bar{\mu}_{\tilde{A}}(x)$ and a lower membership function $\underline{\mu}_{\tilde{A}}(x)$.

A FLS described using at least one type-2 fuzzy set is called a type-2 FLS. Type-1 FLSs are unable to directly handle rule uncertainties, because they use type-1 fuzzy sets that are certain [21]. On the other hand, type-2 FLSs, are very useful in circumstances where it is difficult to determine an exact membership function, and there are measurement uncertainties [23].

It is known that type-2 fuzzy sets enable modeling and minimizing the effects of uncertainties in rule-based FLS. Unfortunately, type-2 fuzzy sets are more difficult to use and understand than type-1 fuzzy sets; hence, their use is not widespread yet. As a justification for the use of type-2 fuzzy sets, in [22] are mentioned at least four sources of uncertainties not considered in type-1 FLSs:

1. The meanings of the words that are used in the antecedents and consequents of rules can be uncertain (words mean different things to different people).
2. Consequents may have histogram of values associated with them, especially when knowledge is extracted from a group of experts who do not all agree.
3. Measurements that activate a type-1 FLS may be noisy and therefore uncertain.
4. The data used to tune the parameters of a type-1 FLS may also be noisy.

All of these uncertainties translate into uncertainties about fuzzy set membership functions. Type-1 fuzzy sets are not able to directly model such uncertainties because their membership functions are totally crisp. On the other hand, type-2 fuzzy sets are able to model such uncertainties because their membership functions are themselves fuzzy. A type-1 fuzzy set is a special case of a type-2 fuzzy set; its secondary membership function is a subset with only one element, unity.

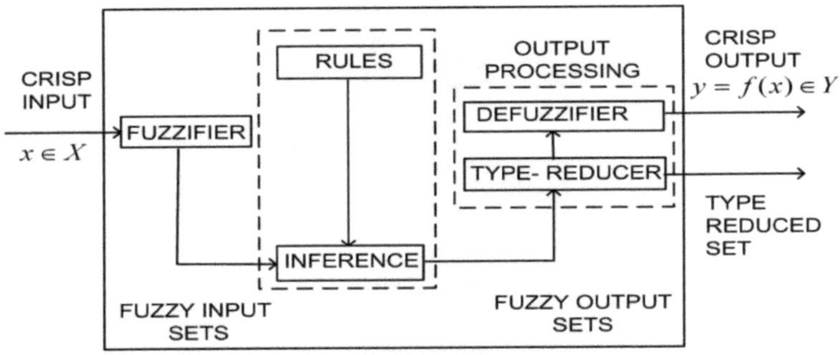

Fig. 3. Type-2 Fuzzy Logic System.

A type-2 FLS is again characterized by IF-THEN rules, but its antecedent or consequent sets are now of type-2. Type-2 FLSs, can be used when the circumstances are too uncertain to determine exact membership grades such as when the training data is corrupted by noise. Similar to a type-1 FLS, a type-2 FLS includes a fuzzifier, a rule base, fuzzy inference engine, and an output processor, as we can see in Figure 3. The output processor includes type-reducer and defuzzifier; it generates a type-1 fuzzy set output (from the type-reducer) or a crisp number (from the defuzzifier) [6,5]. Next we will explain each of the blocks of Figure 3.

2.1.1 Fuzzifier

The fuzzifier maps a crisp point $\mathbf{x}=(x_1,\ldots,x_p)^T \in X_1 x X_2 x \ldots x X_p \equiv \mathbf{X}$ into a type-2 fuzzy set $\tilde{A}_x$ in $\mathbf{X}$ [12], interval type-2 fuzzy sets in this case. We will use type-2 singleton fuzzifier, in a singleton fuzzification, the input fuzzy set has only a single point on nonzero membership [31, 34]. $\tilde{A}_x$ is a type-2 fuzzy singleton if $\mu_{\tilde{A}_x}(x) = 1/1$ for $\mathbf{x}=\mathbf{x}'$ and $\mu_{\tilde{A}_x}(x) = 1/0$ for all other $\mathbf{x} \neq \mathbf{x}'$[23].

2.1.2 Rules

The structure of rules in a type-1 FLS and a type-2 FLS is the same, but in the latter the antecedents and the consequents will be represented by type-2 fuzzy sets. So for a type-2 FLS with p inputs $x_1 \in X_1,\ldots,x_p \in X_p$ and one output $y \in Y$, Multiple Input Single Output (MISO), if we assume there are M rules, the lth rule in the type-2 FLS can be written as follows [23]:

$$R^{l}\text{: IF } x_1 \text{ is } \tilde{F}_1^{\,l} \text{ and } \cdots \text{and } x_p \text{ is } \tilde{F}_p^{\,l}\text{ , THEN } y \text{ is } \tilde{G}^{l} \qquad l=1,\ldots,M \tag{3}$$

2.1.3 Inference

In the type-2 FLS, the inference engine combines rules and gives a mapping from input type-2 fuzzy sets to output type-2 fuzzy sets. It is necessary to compute the join $\sqcup$, (unions) and the meet Π (intersections), as well as extended sup-star compositions (sup star compositions) of type-2 relations [23].If $\tilde{F}^{l}{}_{1}\times\cdots\times\tilde{F}^{l}{}_{p}=\tilde{A}^{l}$, equation (3) can be re-written as

$$R^{l}:\tilde{F}^{l}{}_{1}\times\cdots\times\tilde{F}^{l}{}_{p}\rightarrow\tilde{G}^{l}=\tilde{A}^{l}\rightarrow\tilde{G}^{l} \qquad l=1,\ldots,M \tag{4}$$

R^l is described by the membership function $\mu_{R^l}(\mathbf{x},y)=\mu_{R^l}(x_1,\ldots,x_p,y)$, where

$$\mu_{R^l}(\mathbf{x},y)=\mu_{\tilde{A}^l\rightarrow\tilde{G}^l}(\mathbf{x},y) \tag{5}$$

can be written as [23]:

$$\mu_{R^l}(\mathbf{x},y)=\mu_{\tilde{A}^l\rightarrow\tilde{G}^l}(\mathbf{x},y)=\mu_{\tilde{F}_1^l}(x_1)\,\Pi\cdots\Pi\,\mu_{\tilde{F}_p^{\,l}}(x_p)\,\Pi\,\mu_{\tilde{G}^l}(y)=$$
$$[\Pi_{i=1}^{p}\,\mu_{\tilde{F}^{l}{}_{i}}(x_i)\,]\Pi\,\mu_{\tilde{G}^l}(y) \tag{6}$$

In general, the p-dimensional input to R^l is given by the type-2 fuzzy set $\tilde{A}_x$ whose membership function is

$$\mu_{\tilde{A}_x}(\mathbf{x})=\mu_{\tilde{x}_1}(x_1)\,\Pi\cdots\Pi\,\mu_{\tilde{x}p}(x_p)=\Pi_{i=1}^{p}\,\mu_{\tilde{x}i}(x_i) \tag{7}$$

where $\tilde{X}_i(i=1,\ldots,p)$ are the labels of the fuzzy sets describing the inputs. Each rule R^l determines a type-2 fuzzy set $\tilde{B}^{l}=\tilde{A}_x\circ R^{l}$ such that [23]:

$$\mu_{\tilde{B}^l}(y)=\mu_{\tilde{A}_x\circ R^l}=\sqcup_{x\in\mathbf{X}}\left[\mu_{\tilde{A}_x}(\mathbf{x})\,\Pi\,\mu_{R^l}(\mathbf{x},y)\right] \qquad y\in Y\;\; l=1,\ldots,M \tag{8}$$

This equation is the input/output relation in Figure 3 between the type-2 fuzzy set that activates one rule in the inference engine and the type-2 fuzzy set at the output of that engine [23].

In the FLS we used interval type-2 fuzzy sets and meet under product t-norm, so the result of the input and antecedent operations, which are contained in the firing set $\Pi_{i=1}^{p}\mu_{\tilde{F}_{i_i}}(x'_i\equiv F^{l}(\mathbf{x}')$, is an interval type-1 set [23],

$$F^{l}(\mathbf{x}')=\left[\underline{f}^{l}(\mathbf{x}'),\bar{f}^{l}(\mathbf{x}')\right]\equiv\left[\underline{f}^{l},\bar{f}^{l}\right] \tag{9}$$

where

$$\underline{f}^{l}(\mathbf{x}') = \underline{\mu}_{\tilde{F}_1^l}(x_1') * \cdots * \underline{\mu}_{\tilde{F}_p^l}(x_p') \tag{10}$$

and

$$\bar{f}^{l}(\mathbf{x}') = \bar{\mu}_{\tilde{F}_1^l}(x_1') * \cdots * \bar{\mu}_{\tilde{F}_p^l}(x_p') \tag{11}$$

where * is the product operation.

2.1.4 Type Reducer

The type-reducer generates a type-1 fuzzy set output which is then converted in a crisp output through the defuzzifier. This type-1 fuzzy set is also an interval set, for the case of our FLS we used center of sets (cos) type reduction, $Y_{\cos}$ which is expressed as [23]

$$Y_{\cos}(\mathbf{x}) = [y_l, y_r] = \int_{y^1 \in [y_l^1, y_r^1]} \cdots \int_{y^M \in [y_l^M, y_r^M]} \int_{f^1 \in [\underline{f}^1, \bar{f}^1]} \cdots \int_{f^M \in [\underline{f}^M, \bar{f}^M]} 1 \Big/ \frac{\sum_{i=1}^{M} f^i y^i}{\sum_{i=1}^{M} f^i} \tag{12}$$

this interval set is determined by its two end points, y_l and y_r, which corresponds to the centroid of the type-2 interval consequent set $\tilde{G}^i$ [23],

$$C_{\tilde{G}^i} = \int_{\theta_1 \in J_{y1}} \cdots \int_{\theta_N \in J_{yN}} 1 \Big/ \frac{\sum_{i=1}^{N} y_i \theta_i}{\sum_{i=1}^{N} \theta_i} = [y_l^i, y_r^i] \tag{13}$$

before the computation of $Y_{\cos}(\mathbf{x})$, we must evaluate equation 13, and its two end points, y_l and y_r. If the values of f_i and y_i that are associated with y_l are denoted f_l^i and y_l^i, respectively, and the values of f_i and y_i that are associated with y_r are denoted f_r^i and y_r^i, respectively, from 12, we have [23]

$$y_l = \frac{\sum_{i=1}^{M} f_l^i y_l^i}{\sum_{i=1}^{M} f_l^i} \tag{14}$$

$$y_r = \frac{\sum_{i=1}^{M} f_r^i y_r^i}{\sum_{i=1}^{M} f_r^i} \tag{15}$$

2.1.5 Defuzzifier

From the type-reducer we obtain an interval set $Y_{\cos}$, to defuzzify it we use the average of y_l and y_r, so the defuzzified output of an interval singleton type-2 FLS is [23]

$$y(\mathbf{x}) = \frac{y_l + y_r}{2} \tag{16}$$

2.2 Evolutionary Methods Applied to Fuzzy Systems

2.2.1 Genetic Fuzzy Systems

Fuzzy systems have been successfully applied to problems in classification [8], modeling [24] control [11], and in a considerable number of applications. In most cases, the key for success was the ability of fuzzy systems to incorporate human expert knowledge. In the 1990s, despite the previous successful history, the lack of learning capabilities characterizing most of the works in the field generated a certain interest for the study of fuzzy systems with added learning capabilities.

Two of the most successful approaches have been the hybridization attempts made in the framework of soft computing, were different techniques, such as neural and evolutionary; provide fuzzy systems with learning capabilities. Neuro-fuzzy systems are one of the most successful and visible directions of that effort. A different approach to achieve hybridization has lead to genetic fuzzy systems (GFSs). A GFS is basically a fuzzy system augmented by a learning process based on a genetic algorithm (GA) [9].

GAs are search algorithms, based on natural genetics, that provide robust search capabilities in complex spaces, and thereby other a valid approach to problems requiring efficient and effective search processes [16,19,20]. Genetic learning processes cover different levels of complexity according to the structural changes produced by the algorithm [10], from the simplest case of parameter optimization to the highest level of complexity of learning the rule set of a rule based system. Parameter optimization has been the approach utilized to adapt a wide range of different fuzzy systems, as in genetic fuzzy clustering or genetic neuro-fuzzy systems.

An analysis of the literature shows that the most prominent types of GFSs are genetic fuzzy rule-based systems (GFRBSs) [9], whose genetic process learns or tunes different components of a fuzzy rule-based system (FRBS). Inside GFRBSs it is possible to distinguish between either parameter optimization or rule generation processes, that is, adaptation and learning.

It is important to distinguish between tuning (alternatively, adaptation) and learning problems:

- Tuning is concerned with optimization of an existing FRBS, whereas learning constitutes an automated design method for fuzzy rule sets that starts from scratch. Tuning processes assume a predefined RB and have the objective to find a set of optimal parameters for the membership and or the scaling functions, DB parameters.
- Learning processes perform a more elaborated search in the space of possible RBs or whole KBs and do not depend on a predefined set of rules.
- They are different kind of genetic fuzzy systems applications, but we focused on genetic tuning for optimization parameters of membership functions [2] [1] [7].

2.2.2 Particle Swarm Optimization

Particle swarm optimization (PSO) is a population based stochastic optimization technique developed by Eberhart and Kennedy in 1995, inspired by social behavior of bird flocking or fish schooling [13].

PSO shares many similarities with evolutionary computation techniques such as Genetic Algorithms (GA) [15]. The system is initialized with a population of random solutions and searches for optima by updating generations. However, unlike the GA, the PSO has no evolution operators such as crossover and mutation. In the PSO, the potential solutions, called particles, fly through the problem space by following the current optimum particles [4].

Each particle keeps track of its coordinates in the problem space, which are associated with the best solution (fitness) it has achieved so far (The fitness value is also stored). This value is called *pbest*. Another "best" value that is tracked by the particle swarm optimizer is the best value, obtained so far by any particle in the neighbors of the particle. This location is called *lbest*. When a particle takes all the population as its topological neighbors, the best value is a global best and is called *gbest* [17].

The particle swarm optimization concept consists of, at each time step, changing the velocity of (accelerating) each particle toward its p*best* and *lbest* locations (local version of PSO). Acceleration is weighted by a random term, with separate random numbers being generated for acceleration toward *pbest* and *lbest* locations [27].

In the past several years, PSO has been successfully applied in many research and application areas. It is demonstrated that PSO gets better results in a faster, cheaper way compared with other methods [3] [18].

Another reason that PSO is attractive is that there are few parameters to adjust. One version, with slight variations, works well in a wide variety of applications. Particle swarm optimization has been used for approaches that can be used across a wide range of applications, as well as for specific applications focused on a specific requirement [28].

3 Fuzzy Logic Controller Design

In this section we design a fuzzy logic controller (FLC) where the optimal controller was found with first evolutionary method, which in this case is the genetic algorithm.

The FLC a Takagi-Sugeno type of fuzzy systems is used with two inputs a) error, and b) error change, with three membership functions each input, "Negative, Zero and Positive" (Gaussian and triangular), and one output which are constant

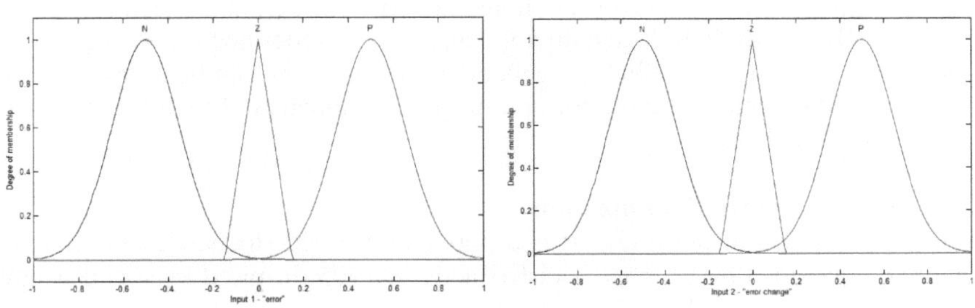

Fig. 4. a) input 1 "error", b) input 2 "error change".

Table 1. Fuzzy Rules of the FLC.

	N	**Z**	**P**
N	N	N	Z
Z	N	Z	P
P	Z	P	P

values. The fuzzy rules "If-Then" type. Figure 5 shows the FLC base for the plant control and Table 1 show the Fuzzy Rules.

Once we obtained the FLC design, we used an evolutionary method (Genetic Algorithm) to find the optimal Controller. The genetic algorithm chromosome has 17 genes of real values and they represent the two inputs, error and error change and one output constant values. Figure 6 show the chromosome representation for the FLC and Table 2 shows the parameters of the membership functions, the minimal and the maximum values in the search range for the genetic algorithm to find the best fuzzy controller system.

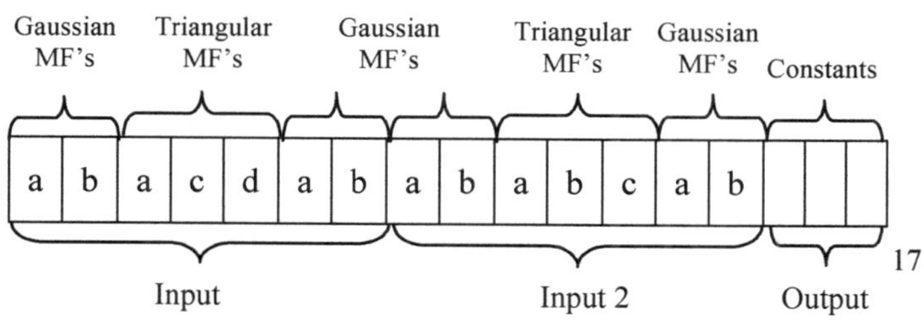

Fig. 5. Chromosome representation for the fuzzy logic controller.

4 Simulations Results

In this section, we evaluate, through computer simulations performed in MATLAB® and SIMULINK®, the designed FLC for two plants.

4.1 Plant 1

Plant 1 is given by the following transfer function:

$$g(s) = \frac{w_n^2}{s^2 + 2\varepsilon\, w_n\, s + w_n^2} \qquad \varepsilon = 0.5, \quad w_n = 2 \tag{17}$$

Table 2 presents the parameters of the membership functions used in the genetic algorithm for the plant 1.

Table 3 present the main results of the FLC obtained by genetic algorithms showing in the result 3 our best result.

Table 2. Parameters of the membership functions

MF Type	Point	Minimum Value	Maximum Value
Gaussian	a	0.3	0.6
	b	-1	-1
Triangular	a	-0.3	-0.8
	b	0	0
	c	0.3	0.8
Gaussian	a	0.3	0.6
	b	1	1

Table 3. Results of the FLC obtained by genetic algorithms.

No.	Indiv.	Gen.	% Remp.	Cross.	Mut.	GA Time	Average error
1	90	35	0.7	0.6	0.3	00:18:06	0.050870
2	150	80	0.7	0.5	0.2	01:19:13	0.044310
3	**80**	**50**	**0.7**	**0.5**	**0.2**	**00:23:01**	**0.071366**
4	45	60	0.7	0.6	0.3	00:16:03	0.068477
5	75	50	0.7	0.6	0.1	00:21:37	0.068158
6	100	40	0.7	0.6	0.1	00:24:08	0.067052
7	65	35	0.7	0.7	0.2	00:15:35	0.069994
8	200	70	0.7	0.4	0.1	01:30:02	0.072356
9	25	15	0.7	0.8	0.3	00:02:58	0.129872
10	50	45	0.7	0.5	0.2	00:13:25	0.068855
11	90	35	0.7	0.6	0.2	00:19:15	0.065290
12	40	25	0.7	0.7	0.4	00:06:48	0.175755
13	120	454	0.7	0.4	0.1	00:29:51	0.065761

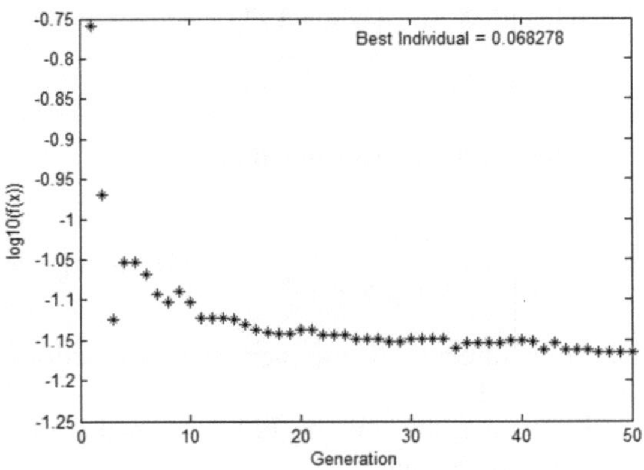

Fig. 6. Evolution of the GA for the FLC optimization.

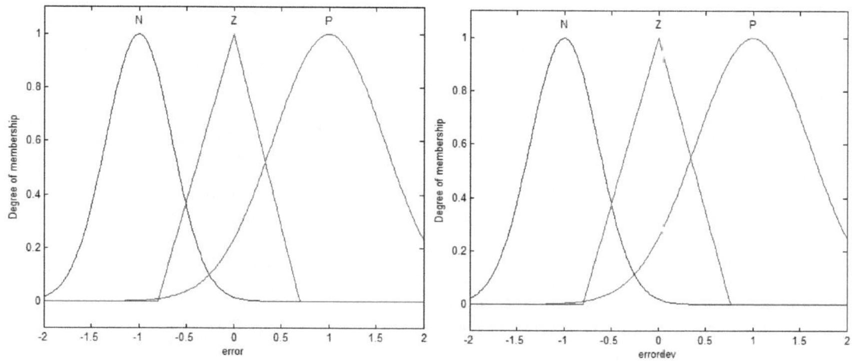

Fig. 7. Membership functions of the FLC obtained by GA.

Figure 6 shows the evolution of the genetic algorithm giving the best FLC for control the plant and figure 7 show the membership functions of the FLC obtained by the GA.

Figure 8 show the control result of the plant 1 using the FLC optimized obtained by GA.

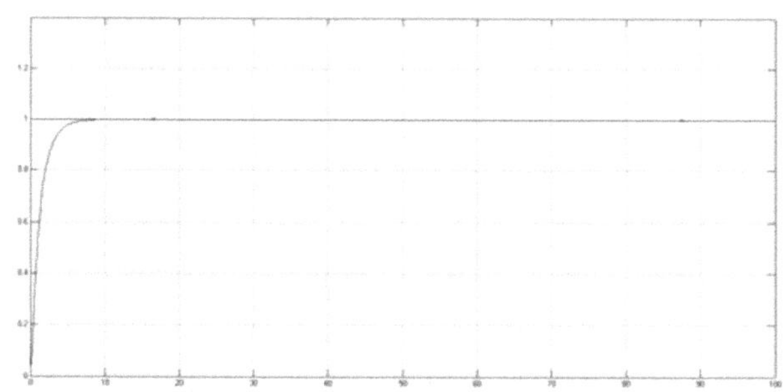

Fig. 8. Simulation result of the control of plant 1.

4.2 Plant 2

Plant 2 is given by the following equation:

$$g(s) = \frac{1}{s^2 + 4} \tag{18}$$

For the experiment with plant 2 we test different range of the membership function finding the best result in the range of the -10 to 10. Table 4 present the main results of the FLC obtained by genetic algorithms.

Table 4. Simulation results of plant 2.

No.	Ind.	Gen.	% Remp.	Cross	Mut.	GA Time	Average error	Param. Range of MF
1	45	20	0.7	0.4	0.1	00:05:10	0.889298	from 0.5 to 1
2	160	75	0.5	0.6	0.1	01:10:43	0.889298	from -1 to 1
3	80	30	0.7	0.6	0.1	00:14:01	0.889302	
4	50	20	0.7	0.5	0.1	00:05:13	0.889784	
5	50	20	0.7	0.5	0.1	00:05:11	0.894638	
6	55	45	0.5	0.6	0.1	00:12:45	0.574186	from -2 to 2
7	95	65	0.5	0.5	0.1	00:32:13	0.569497	
8	150	70	0.5	0.4	0.1	00:53:32	0.569667	
9	55	45	0.5	0.6	0.1	00:12:45	0.473400	from -4 to 4
10	150	70	0.5	0.4	0.1	00:53:52	0.447686	
11	95	65	0.5	0.5	0.1	00:30:50	0.461219	
12	150	100	0.5	0.5	0.1	01:20:24	0.064981	from -8 to 8
13	95	65	0.5	0.5	0.1	00:32:07	0.065394	
14	85	65	0.5	0.5	0.1	00:29:11	0.066594	
15	150	70	0.5	0.4	0.1	00:54:30	0.068396	
16	55	45	0.5	0.6	0.1	00:12:50	0.073958	
17	80	50	0.5	0.6	0.1	00:22:18	0.076842	
18	150	70	0.5	0.4	0.1	00:54:48	0.196999	
19	95	65	0.5	0.5	0.1	00:30:56	0.197443	
20	55	45	0.5	0.6	0.1	00:37:22	0.203600	
21	55	45	0.5	0.6	0.1	00:12:16	0.213285	
22	200	90	0.5	0.4	0.1	01:33:19	0.065268	from -10 to 10
23	**120**	**85**	**0.5**	**0.4**	**0.1**	**00:52:13**	**0.070636**	
24	90	35	0.5	0.5	0.2	00:16:26	0.074058	
25	65	35	0.5	0.4	0.2	00:11:53	0.076711	
26	55	45	0.5	0.6	0.1	00:39:01	0.077597	
27	100	40	0.5	0.3	0.1	00:20:41	0.078355	
28	40	25	0.5	0.7	0.1	00:05:12	0.134507	
29	25	15	0.5	0.8	0.3	00:02:15	0.261787	

Figure 9 shows the evolution of the genetic algorithm and figure 10 shows the membership functions of FLC obtained by GA.

Figure 11 show the control result of the plant 2 using the FLC optimized obtained by GA.

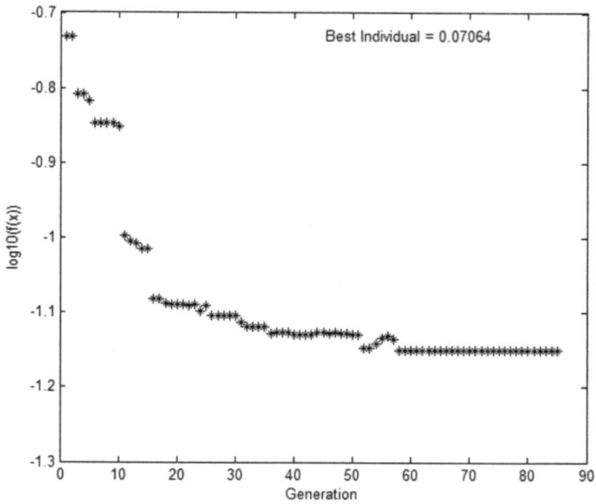

Fig. 9. Evolution of the GA for the FLC optimization.

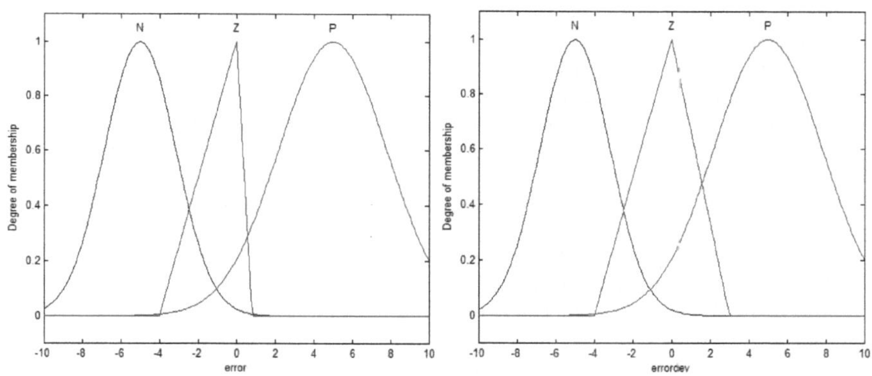

Fig. 10. Membership functions of the FLC obtained with the GA.

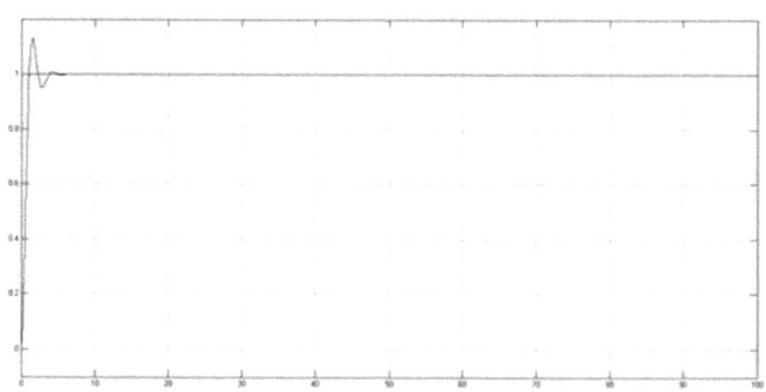

Fig. 11. Simulation result of the control Plant 2.

5 Conclusions

We described the use of evolutionary methods for the design of optimized FLC's; in particular we present results of a genetic algorithm in FLC optimization for linear plants. The results of first plant are satisfactory controlling the plant and getting stability in less than 10 seconds using a fuzzy logic controller with three membership functions and nine fuzzy rules. On the other hand, in second plant we can get stability faster that the first plant close to five seconds, but with a high overshoot and undershoot for the plant. We have achieved satisfactory results with genetic algorithms and the next step is to solve the problem using multiple objective optimizations to obtain better results. Moreover, we will extend the results for nonlinear systems like autonomous mobile robots.

References

1. Alcalá, R., Alcalá-Fdez, J., Herrera, F.: A proposal for the Genetic Lateral Tuning of Linguistic Fuzzy Systems and its Interaction with Rule Selection. IEEE Transactions on Fuzzy Systems 15(4), 616–635 (2007)
2. Alcalá, R., Gacto, M.J., Herrera, F., Alcalá-Fdez, J.: A Multi-objective Genetic Algorithm for Tuning and Rule Selection to Obtain Achúrate and Compact Linguistic Fuzzy Rule-Based Systems. International Journal of Uncertainty, Fuzziness and Knowledge-Based Systems 15(5), 539–557 (2007)
3. Angeline, P.J.: Using Selection to Improve Particle Swarm Optimization. In: Proceedings 1998 IEEE World Congress on Computational Intelligence, Anchorage, Alaska, pp. 84–89. IEEE, Los Alamitos (1998)
4. Angeline, P.J.: Evolutionary Optimization Versus Particle Swarm Optimization: Philosophy and Performance Differences. In: Porto, V.W., Waagen, D. (eds.) EP 1998. LNCS, vol. 1447, pp. 601–610. Springer, Heidelberg (1998)
5. Bentalba, S., El Hajjaji, A., Rachid, A.: Fuzzy Control of a Mobile Robot: A New Approach. In: Proc. IEEE Int. Conf. on Control Applications, Hartford, CT, October 1997, pp. 69–72 (1997)
6. Brockett, R.W.: Asymptotic stability and feedback stabilization. In: Millman, R.S., Sussman, H.J. (eds.) Differential Geometric Control Theory, pp. 181–191. Birkhauser, Boston (1983)
7. Casillas, J., Cordon, O., del Jesús, M.J., Herrera, F.: Genetic Tuning of Fuzzy Rule Deep Structures Preserving Interpretability and its Interaction with Fuzzy Rule Set Reduction. IEEE Transaction on Fuzzy Systems 13(1), 13–29 (2005)
8. Chi, Z., Yan, H., Pham, T.: Fuzzy Algorithms: With Applications to Image Processing and Pattern recognition. World Scientific, Singapore (1996)
9. Cordon, O., Gomide, F., Herrera, F., Hoffmann, F., Magdalena, L.: Ten Years of Genetic Fuzzy Systems: Current Framework and New trends. Fuzzy Sets and Systems 141(1), 5–31 (2004)
10. DeJong, K.: Learning with genetic algorithms: an overview. Mach. Learning 3(3), 121–138 (1988)
11. Driankov, D., Hellendoorn, H., Reinfrank, M.: An Introduction to Fuzzy Control. Springer, Berlin (1993)
12. Duc Do, K., Zhong-Ping, J., Pan, J.: A global output-feedback controller for simultaneous tracking and stabilizations of unicycle-type mobile robots. IEEE Trans. Automat. Contr. 30, 589–594 (2004)

13. Eberhart, R.C., Kennedy, J.: A new optimizer using particle swarm theory. In: Proceedings of the Sixth International Symposium on Micromachine and Human Science, Nagoya, Japan, pp. 39–43 (1995); Lu, J.-G.: Title of paper with only the first word capitalized. J. Name Stand. Abbrev. (in press)
14. Fierro, R., Lewis, F.L.: Control of a Nonholonomic Mobile Robot Using Neural Networks. IEEE Trans. on Neural Networks 9(4), 589–600 (1998)
15. Fogel, D.B.: An introduction to simulated evolutionary optimization. IEEE transactions on neural networks 5(1), 3–14 (1994)
16. Fukao, T., Nakagawa, H., Adachi, N.: Adaptive Tracking Control of a NonHolonomic Mobile Robot. IEEE Trans. on Robotics and Automation 16(5), 609–615 (2000)
17. Kennedy, J., Mendes, R.: Population structure and particle swarm performance. In: Proceeding of IEEE conference on Evolutionary Computation, pp. 1671–1676 (2002)
18. Kennedy, J., Mendes: The particle swarm-explosion, stability, and convergence in a multidimensional complex space. IEEE Transactions on Evolutionary Computation 6(1), 58–73 (2002)
19. Lee, T.H., Leung, F.H.F., Tam, P.K.S.: Position Control for Wheeled Mobile Robot Using a Fuzzy Controller, pp. 525–528. IEEE, Los Alamitos (1999)
20. Liang, Q., Mendel, J.M.: Interval Type-2 Fuzzy Logic Systems: Theory and Design. IEEE Trans. on Fuzzy Systems 8(5), 535–550 (2000)
21. Mendel, J., Mouzouris George, C.: Type-2 fuzzy logic systems. IEEE Trans. Fuzzy Systems 7, 643–658 (1999)
22. Mendel, J., Bob John, R.I.: Type-2 Fuzzy Sets Made Simple. IEEE Transactions on Fuzzy Systems 10(2) (April 2002)
23. Mendel, J.: Uncertain Rule-Based Fuzzy Logic Systems: Introduction and new directions. Prentice Hall, USA (2000)
24. Pedrycz, W. (ed.): Fuzzy Modelling: Paradigms and Practice. Kluwer Academic Press, Dordrecht (1996)
25. Sepulveda, R., Castillo, O., Melin, P., Montiel, O.: An Efficient Computational Method to Implement Type-2 Fuzzy Logic in Control Applications, Analysis and Design of Intelligent Systems Using Soft Computing Techniques. Advances in Soft Computing 41, 45–52 (2007)
26. Takagi, T., Sugeno, M.: Fuzzy Identification of Systems and its application to modeling and control. IEEE Transactions on Systems, Man, and Cybernetics 15(1) (1985)
27. Veeramachaneni, K., Osadciw, L., Yan, W.: Improving Classifier Fusion Using Particle Swarm Optimization. In: IEEE Fusion Conference, Italy (July 2006)
28. Wei, W., Jiatao, S., Zhongzxiu, Y., Zheru: Wavelet-based Illumination Compensation for Face Recognition using Eigenface Method Intelligent Control and Automation. In: WCICA 2006, June 21-23, vol. 2, pp. 10356–10360 (2006)

Multi-Agent System with Fuzzy Logic Control for Autonomous Mobile Robots in Known Environments

Cinthya Solano-Aragón[1] and Arnulfo Alanis[2]

[1] Student of Master of Science in Computer Science, Instituto Tecnologico de Tijuana de Tijuana, Calzada Tecnologico S / N, Tijuana, Mexico
cinthyasolanoa@gmail.com

[2] Division of Graduate Studies and Research, Instituto Tecnologico de Tijuana, Calzada Tecnologico, S/N, Tijuana, Mexico
alanis@tectijuana.edu.mx

Abstract. This paper describes the development of a Multi-Agent System (MAS), which is supported with fuzzy logic (to control the robots movements in a reactive path) and vision, which controls an autonomous mobile robot to exit a maze. The research consists of two stages. In the first stage the problem is to be able to make the robot exit a maze, the mobile robot is positioned at the entrance (point A) and should reach an output (B). It should be noted that we are working with a NXT Lego MINDSTORMS robot. In its second phase the problem is to make the robot search for a recognized object, for this, a camera is used to capture images, which will be processed with vision techniques, for their identification, and after that, the SMA takes the decision to evade or take the object as appropriate.

1 Introduction

Making smarter robots is a problem that has captivated the scientific community for several years now, this being a big challenge to overcome even for man. To move from one place to another from an initial to a final point with only the information of what we want to reach at the end is a complicated task. The problem has been solved with soft computing methods, such as fuzzy logic, neural networks, genetic algorithms, hybrid systems, among others. Navigation of the robot has been achieved taking a completely controlled environment where we have total knowledge of the world [8, 11, 15, 14].

Human beings have the ability to react to unexpected situations and the ability to use their reasoning to react to situations that arise, an example is a situation of traffic management, to which we think that we have the best route and with this the rest of the trip is done reacting to what was observed. This could be, at a traffic light, when it changes from green to yellow, the reaction will be learning, acceleration or deceleration, a situation that when you start, it has to be learned.

O. Castillo et al. (Eds.): Evolutionary Design of Intelligent Systems, SCI 257, pp. 33–52.
springerlink.com

Knowing that we have this ability to react, we will use fuzzy logic to transfer the knowledge we use to carry out this process by establishing computer-type fuzzy "if-then" rules and exploiting the advantages offered to us by fuzzy logic, which is the use of linguistic variables [26].

A man driving a car, as in many other activities, derives its information by looking at distances, clearances and other identifying information that we collect through the light. In this research we use vision techniques to draw from that vital information for navigation control, supported with special sensors. An example is the ultrasonic sensors that are used to measure distance and the light sensors that we can use to measure the change in light intensity, all of these help to extract environmental information necessary for decision making.

2 Agents

Let's first deal with the notion of intelligent agents. These are generally defined as "software entities", which assist their users and act on their behalf. Agents make your life easier, save you time, and simplify the growing complexity of the world, acting like a personal secretary, assistant, or personal advisor, who learns what you like and can anticipate what you want or need. The principle of such intelligence is practically the same of human intelligence. Through a relation of collaboration-interaction with its user, the agent is able to learn from himself, from the external world and even from other agents, and consequently act autonomously from the user, adapt itself to the multiplicity of experiences and change its behavior according to them. The possibilities offered for humans, in a world whose complexity is growing exponentially, are enormous [20][19][7][18].

We need to be careful to distinguish between rationality and omniscience. An omniscient agent knows the actual outcome of its actions, and can act accordingly; but omniscience is impossible in reality. Consider the following example: I am walking along the Champs Elys´ees one day and I see an old friend across the street. There is no traffic nearby and I'm not otherwise engaged, so, being rational, I start to cross the street. Meanwhile, at 33,000 feet, a cargo door falls off a passing airliner, and before I make it to the other side of the street I am flattened. Was I irrational to cross the street? It is unlikely that my obituary would read "Idiot attempts to cross street."Rather, this point out that rationality is concerned with expected success given what has been perceived. Crossing the street was rational because most of the time the crossing would be successful, and there was no way I could have foreseen the falling door. Note that another agent that was equipped with radar for detecting falling doors or a steel cage strong enough to repel them would be more successful, but it would not be any more rational [23].

3 FIPA

FIPA (The Foundation of Intelligence Physical Agents) specifications represent a collection of standards, which are intended to promote the interoperation of heterogeneous agents and the services that they can represent. The life cycle [46] of specifications details what stages a specification can attain while it is part of the FIPA standards process. Each specification is assigned a specification identifier [47] as it enters the FIPA specification life cycle. The specifications themselves can be found in a Repository [11]. The Foundation of Intelligent Physical Agents (FIPA) is now an official IEEE Standards Committee. The UML modeling language is the most popular system used today and is a graphic language to visualize, specify, build and document a system. Gaia is a methodology for application development in the paradigm of agents, one of the most curious aspects of Gaia, is the fact that the requirements specification is completely independent of the analysis and design.

4 Proposed Method

To realize this work, the creation of a SMA is needed, which is composed of three agents, an agent for the management of sensors, an agent for handling the drive and a last agent to act as coordinator of the multi-agent system.

Summarizing the SMA is composed of the following agents:

Node Agent (NA), a set of sensors.
Task Agent (TA), a set of servo motors.
Agent System (AS), coordinator of the system [1].

The Node Agent (NA) will be responsible for the sensors used by the robot, which are:

- Ultrasonic Sensor
- Light Sensor left
- Light Sensor right
- Camera

These sensors are used to interact with the world to achieve our goal.

The Task Agent (TA) will be used to drive to move forward, rewind, or make some movement to escape.

The System Agent (SA) will be responsible for coordinating the other actors, the NA and TA, and will also have activities not only the coordination, which will be to recognize the object that is opposite, calculate the angle of the object, which will help us to know that will guide the robot. Depending on the recognition that was obtained, whether or not the object, which will tell us that there is no escape or the object based on this decision was to trigger the switch which we will control the outcome to be used, reactive or path to better performance of the controllers.

In Fig. 1 we show graphically the architecture of the complete system.

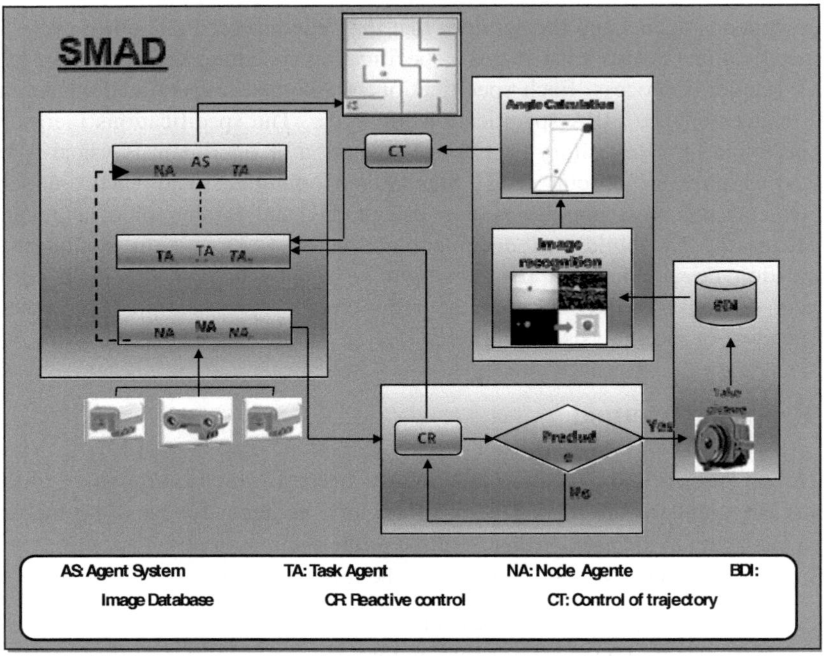

Fig. 1. Architecture of the complete system.

We can describe the proposed method as follows:

In the operation of the model, we have 2 main blocks, which are responsible for knowledge and learning (with the paradigm of intelligent agents) and the vision and control (using fuzzy logic).

These modules are described below, the first module contains 3 agents, NA [Node Agent], TA [Task Agent] SA [System Agent] Agent System (AS), and will know every time the operation of the other agents.

To detect a change in the environment the Agent Node [that is in charge of the sensors (ultrasonic, camera and two light sensors)] with the ultrasonic sensor and two light sensors, starts its operation, and we start at the state called reactive control; this because you have to move and sense all the time until it finds an obstacle, this can happen once a photo is taken, which will be sent to the database of images (BDI), the image will be applied a pre-processing for recognition in order to know whether the object is found, this decision was taken on the angle (angle to take the decision to bypass the object), able to escape and move forward, trend data, they are caught in the trajectory control process, to observe the speed values, the Task Agent (which is responsible for the drive), once all the parameters necessary for the performance of the robot agent system (AS) who is the coordinator, and executes instructions in the robot.

For the development of the two key tools we used Gaia and UML for the analysis and design of agents. The completion of the Multi-Agent System (MAS) is based on the FIPA standards for better utilization and greater possibility of extension.

5 Analysis and Design of the Multi-Agent System (MAS)

The first activity was the analysis and design of the multi-agent system as mentioned above and to develop the two tools and we describe as follows.

5.1 UML

Use Case Diagrams

The Use Case diagrams show the granularity of the system into reusable pieces of functionality, interaction of players with the functionality of the system, visually organize user requirements and allow the contract to certify the functionality, formalize the process map.

Fig. 2 presents the use case of the sense in which the distance is appreciated that all other processes must be carried out before and after to complete. In Fig. 3 we show the use case of the motor A move, which is seen in all other processes that must be carried out before and after to complete [2][3][4][5][6].

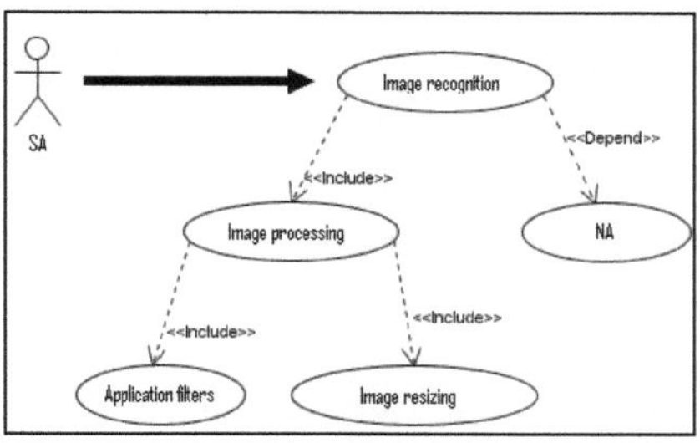

Fig. 2. Use case diagram of the process of image recognition.

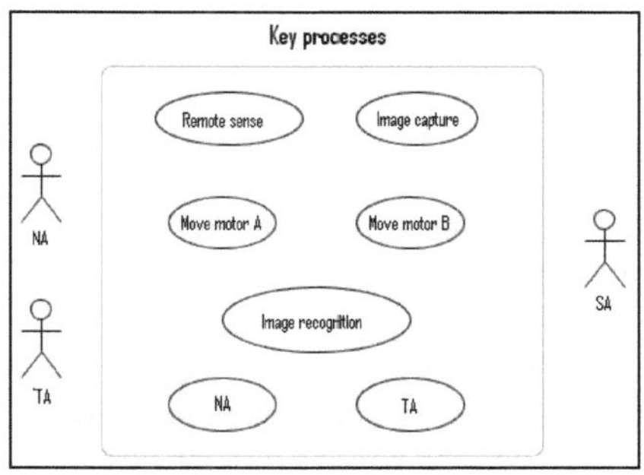

Fig. 3. Main processes of use cases.

Fig. 4 presents the case for use of the remote sensing process, which shows that other processes must be carried out before and after to complete. Fig. 5 shows the use case of the motor A move, which is seen in all other processes must be carried out before and after to complete.

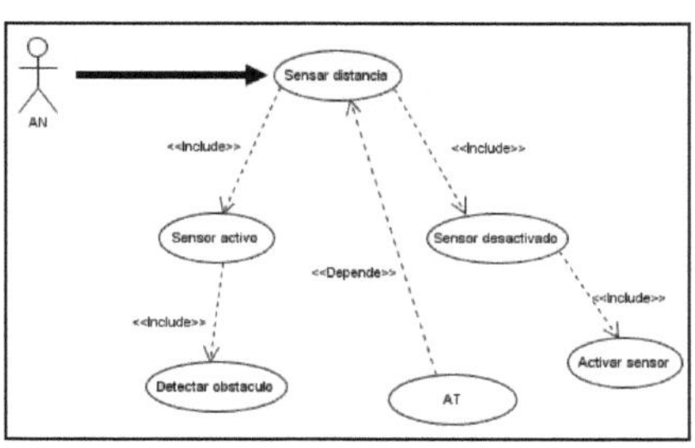

Fig. 4. Use case diagram of the remote sense.

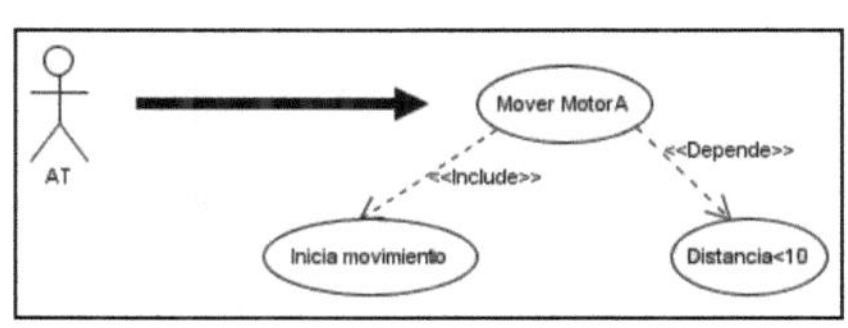

Fig. 5. Use case diagram of the process moving motor A.

Sequence diagrams describe the interaction of objects that require the functionality of the different scenarios of a use case, objects are represented with their life cycle within a time series, and each possible scenario of a use case can be represented as a sequence diagram.

Below in Fig. 6 we show the sequence diagram of the use case moving motor A. In Fig. 7 presents the sequence diagram of use case to capture the image which is carried out by the agent node (AN).

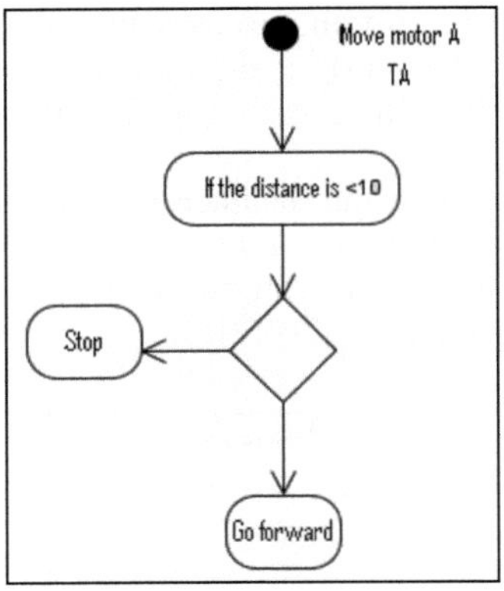

Fig. 6. Sequence diagram use case move motor A.

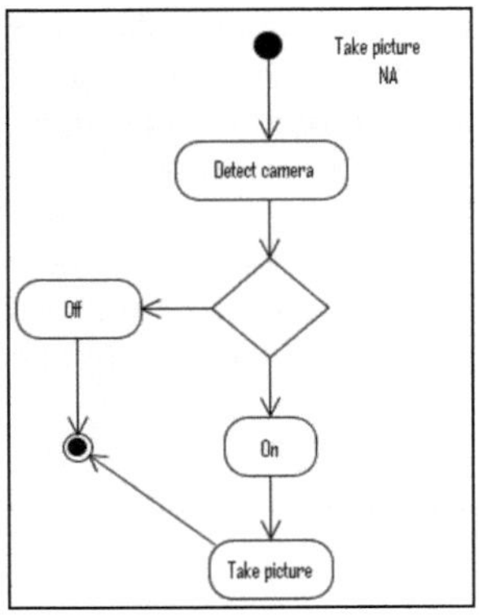

Fig. 7. Sequence diagram use case capture image.

In Fig. 8 presents the sequence diagram used for the case sense of distance, which is carried out by the agent node (AN).

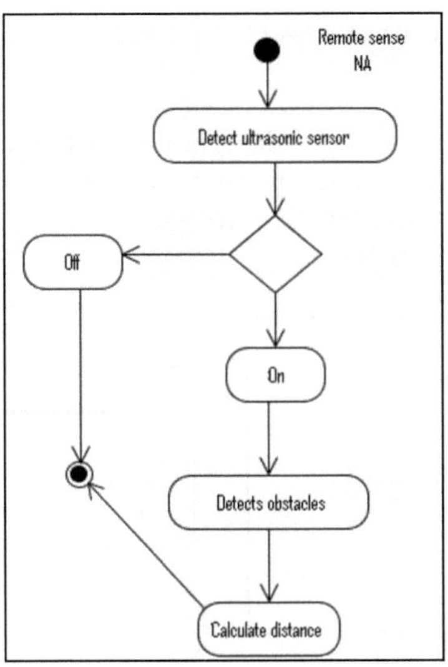

Fig. 8. Sequence diagram use case sense distance.

5.2 Gaia

Below we show the diagrams in which we can observe the responsibilities, permissions, activities and protocols to be followed by each agent in the Multi-Agent System (MAS).

Agent node

Responsibilities:

Sense:

Active

Inactive

Permissions:

Ultrasonic sensor: measures distances from 0 cm to 255 cm, with a delay of one millisecond signal.

Camera Sensor: take pictures in an estimated time of seconds, saves all images in a folder for processing. It has a degree of vision approximately of 50 degrees.

Light sensors: they measure the intensity with which an object reflects light. This makes a light emitting and measuring the portion of the return is received, the table 3.2 shows the ranges of values.

Activities:

The only activities are sensing and taking pictures of the world.

Protocols:

Ultrasound: This is responsible for sensing the distance to get to an obstacle. The initiative for this would be the agent system, which indicates the time of initiation.

Chamber is responsible for obtaining the images within the scene, and as for the ultrasonic sensor system depends for its initiation.

Light sensors: the role of these sensors is to measure the intensity with which an object reflects light (walls, diagrams). This makes a light emitting and measuring the portion of the return is received and handled the ranges are 0 to 1023 RAW.

Agent system

Responsibilities:

It is responsible for image processing and decision making, and sends a message to agent communication process completed?

Permissions:

This is the one that has full access, sensors, communication, engine and handling agents and NXT in general.

Activities:

Making decisions based on the behavior and implements what is needed to finish the job.

Protocols:

Full Access, sensors, communication, engine and handling agents and NXT in general, the entry for this information would be provided by the agents, getting a response from these so they generate an output and generate interaction among them.

Task agent

Responsibilities:

Motor movement

Permission:

Depends entirely on the agent system to perform its task.

Activities:

Sensors for the engine: they measure in degrees is given for each engine is 360 degrees for one revolution, a revolution for the tire of each motor, the motors can rotate independently, can also be used with constant speed motors.

Protocols:

The system is the initiator, resulting in a movement or departure, which can be any direction, depending on the decision taken by the agent system.

Figure 9 shows the three models that form the multi-agent system (MAS) using Gaia.

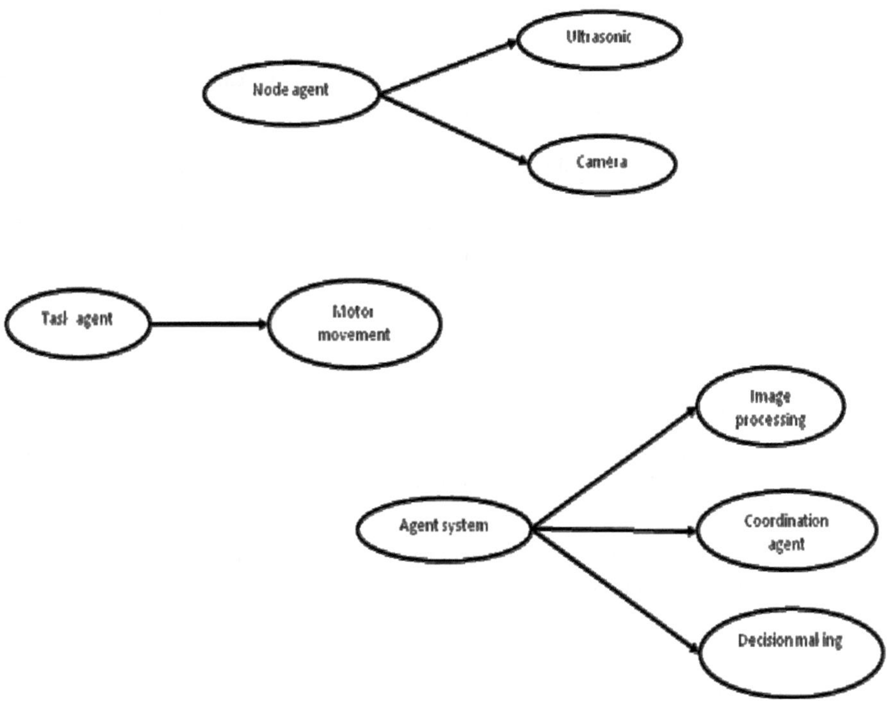

Fig. 9. Shows the diagrams of the Agent Node (AN), Task Agent (TA), and Agent System (AS) in Gaia.

Fig. 10. Diagram of the relationship model of the SMA.

Fig. 10 presents the relationship between the agents of the multi-agent system (MAS).

5.3 Knowledge Base

Below we show the knowledge base that the MAS uses to function properly.

Facts

NA_i.SU. STATE =(X), where X can have 2 states (ON, OFF)
NA_i.SU.OBJ=(Y) where Y can have 2 states (DETECTS, NO DETECTS)
NA_i.SC. STATE = (Z) where Z can have 2 states (ON, OFF)
NA_i.SC.TIMAGE
NA_i.SLI. STATE = (W) where W can have 2 states (ON, OFF)
NA_i.SLD. STATE = (V) where V can have 2 states (ON, OFF)
TA_j.M.GO(evaluationFIS[SU, SLI, SLD]) where SU,SLI,SLD FIS are entries for return rates of both servomotors (MA, MB)
TA_j.M.STOP
TA_j.MA.GO(evaluationFIS[SU, SLI, SLD]) where SU,SLI,SLD FIS are entries for return rates of both servomotors (MA, MB)
TA_j.MA.DECREASEV
TA_j.MB.GO(evaluationFIS[SU, SLI, SLD]) where SU,SLI,SLD FIS are entries for return rates of both servomotors (MA, MB)
TA_j.MB. DECREASEV
SA. BEGINTASK
SA.RECONOCIMAGE.STATE = (N) where N can have 2 states (YES, NO)
SA.ACTION.TAKEFIGURE
SA.ACTION.AVOID
SA.FIGUREFOUND
SA.ENDMAZE = (M) where M can have 2 states (YES, NO)
SA.ENDTASK

Rules

If SA. BEGINTASK and NA_i.*SU.STATE=OFF* then NA_i.*SU.STATE=ON*
If SA. BEGINTASK and NA_i.*SC.STATE=OFF* then NA_i.*SC.STATE=ON*
If SA. BEGINTASK and NA_i.*SLI.STATE=OFF* then NA_i.*SLI.STATE=ON*
If SA. BEGINTASK and NA_i.*SLD.STATE=OFF* then NAi.*SLD.STATE=ON*
If SA.ENDTASK and NA_i.*SU.STATE=ON* then NA_i.*SU.STATE=OFF*
If SA.ENDTASK and NA_i.*SC.STATE=ON* then NA_i.*SC.STATE=OFF*
If SA.ENDTASK and NA_i.*SLI.STATE=ON* then NA_i.*SLI.STATE=OFF*

If SA.ENDTASK and NA$_i$.*SLD.STATE=ON* then NA$_i$.*SLD.STATE=OFF*

If SA. BEGINTASK and NA$_i$.*SU.STATE=ON* then TA$_j$.M.GO(evaluationFIS[SU, SLI, SLD])

If TA$_j$.M.GO(evaluationFIS[SU, SLI, SLD]) and NA$_i$.*SU.OBJ=* DETECTS then NA$_i$.*SC.TIMAGE*

If NA$_i$.*SC.TIMAGE* then SA.RECONOCIMAGE

If SA.RECONOCIMAGE.STATE = SI then SA.ACTION.TOMAFIG

If SA.RECONOCIMAGE.STATE = NO then SA.ACTION.EVADIR

If SA.ACTION.TAKEFIGURE then TA$_j$.MA.GO(evaluationFIS[SU, SLI, SLD]) and TA$_j$.MB.GO(evaluationFIS[SU, SLI, SLD])

If SA.ACTION.AVOID then TA$_j$.MA.GO(evaluationFIS[SU, SLI, SLD]) and TA$_j$.MB.GO(evaluationFIS[SU, SLI, SLD])

If SA.ACTION.TAKEFIGURE and SA.ENDMAZE = YES then SA.ENDTASK

If SA.ACTION.TAKEFIGURE and SA.ENDMAZE = NO then TA$_j$.MA.GO(evaluationFIS[SU, SLI, SLD]) and TA$_j$.MB.GO(evaluationFIS[SU, SLI, SLD])

If SA.ACTION.AVOID and SA.ENDMAZE = YES then TA$_j$.MA.GO(evaluationFIS[SU, SLI, SLD]) and TA$_j$.MB.GO(evaluationFIS[SU, SLI, SLD])

6 Simulation

Figure 11 shows the plant and the controller for the simulations that were carried out in Simulink of Matlab 2007b.

The controller is based on the following fuzzy system, which contains the rules that work with the robot.

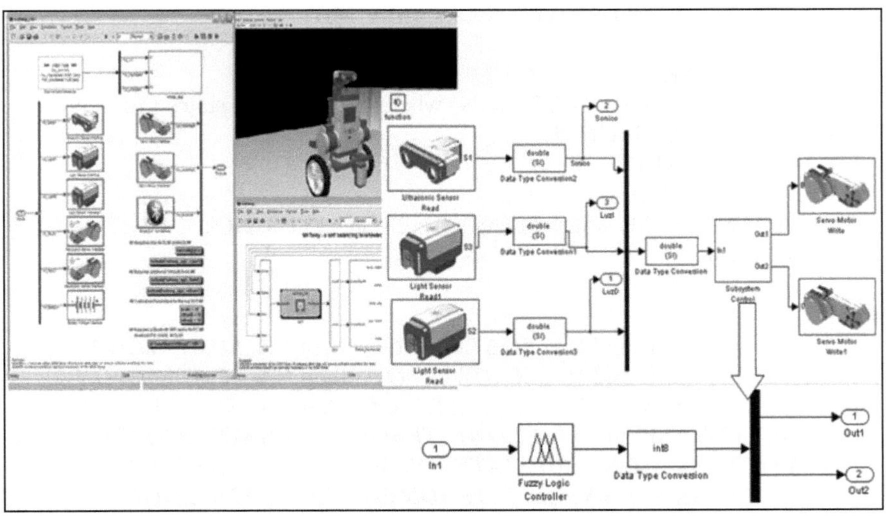

Fig. 11. Plant for performing simulations.

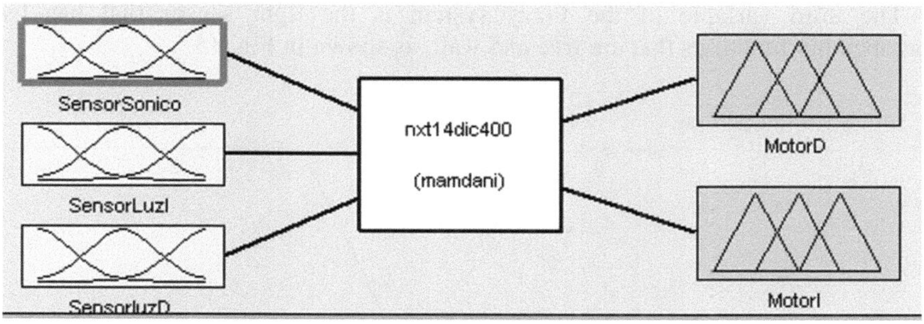

Fig. 12. Structure of the fuzzy system that works for the simulations.

The fuzzy system has three inputs, two outputs and consists of ten rules for inference.

The first input variable of the fuzzy system is the ultrasonic sensor, which has three membership functions, which are linguistic (close, near, far), as shown in Fig. 13.

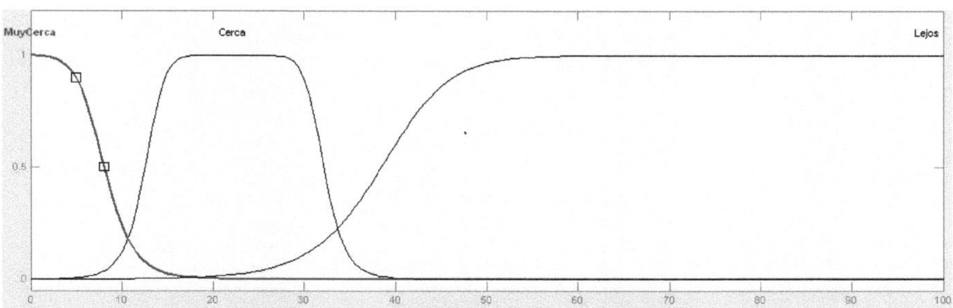

Fig. 13. Membership functions for the ultrasonic sensor.

The second variable of the fuzzy system is the light sensor that has two membership functions that are free and wall, as shown in Fig. 14.

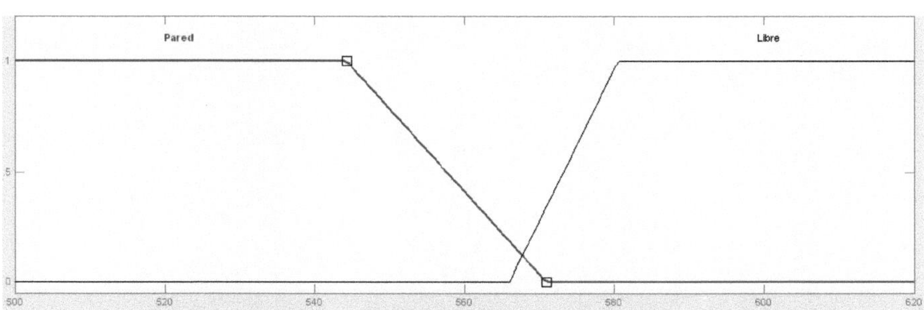

Fig. 14. Membership functions of the left light sensor.

The third variable of the fuzzy system is the light sensor that has two membership functions that are free and wall, as shown in Fig. 15.

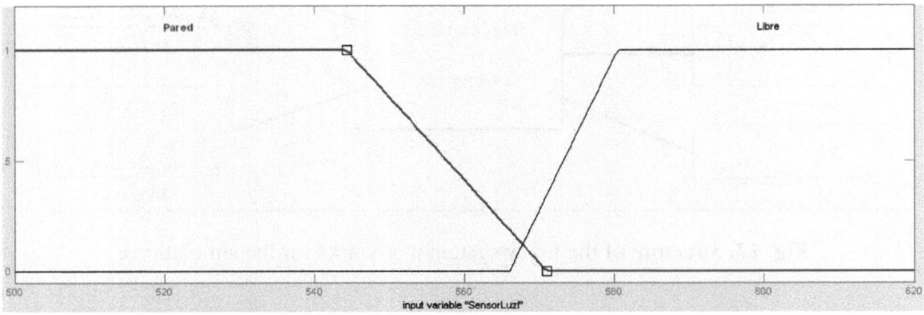

Fig. 15. Membership functions of the right light sensor.

This is the first output variable is the speed of the left engine, has five membership functions as shown in Fig. 16.

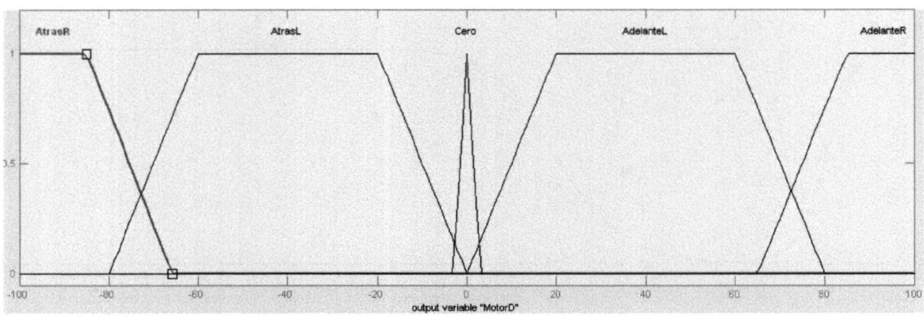

Fig. 16. Membership functions of the left motor.

This is the second output variable is the speed of the right engine, has five membership functions as shown in Fig. 17.

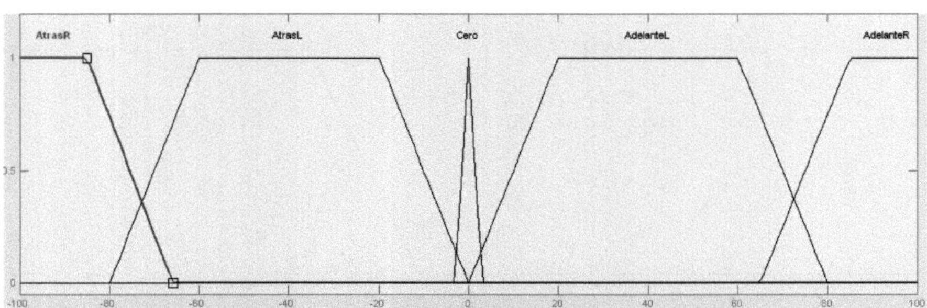

Fig. 17. Membership functions of the right motor.

In Fig. 18 we show the rules that are used in the system, which are 10 fuzzy rules.

1. If (SensorSonico is Lejos) and (SensorLuzI is Pared) and (SensorluzD is Pared) then (MotorD is AdelanteR)(MotorI is AdelanteR) (1)
2. If (SensorSonico is Cerca) and (SensorLuzI is Pared) and (SensorluzD is Pared) then (MotorD is AdelanteL)(MotorI is AdelanteL) (1)
3. If (SensorSonico is MuyCerca) and (SensorLuzI is Pared) and (SensorluzD is Pared) then (MotorD is AdelanteL)(MotorI is AdelanteL) (1)
4. If (SensorSonico is Lejos) and (SensorLuzI is Libre) and (SensorluzD is Pared) then (MotorD is AdelanteR)(MotorI is AtrasL) (1)
5. If (SensorSonico is Cerca) and (SensorLuzI is Pared) and (SensorluzD is Libre) then (MotorD is AtrasL)(MotorI is AdelanteR) (1)
6. If (SensorSonico is MuyCerca) and (SensorLuzI is Pared) and (SensorluzD is Libre) then (MotorD is AtrasL)(MotorI is AdelanteR) (1)
7. If (SensorSonico is Lejos) and (SensorLuzI is Pared) and (SensorluzD is Libre) then (MotorD is AtrasL)(MotorI is AdelanteR) (1)
8. If (SensorSonico is Cerca) and (SensorLuzI is Libre) and (SensorluzD is Libre) then (MotorD is AdelanteR)(MotorI is Cero) (1)
9. If (SensorSonico is Lejos) and (SensorluzD is Pared) then (MotorD is AdelanteL)(MotorI is AdelanteL) (1)
10. If (SensorSonico is MuyCerca) and (SensorLuzI is Libre) and (SensorluzD is Pared) then (MotorD is AdelanteR)(MotorI is AtrasL) (1)

Fig. 18. Rules of the fuzzy system (FIS).

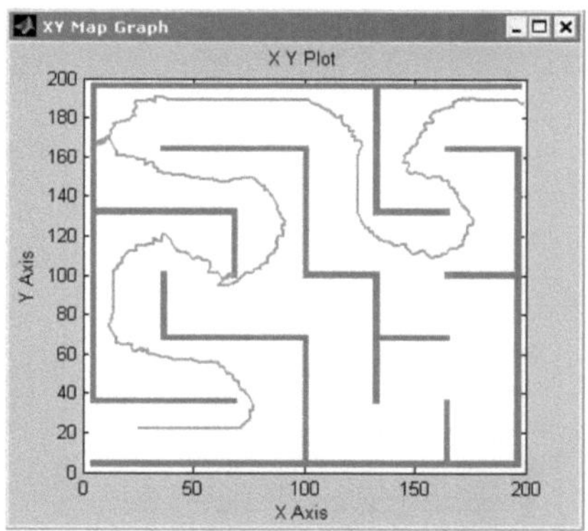

Fig. 19. Simulation1 with a duration of 00:63 minutes.

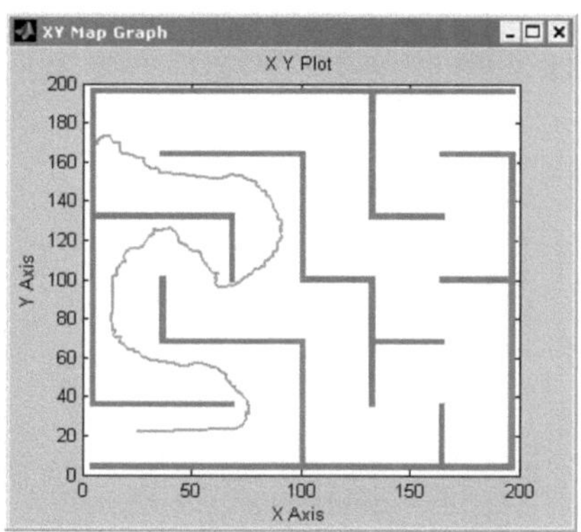

Fig. 20. Simulation2 with a duration of 01:30 minutes.

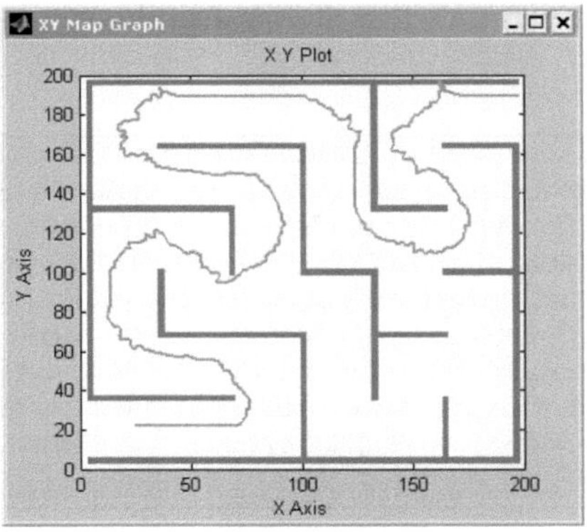

Fig. 21. Simulation3 with a duration of 00:51 minutes.

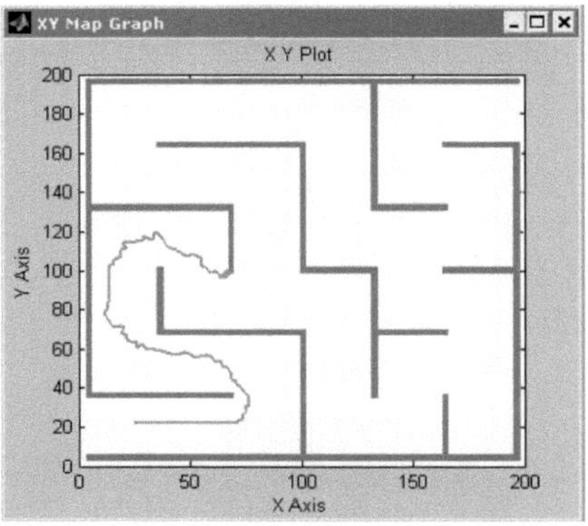

Fig. 22. Simulation4 with a duration of 00:50 minutes.

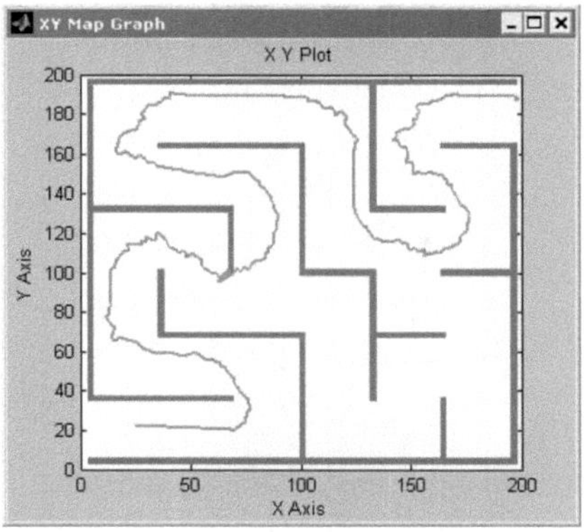

Fig. 23. Simulation5 with a duration 00:50 minutes.

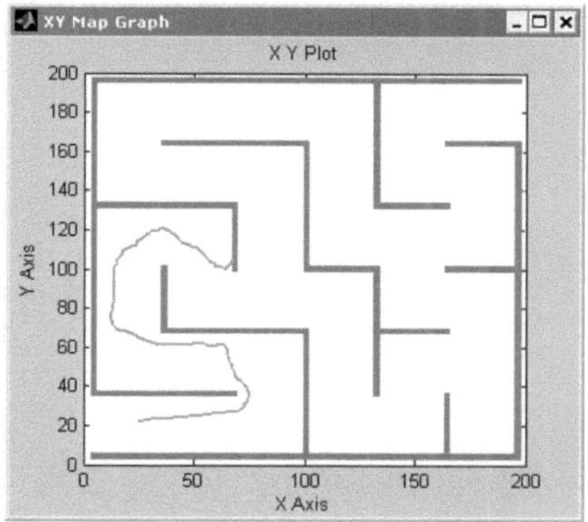

Fig. 24. Simulation6 with a duration of 00:17 minutes.

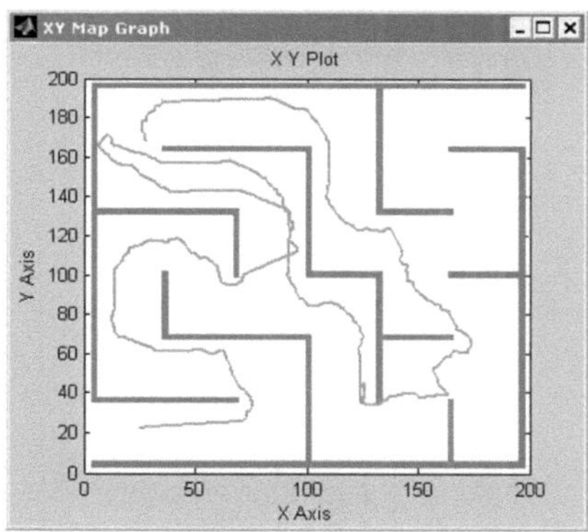

Fig. 25. Simulation7 with a duration of 00:57 minutes.

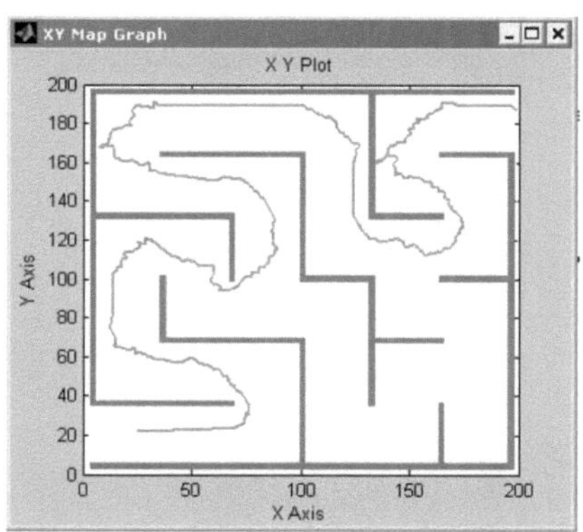

Fig. 26. Simulation8 with a duration of 00:52 minutes.

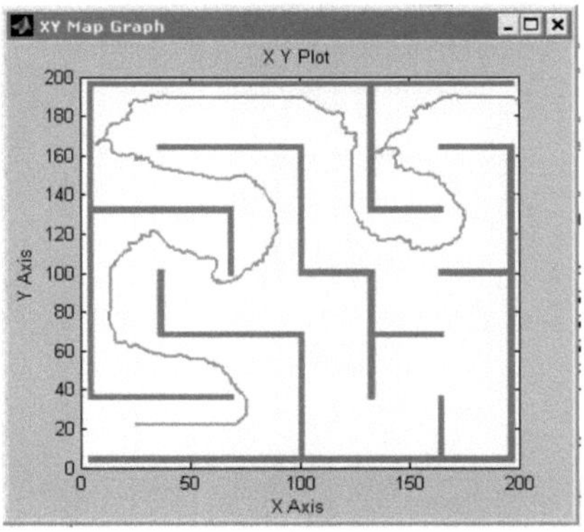

Fig. 27. Simulation9 with a duration of 00:52 minutes.

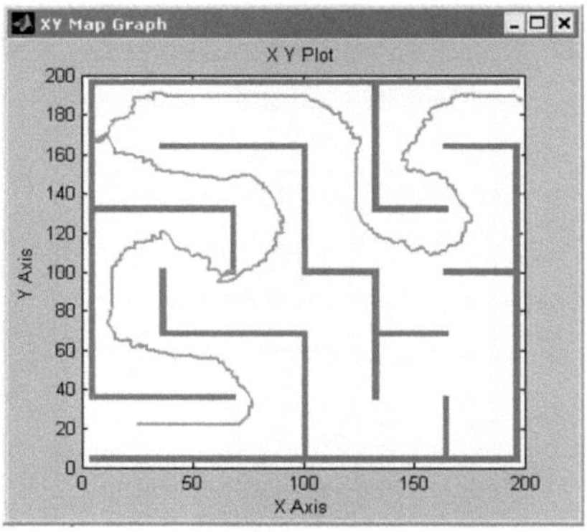

Fig. 28. Simulation10 with a duration of 00:51 minutes.

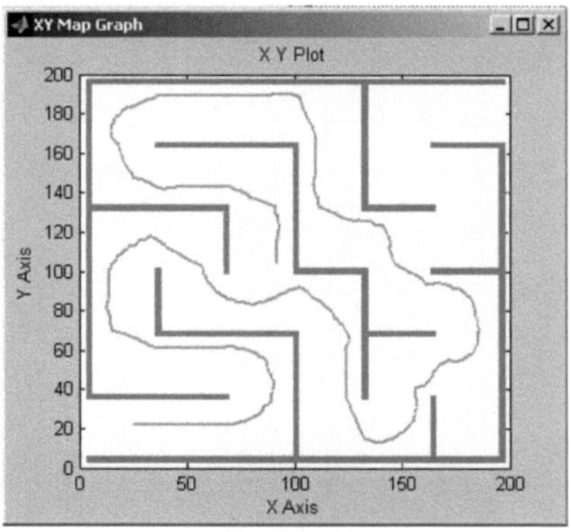

Fig. 29. Simulation11 with a duration of 01:25 minutes.

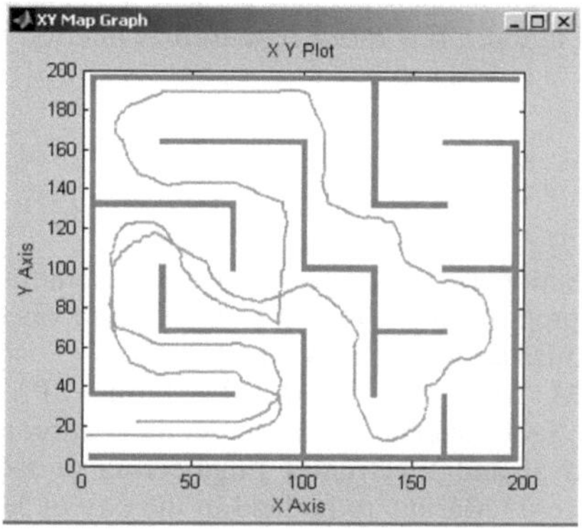

Fig. 30. Simulation12 with a duration 01:46 minutes.

Table 1 contains the information from the simulations regarding the duration that each one had, and whether or not out the robot was able to get out of the labyrinth, it also highlights the best simulation time.

Table 1. A comparison between the simulations.

Number of Simulation	Duration/minutes	Exit the maze
1	00:63	Yes
2	1:30	No
3	00:51	Yes
4	00:50	No
5	00:50	Yes
6	00:17	No
7	00:57	No
8	00:52	Yes
9	00:52	Yes
10	00:51	Yes
11	1:25	No
12	1:46	No

7 Conclusions

There is a big diversity of intelligent applications, in this particular case, the implementation of multi-agent systems is complex because it is short, and it incorporates other techniques, making it even more complex.

One of the main objectives of this research initially was to open a new view of intelligent agents using fuzzy logic, which hitherto has not been given the approach that is intended here. Fuzzy logic gives a degree of utilization of this paradigm of intelligent agents, but more research is needed in order to defend this idea as innovative.

References

[1] Alanis, A.: Contribution to the design of Fault Tolerant Distributed Systems for industrial control: proposal for a new paradigm based on Intelligent Agents, PhD, Universidad Politecnica de Valencia (November 2007)

[2] Alanis, A., López, M., Muñoz, R., Pulido, M., Martínez, P., Beltrán, M.: Multi-agent System for a Lego NXT robot to access the collaboration object 1.8 °. In: National Congress of Electrical and Electronics Engineering of the Mayab conieem 2008 Instituto Tecnológico de Mérida, to be held in the city of Merida, Yucatan, Mexico, April 21-25 (2008)

[3] Alanis, A., López, M., Muñoz, R., Pulido, M., Martínez, P., Beltrán, M.: Multi-agent system for a Lego NXT robot to access the collaboration of objects 1. In: XXIX International Congress of Engineering in electronics, electro 2008, Instituto Tecnológico de Chihuahua, division of graduate studies and research, Chihuahua, Chih. Mexico, October 29-31 (2008)

[4] Alanis, A., López, M., Parra, B., Serrano, M., Ayala, E., Solano, C.: Multi-agent system for search and object recognition using vision, in a Lego NXT robot 1.8 °. In: National Congress of Electrical and Electronics Engineering of the Mayab conieem 2008 Instituto Tecnológico de Mérida, to be held in the city of Merida, Yucatan, Mexico, April 21-25 (2008)

[5] Alanis, A., López, M., Parra, B., Serrano, M., Ayala, E., Solano, C.: Multi-agent system for search and object recognition using vision, in a Lego NXT robot 1. In: XXIX International Congress of Engineering in electronics, electro 2008 Instituto Tecnologico de Chihuahua, division of graduate studies and research, Chihuahua, Chih. Mexico, October 29, 31 (2008)

[6] Antoquia University, Faculty of Engineering, "Neural Networks" (June 2003)

[7] Bratko, I.: Prolog for Programming Artificial Intelligence. Addison-Wesley, Reading (1986)

[8] Cupertino, F., Giordano, V., Naso, D., Delfine, L.: Fuzzy control of a mobile robot. IEEE Robotics & Automation Magazine, 74–81

[9] Galván, R., Monsivais, A.: Resolution of mazes with Lego Mindstorms NXT with LabVIEW 7.1, thesis work ITT

[10] Dhananjay, G.S.: ANN Models for Speaker Recognition Based on Difference Cepstrals. Indian Institute of Technology, Madras (2000)

[11] Jang, J.-S.R., Sun, C.-T., Mizutani, E.: Neuro-Fuzzy and Soft Computing for computational approach to learning and machine intelligence. Prentice Hall, Englewood Cliffs (1997)

[12] Jang, Sun, J., Mizutani, J.: Neuro-Fuzzy and soft computing: a computational approach to learning and machine intelligence. Prentice-Hall, Englewood Cliffs (1997)

[13] López, A.: Program Development in Matlab, Matlab. no. Bulletin of the Cuban Society of Mathematics and Computer Science 2(2), 105–112 (2004)

[14] Maček, K., Petrović, I., Siegwart, R.: A control method for stable and smooth path following of mobile robots. In: Proceedings of the 2nd European Conference on Mobile Robots - ECMR 2005, Ancona, Italy, September 7-10, pp. 128–133 (2005)

[15] Leyden, M., Toal, D., Flanagan, C.: A Fuzzy Logic Based Navigation System for a Mobile Robot. In: Proceedings of Automatisierungs Symposium, Wismar, Germany (1999)
[16] Melendez, A.: Control and monitoring reagent for a mobile robot using fuzzy logic, thesis work, ITT (August 2008)
[17] Morgan, D.P., Scofield, C.L.: Neural Networks and Speech Processing. Kluwer Academic Publishers, Dordrecht (1991)
[18] Norving, P., Russell, S.: Artificial intelligence a modern approach. Prentice Hall, Australia (1996)
[19] Etzioni, O., Lesh, N., Segal, R.: Bulding for Softbots UNIX (preliminary report). Tech. Report 93-09-01. Univ. of Washington, Seattle (1993)
[20] Cohen, P.R., et al.: An Open Agent Architecture. In: Working Notes of the AAAI Spring symp.: Software Agent, pp. 1–8. AAAI Press, Cambridge (1994)
[21] Russell, S., Norvig, P.: Intelligent Agent. In: Artificial Intelligence to Modern Aproach. Pretence artificial Hall series in intelligence, pp. 31–52 (1994)
[22] Sánchez, O.: Voice and image recognition by neural networks to guide a robot, thesis work ITT (October 2008)
[23] Suárez, J.: Location and tracking of trajectories with robots Wayfarer in natural surroundings, thesis work Universidad Complutense de Madrid
[24] University of Vigo, Department of Computer Science, Problems of search and optimization (October 2004)
[25] Zadeh, L.A.: Outline of a new approach to the analysis of complex systems and decision processes. IEEE Transactions on Systems, Man, and Cybernetics 31(6), 891–901 (2001)
[26] Zadeh, L.A.: The concept of a linguistic variable and its application to approximate reasoning, Parts 1, 2, and 3, Information Sciences (1975)
[27] Intelligent Autonomous Robots are the new generation, September 3 (2008), http://www.tendencias21.net/Los-Robots-Inteligentes-Autonomos-son-la-nueva-generacion_a744.html
[28] Introduction to intelligent control, September 3 (2008), http://isa.umh.es/isa/es/asignaturas/cas/TRANSPCI.PDF
[29] What is Simulink?, September 4 (2008), http://voltio.ujaen.es/jaguilar/matlab/Manual20Matlab_Simulink/manual%%20simulink/SIM_01%20-%20Que_%20es.htm
[30] Simulation of a fuzzy system for controlling engine speed of a CD, September 8 (2008), http://www.uaem.mx/cicos/memorias/3ercic2004/Articulos/articulo3.pdf
[31] fuzzy systems, September 8 (2008), http://ants.dif.um.es/staff/juanbot/ml/files/20022003/fuzzy.pdf
[32] Control of local mobile robots based on statistical methods and genetic algorithms, October 1 (2008), http://www.dccia.ua.es/ria/artics/caepia97.pdf
[33] The NXT Bluetooth C + + library, October 6 (2008), http://www.norgesgade14.dk/bluetoothlibrary.php
[34] Embedded Coder Robot NXT, October 6 (2008), http://www.mathworks.com/matlabcentral/fileexchange/loadFile.do?objectId=13399
[35] Tools for programming free agents, October 13 (2008), http://torio.unileon.es/~vmo/pubs/waf00.pdf

[36] Analysis and design of reinforced learning agents, October 13 (2008), http://www.itcm.edu.mx/mccc06/publicaciones/Analisisydise%F1odeagentes.pdf
[37] An architecture for mobile robots small, October 13 (2008), http://igarrido.vtrbandaancha.net/papers/gbot_arquitectura.pdf
[38] Class diagrams, UML artifacts, October 22 (2008), http://www.vico.org/aRecursosPrivats/UML_TRAD/talleres/mapas/UMLTRAD_101A/LinkedDocuments/UML_diagClases.pdf
[39] Altamirano, R.L.: Laboratory of Computer Vision, http://ccc.inaoep.mx/labvision~/vcomp.htm
[40] Maldonado, O.: Republic of Knowledge, Computer Vision. p. 3, http://www.depi.itch.edu.mx/apacheco/expo/html/ai11/vision.html
[41] Salinas, R.: Neural Network Architecture Parametric Face Recognition, University of Santiago de Chile, http://cabierta.uchile.cl/revista/17/articulos/pdf/paper4.pdf
[42] http://www.fipa.org/specifications/lifecycle.html
[43] http://www.fipa.org/specifications/identifiers.html

Hybrid Interval Type-1 Non-singleton Type-2 Fuzzy Logic Systems Are Type-2 Adaptive Neuro-fuzzy Inference Systems

Gerardo M. Mendez[1] and Ma. De Los Angeles Hernandez[2]

[1] Instituto Tecnologico de Nuevo Leon Department of Electronics and Electrical Engineering Av. Eloy Cavazos 2001, Cd. Guadalupe, NL, Mexico, CP 67170
gmm_paper@yahoo.com.mx

[2] Instituto Tecnologico de Nuevo Leon Department of Economic Sciences Av. Eloy Cavazos 2001, Cd. Guadalupe, NL, Mexico, CP 67170
ahernandez@yturria.com.mx

Abstract. This article presents a new learning methodology based on a hybrid algorithm for interval type-1 non-singleton type-2 TSK fuzzy logic systems (FLS). Using input-output data pairs during the forward pass of the training process, the interval type-1 non-singleton type-2 TSK FLS output is calculated and the consequent parameters are estimated by the recursive least-squares (RLS) method. In the backward pass, the error propagates backward, and the antecedent parameters are estimated by the back-propagation (BP) method. The proposed hybrid methodology was used to construct an interval type-1 non-singleton type-2 TSK fuzzy model capable of approximating the behavior of the steel strip temperature as it is being rolled in an industrial Hot Strip Mill (HSM) and used to predict the transfer bar surface temperature at finishing Scale Breaker (SB) entry zone. Comparative results show the performance of the hybrid learning method (RLS-BP) against the only BP learning.

Keywords: Interval type-2 fuzzy logic systems, ANFIS, neuro-fuzzy systems, hybrid learning.

1 Introduction

Interval type-2 (IT2) fuzzy logic systems (FLS) constitute an emerging technology. In [1] both, one-pass and back-propagation (BP) methods are presented as IT2 Mamdani FLS learning methods, but only BP is presented for IT2 Takagi-Sugeno-Kang (TSK) FLS systems. The one-pass method generates a set of IF-THEN rules by using the given training data one time, and combines the rules to construct the final FLS. When BP method is used in both Mamdani and TSK FLS, none of antecedent and consequent parameters of the IT2 FLS are fixed at starting of training process; they are tuned using exclusively steepest descent method. In [1] recursive least-squares (RLS) and recursive filter (REFIL) algorithms are not presented as IT2 FLS learning methods.

O. Castillo et al. (Eds.): Evolutionary Design of Intelligent Systems, SCI 257, pp. 53–61.
springerlink.com

The aim of this work is to present and discuss a new hybrid learning algorithm for antecedent and consequent parameters tuning during training process for interval type-1 non-singleton type-2 TSK FLS (IT2 NSFLS-1 or IT2 NS-1 ANFIS). In the forward pass, the FLS output is calculated and the consequent parameters are tuned using RLS method. In the backward pass, the error propagates backward, and the antecedent parameters are tuned using the BP method.

The hybrid algorithm for IT2 Mamdani FLS has been already presented elsewhere [2, 3, 4] with three combinations of the learning method: RLS-BP, REFIL-BP and orthogonal least-squares-BP (OLS-BP), whilst the hybrid algorithm for singleton IT2 TSK FLS (IT2 ANFIS) has been presented elsewhere [5] with two combinations of the learning method: RLS-BP and REFIL-BP.

Since in the literature, only the BP learning method for IT2 FLS has been proposed, in this work the IT2 TSK NSFLS-1 system that uses the hybrid learning methodology has been developed and implemented for temperature prediction at hot strip mill (HSM) finishing scale breaker (SB) entry zone. This motivated by the success of the hybrid learning method in type-1 (T1) FLS (ANFIS) over BP only method.

Convergence has been practically tested for particular conditions; it is no the purpose of this work the generalization of the algorithm developed here, but only to show comparative results and feasibility of application. Mathematical proof is still to be done in general for hybrid learning algorithms.

2 Problem Formulation

Most of the hot strip mill processes are highly uncertain, non-linear, time varying and non-stationary [2, 6], having very complex mathematical representations. IT2 NS-1 ANFIS takes easily the random and systematic components of type A or B standard uncertainty [7] of industrial measurements. The non-linearities are handled by FLS as identifiers and universal approximators of nonlinear dynamic systems [8, 9, 10, 11]. Stationary and non-stationary additive noise is modeled as a Gaussian function centered at the measurement value. In stationary additive noise the standard deviation takes a single value, whereas in non-stationary additive noise the standard deviation varies over an interval of values [1]. Such characteristics make IT2 NS-1 ANFIS a powerful inference system to model and control industrial processes.

Only the BP learning method for IT2 TSK SFLS has been proposed in the literature and it is used as a benchmark algorithm for parameter estimation or systems identification [1]. To the best knowledge of the authors, IT2 NS-1 ANFIS has not been reported in the literature [1, 12].

One of the main contributions of this work is to implement an application of the IT2 NS-1 ANFIS using the hybrid RLS-BP learning algorithm, capable of compensates for uncertain measurements.

3 Problem Solution

3.1 Using Hybrid RLS-BP Method in IT2 NS-1 ANFIS Training

The Table 1 shows the activities of the one pass learning algorithm of BP method. Both, IT2 TSK SFLS (BP) and IT2 TSK NSFLS-1 (BP) outputs are calculated during forward pass. During the backward pass, the error propagates backward and the antecedent and consequent parameters are estimated using only the BP method.

Table 1. One pass in learning procedure for IT2 TSK (BP)

	Forward Pass	Backward Pass
Antecedent Parameters	Fixed	BP
Consequent Parameters	Fixed	BP

The proposed hybrid algorithm (IT2 NS-1 ANFIS) uses RLS during forward pass for tuning of consequent parameters as well as the BP method for tuning of antecedent parameters, as shown in Table 2. It looks like Sugeno type-1 ANFIS [13, 14], which uses RLS-BP hybrid learning rule for type-1 FLS systems.

Table 2. Two passes in hybrid learning procedure for IT2 NS-1 ANFIS

	Forward Pass	Backward Pass
Antecedent Parameters	Fixed	BP
Consequent Parameters	RLS	Fixed

3.2 Adaptive BP Learning Algorithm

The training method is presented as in [1]: Given N input-output training data pairs, the training algorithm for E training epochs, should minimize the error function:

$$e^{(t)} = \frac{1}{2}\left[f_{IT2-FLS}\left(\mathbf{x}^{(t)}\right) - y^{(t)}\right]^2 . \tag{1}$$

where e^t is the error function at time t, $f_{IT2-FLS}\left(\mathbf{x}^{(t)}\right)$ is the output of the IT2 FLS using the input vector $\mathbf{x}^{(t)}$ from the input-output data pairs, and $y^{(t)}$ is the output from the input-output data pairs.

4 Application to Transfer Bar Surface Temperature Prediction

4.1 Hot Strip Mill

Because of the complexities and uncertainties involved in rolling operations, the development of mathematical theories has been largely restricted to two-dimensional models applicable to heat losing in flat rolling operations.

Fig. 1, shows a simplified diagram of a HSM, from the initial point of the process at the reheat furnace entry to its end at the coilers.

Besides the mechanical, electrical and electronic equipment, a big potential for ensuring good quality lies in the automation systems and the used control techniques. The most critical process in the HSM occurs in the Finishing Mill (FM). There are several mathematical model based systems for setting up the FM. There is a model-based set-up system [15] that calculates the FM working references needed to obtain gauge, width and temperature at the FM exit stands. It takes as inputs: FM exit target gage, target width and target temperature, steel grade, hardness ratio from slab chemistry, load distribution, gauge offset, temperature offset, roll diameters, load distribution, transfer bar gauge, transfer bar width and transfer bar temperature entry.

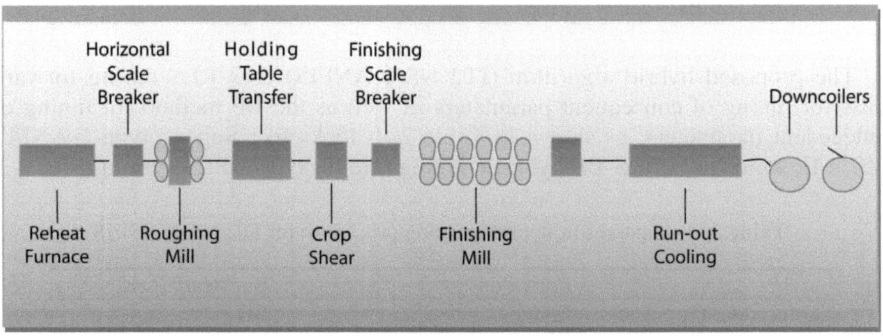

Fig. 1. Typical hot strip mill

The errors in the gauge of the transfer bar are absorbed in the first two FM stands and therefore have a little effect on the target exit gauge. It is very important for the model to know the FM entry temperature accurately. A temperature error will propagate through the entire FM.

4.2 Design of the IT2 NS-1 ANFIS

The architecture of the IT2 NS-1 ANFIS was established in such away that its parameters are continuously optimized. The number of rule-antecedents was fixed to two; one for the Roughing Mill (RM) exit surface temperature and one for transfer bar head traveling time. Each antecedent-input space was divided in three fuzzy sets (FSs), fixing the number of rules to nine. Gaussian primary membership functions (MFs) of uncertain means were chosen for the antecedents. Each rule of the each IT2 NS-1 ANFIS is characterized by six antecedent MFs parameters (two for left-hand and right-hand bounds of the mean and one for standard deviation, for each of the two antecedent Gaussian MFs) and six consequent parameters (one for left-hand and one for right-hand end points of each of the three consequent type-1 FSs), giving a total of twelve parameters per rule. Each input value has one standard deviation parameter, giving two additional parameters.

4.3 Noisy Input-Output Training Data Pairs

From an industrial HSM, noisy input-output pairs of three different product types were collected and used as training and checking data. The inputs are the noisy measured RM exit surface temperature and the measured RM exit to SB entry transfer bar traveling time. The output is the noisy measured SB entry surface temperature.

4.4 Fuzzy Rule Base

The IT2 NS-1 ANFIS fuzzy rule base consists of a set of IF-THEN rules that represents the model of the system. The IT2 ANFIS system has two inputs $x_1 \in X_1$, $x_2 \in X_2$ and one output $y \in Y$. The rule base has $M = 9$ rules of the form:

$$R^i : IF \quad x_1 \quad is \quad \tilde{F}_1^i \quad and \quad x_2 \quad is \quad \tilde{F}_2^i, \quad THEN \quad Y^i = C_0^i + C_1^i x_1 + C_2^i x_2 \ . \tag{2}$$

where Y^i the output of the ith rule is a fuzzy type-1 set, and the parameters C_j^i, with $i = 1,2,3,\ldots,9$ and $j = 0,1,2$, are the consequent type-1 FSs.

4.5 Input Membership Function

The primary MFs for each input of the IT2 NS-1 ANFIS are Gaussians of the form:

$$\mu_{X_k}(x_k) = \exp\left[-\frac{1}{2}\left[\frac{x_k - x_k^{'}}{\sigma_{X_k}}\right]^2\right] . \tag{3}$$

where: $k = 1,2$ (the number of type-2 non-singleton inputs), $\mu_{X_k}(x_k)$ is centered at $x_k = x_k^{'}$ and σ_{X_k} is the standard deviation. The standard deviation of the RM exit surface temperature measurement, σ_{X_1}, was initially set to 13.0 ^oC and the standard deviation head end traveling time measurement, σ_{X_2}, was initially set to 2.41 s. The uncertainty of the input data was modeled as stationary additive noise using type-1 FSs.

4.6 Antecedent Membership Functions

The primary MFs for each antecedent are FSs described by Gaussian with uncertain means:

$$\mu_k^i(x_k) = \exp\left[-\frac{1}{2}\left[\frac{x_k - m_k^i}{\sigma_k^i}\right]^2\right] . \tag{4}$$

where $m_k^i \in \left[m_{k1}^i, m_{k2}^i\right]$ is the uncertain mean, with $k = 1,2$ (the number of antecedents) and $i =$ 1,2,..9 (the number of M rules), and σ_k^i is the standard deviation. The means of the antecedent fuzzy sets are uniformly distributed over the entire input space.

Table 3 shows the calculated interval values of uncertainty of x_1 input, where $\left[m_{11}, m_{12}\right]$ is the uncertain mean and σ_1 is the standard deviation.

Table 4 shows the calculated interval values of uncertainty of x_2 input, where $\left[m_{21}, m_{22}\right]$ is the uncertain mean and σ_2 is the standard deviation for all the 9 rules.

Table 3. Intervals of uncertainty x_1 input

	m_{11} ^oC	m_{12} ^oC	σ_1 ^oC
1	950	952	60
2	1016	1018	60
3	1080	1082	60

Table 4. Intervals of uncertainty of x_2 input

Product Type	m_{21} s	m_{22} s	σ_2 s
1	32	34	10
2	42	44	10
3	56	58	10

The standard deviation of temperature noise σ_{n1} was initially set to 1 ^oC and the standard deviation of time noise σ_{n2} was set to 1 s.

4.7 Consequent Membership Functions

Each consequent is an interval type-1 FS with $Y^i = \left[y_l^i, y_r^i\right]$ where

$$y_l^i = \sum_{j=1}^{p} c_j^i x_j + c_0^i - \sum_{j=1}^{p} \left|x_j\right| s_j^i - s_0^i \tag{5}$$

and

$$y_r^i = \sum_{j=1}^{p} c_j^i x_j + c_0^i + \sum_{j=1}^{p} \left|x_j\right| s_j^i + s_0^i \tag{6}$$

where c_j^i denotes the center (mean) of C_j^i and s_j^i denotes the spread of C_j^i, with $i = 1,2,3,..,9$ and $j = 0,1,2$. Then y_l^i and y_r^i are the consequent parameters. When only the input-output data training pairs $\left(x^{(1)} : y^{(1)}\right),\ldots,\left(x^{(N)} : y^{(N)}\right)$ are available and there is not data information about the consequents, the initial values for the centroid parameters c_j^i and s_j^i can be chosen arbitrarily in the output space [11]. In this work the initial values of c_j^i were set equal to 0.001 and the initial values of s_j^i equal to 0.0001, for $i = 1,2,3,..,9$ and $j = 0,1,2$.

4.8 Results

The IT2 NS-1 ANFIS (RLS-BP) system was trained and used to predict the SB entry temperature, applying the RM exit measured transfer bar surface temperature and RM exit to SB entry zone traveling time as inputs. We ran fifteen epochs of training; one hundred and ten parameters were tuned using eighty seven, sixty-eight and twenty-eight input-output training data pairs per epoch, for type A, type B and type C products respectively.

The performance evaluation for the hybrid IT2 NS-1 ANFIS system was based on root mean-squared error (RMSE) benchmarking criteria as in [1].

Fig. 2 shows the RMSEs of the two IT2 TSK NSFLS-1 systems (one trained using the BP only method, and the new proposed hybrid algorithm using RLS-BP methods) and the base line interval singleton IT2 TSK SFLS (BP), all of them for fifty epochs' of training for the case for type C products. Observe that from epoch 1 to 4 the hybrid IT2 NS-1 ANFIS (RLS-BP) has better performance than both: the singleton IT2 SFLS (BP) and the IT2 NSFLS-1 (BP). From epoch 1 to 4 the RMSE of the IT2 NSFLS-1 has an oscillation, meaning that it is very sensitive to its learning parameters values. At epoch 5, it reaches its minimum RMSE and is stable for the rest of training. The proposed IT2 NS-1 ANFIS system has the best performance and stability after only one epoch of training.

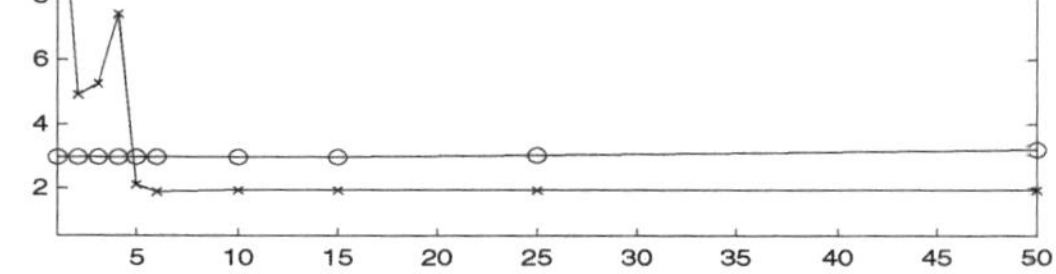

Fig. 2. (*) RMSE $_{\text{TSK 2, SFLS}}$ (BP) (+) RMSE $_{\text{TSK 2, NSFLS-1}}$ (BP) (o) RMSE $_{\text{TSK 2, NSFLS-1}}$ (RLS-BP)

5 Conclusions

An IT2 NS-1 ANFIS using the hybrid RLS-BP training method was tested and compared for predicting the surface temperature of the transfer bar at SB entry. The antecedent MFs and consequent centroids of the IT2 NS-1 ANFIS absorbed the uncertainty introduced by all the factors: the antecedent and consequent initially values, the noisy temperature measurements, and the inaccurate traveling time estimation. The non-singleton type-1 fuzzy inputs are able to compensate the uncertain measurements, expanding the applicability of IT2 NS-1 ANFIS systems.

It has been shown that the proposed IT2 NS-1 ANFIS system can be applied in modeling and control of the steel coil temperature. It has also been envisaged its application in any uncertain and non-linear system prediction and control, as in furnace temperature control, aerospace stability control, turbine trust control, and especially in those applications where there is only one chance of training.

References

1. Mendel, J.M.: Uncertain Rule Based Fuzzy Logic Systems: Introduction and New Directions. Prentice-Hall, Upper Saddle River (2001)
2. Mendez, G., Cavazos, A., Leduc, L., Soto, R.: Hot Strip Mill Temperature Prediction Using Hybrid Learning Interval Singleton Type-2 FLS. In: Proceedings of the IASTED International Conference on Modeling and Simulation, Palm Springs, February 2003, pp. 380–385 (2003)
3. Mendez, G., Cavazos, A., Leduc, L., Soto, R.: Modeling of a Hot Strip Mill Temperature Using Hybrid Learning for Interval Type-1 and Type-2 Non-Singleton Type-2 FLS. In: Proceedings of the IASTED International Conference on Artificial Intelligence and Applications, Benalmádena, Spain, September 2003, pp. 529–533 (2003)
4. Mendez, G., Juarez, I.: Orthogonal-Back Propagation Hybrid Learning Algorithm for Interval Type-1 Non-Singleton Type-2 Fuzzy Logic Systems. WSEAS Transactions on Systems 4(3) (March 2005), ISSN 1109-2777
5. Mendez, G., Castillo, O.: Interval Type-2 TSK Fuzzy Logic Systems Using Hybrid Learning Algorithm. In: FUZZ-IEEE 2005 The international Conference on Fuzzy Systems, Reno Nevada, USA, pp. 230–235 (2005)
6. Lee, D.Y., Cho, H.S.: Neural Network Approach to the Control of the Plate Width in Hot Plate Mills. In: International Joint Conference on Neural Networks, vol. 5, pp. 3391–3396 (1999)
7. Taylor, B.N., Kuyatt, C.E.: Guidelines for Evaluating and Expressing the Uncertainty of NIST Measurement Results, NIST Technical Note 1297 (September 1994)
8. Wang, L.-X.: Fuzzy Systems are Universal Approximators. In: Proceedings of the IEEE Conf. On Fuzzy Systems, San Diego, pp. 1163–1170 (1992)
9. Wang, L.-X., Mendel, J.M.: Back-Propagation Fuzzy Systems as Nonlinear Dynamic System Identifiers. In: Proceedings of the IEEE Conf. On Fuzzy Systems, San Diego, CA, March 1992, pp. 1409–1418 (1992)
10. Wang, L.-X.: Fuzzy Systems are Universal Approximators. In: Proceedings of the IEEE Conf. on Fuzzy Systems, San Diego, pp. 1163–1170 (1992)

11. Wang, L.-X.: A Course in Fuzzy Systems and Control. Prentice Hall PTR, Upper Saddle River (1997)
12. Liang, Q.J., Mendel, J.M.: Interval type-2 fuzzy logic systems: Theory and design. Trans. Fuzzy Sistems. 8, 535–550 (2000)
13. Jang, J.-S.R., Sun, C.-T., Mizutani, E.: Neuro-Fuzzy and Soft Computing: A Computational Approach to Learning and Machine Intelligence. Prentice-Hall, Upper Saddle River (1997)
14. Jang, J.-S.R., Sun, C.-T.: Neuro-Fuzzy Modeling and Control. The Proceedings of the IEEE 3, 378–406 (1995)
15. GE Models, Users reference, Roanoke VA, vol. 1 (1993)

Centralized Direct and Indirect Neural Control of Distributed Parameter Systems

Ieroham S. Baruch and Rosalba Galvan-Guerra

CINVESTAV-IPN, Department of Automatic Control, Ave. IPN No 2508, A.P. 14-470, Mexico D.F., C.P. 07360, México
{baruch,rgalvan}@ctrl.cinvestav.mx

Abstract. The paper proposed to use a Recurrent Neural Network Model (RNNM) for centralized modeling, identification and direct adaptive control of an anaerobic digestion bioprocess, carried out in a fixed bed and a recirculation tank of a wastewater treatment system. The analytical model of the digestion bioprocess represented a distributed parameter system, which is reduced to a lumped system using the orthogonal collocation method, applied in three collocation points plus the recirculation tank. The RNNM learning algorithm is the dynamic backpropagation one. The graphical simulation results of the distributed plant direct and indirect adaptive neural control system, exhibited good convergence and precise reference tracking, outperforming the optimal control.

Keywords: Recurrent neural network model, backpropagation learning, distributed parameter system, system identification, direct and indirect adaptive neural control, anaerobic digestion bioprocess, wastewater treatment bioreactor.

1 Introduction

In the last two decades, a new identification and control tools like Neural Networks (NN), used for biotechnological plants rose fame. Among several possible network architectures the ones most widely used are the Feedforward NN (FFNN) and the Recurrent NN (RNN), [1]. The main NN property namely the ability to approximate complex non-linear relationships without prior knowledge of the model structure makes them a very attractive alternative to the classical modeling and control techniques. This property has been proved for both types of NNs by the universal approximation theorem [1]. The preference given to NN identification with respect to the classical methods of process identification is clearly demonstrated in the solution of the "bias-variance dilemma" [1]. The FFNN and the RNN have been applied for Distributed Parameter Systems (DPS) identification and control too. In [2], [3], [4], an intelligent modeling approach is proposed for Distributed Parameter Systems (DPS). In [5], it is presented a new methodology for the identification of DPS, based on NN architectures, motivated by standard numerical discretization techniques used for the solution of Partial Differential

O. Castillo et al. (Eds.): Evolutionary Design of Intelligent Systems, SCI 257, pp. 63–81.
springerlink.com

Equations (PDE). In [6], an attempt is made to use the philosophy of the NN adaptive-critic design to the optimal control of distributed parameter systems. In [7] the concept of proper orthogonal decomposition is used for the model reduction of DPS to form a reduced order lumped parameter problem. In [8], measurement data of an industrial process are generated by solving the PDE numerically using the finite differences method. Both centralized and decentralized NN models are introduced and constructed based on this data. The models are implemented on FFNN using Backpropagation (BP). Unfortunately, all these works suffered of the same inconvenience that the FFNNs used are multilayer and of higher dimension having great complexity which made difficult their application. In [9] - [13], a new canonical Recurrent Trainable NN (RTNN) architecture, and a dynamic BP learning algorithm has been applied for systems identification and control, obtaining a good results. In the present paper, this RTNN model will be used for identification, direct and indirect adaptive neural control of a digestion anaerobic DPS of wastewater treatment, [14], modeled by PDE/ODE, and simplified using the orthogonal collocation technique.

2 Description of the RTNN Topology and Learning

Block-diagrams of the RTNN topology and its adjoint, are given on Fig. 1, and Fig. 2. Following Fig. 1, and Fig. 2, we could derive the dynamic BP algorithm of its learning based on the RTNN topology using the diagrammatic method of [15]. The RTNN topology and learning are described in vector-matrix form as:

$$X(k+1) = AX(k) + BU(k);\ B = [B_1\ ;\ B_0];\ U^T = [U_1\ ;\ U_2]; \tag{1}$$

$$Z_1(k) = G[X(k)]; \tag{2}$$

$$V(k) = CZ(k);\ C = [C_1\ ;\ C_0];\ Z^T = [Z_1\ ;\ Z_2]; \tag{3}$$

$$Y(k) = F[V(k)]; \tag{4}$$

$$A = \text{block-diag}\ (Ai),\ |Ai\,| < 1; \tag{5}$$

$$W(k+1) = W(k) + \eta\ \Delta W(k) + \alpha\ \Delta W_{ij}(k-1); \tag{6}$$

$$E(k) = T(k)-Y(k); \tag{7}$$

$$E_1(k) = F'[Y(k)]\ E(k);\ F'[Y(k)] = [1-Y^2(k)]; \tag{8}$$

$$\Delta C(k) = E_1(k)\ Z^T(k); \tag{9}$$

$$E_3(k) = G'[Z(k)]\ E_2(k);\ E_2(k) = C^T(k)\ E_1(k);\ G'[Z(k)] = [1-Z^2(k)]; \tag{10}$$

$$\Delta B(k) = E_3(k)\ U^T(k); \tag{11}$$

$$\Delta A(k) = E_3(k)\ X^T(k); \tag{12}$$

$$Vec(\Delta A(k)) = E_3(k) \square X(k); \tag{13}$$

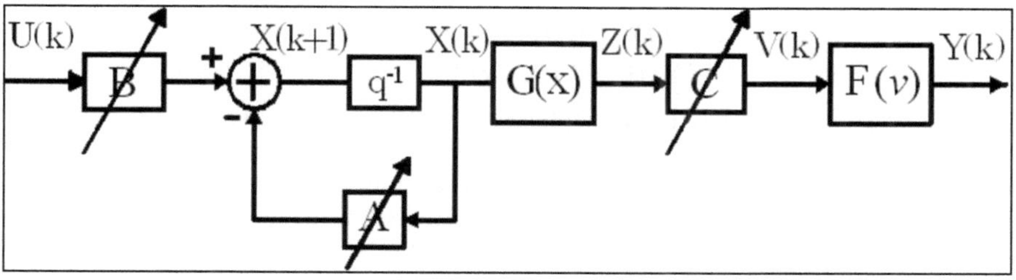

Fig. 1. Block diagram of the RTNN model

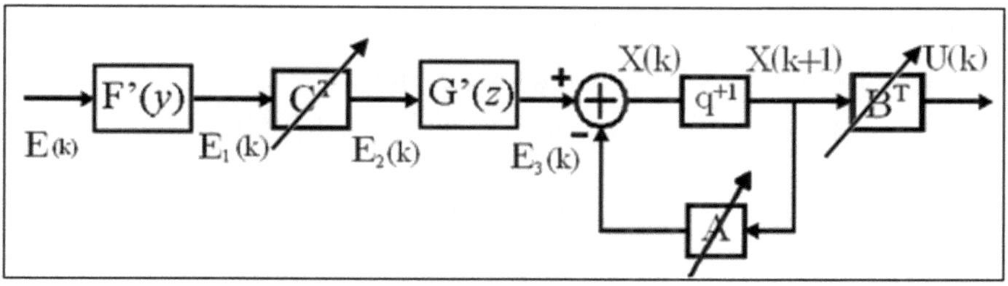

Fig. 2. Block diagram of the adjoint RTNN model

Where: X, Y, U are state, augmented output, and input vectors with dimensions n, (l+1), (m+1), respectively, where Z_1 and U_1 are the (nx1) output and (mx1) input of the hidden layer; the constant scalar threshold entries are Z_2 = -1, U_2 = -1, respectively; V is a (lx1) pre-synaptic activity of the output layer; T is the (lx1) plant output vector, considered as a RNN reference; A is (nxn) block-diagonal weight matrix; B and C are [nx(m+1)] and [lx(n+1)]- augmented weight matrices; B_0 and C_0 are (nx1) and (lx1) threshold weights of the hidden and output layers; F[.], G[.] are vector-valued tanh(.)-activation functions with corresponding dimensions; F'[.], G'[.] are the derivatives of these tanh(.) functions; W is a general weight, denoting each weight matrix (C, A, B) in the RTNN model, to be updated; ΔW (ΔC, ΔA, ΔB), is the weight correction of W; η, α are learning rate parameters; ΔC is an weight correction of the learned matrix C; ΔB is an weight correction of the learned matrix B; ΔA is an weight correction of the learned matrix A; the diagonal of the matrix A is denoted by Vec(.) and equation (13) represents its learning as an element-by-element vector products; E, E_1, E_2, E_3, are error vectors with appropriate dimensions, predicted by the adjoint RTNN model, given on Fig.2. The stability of the RTNN model is assured by the activation functions (-1, 1) bounds and by the local stability weight bound condition, given by (5). Below a theorem of RTNN stability which represented an extended version of Nava's theorem, [9], [10] is given.

Theorem of stability of the RTNN: Let the RTNN with Jordan Canonical Structure is given by equations (1)-(5) (see Fig.1) and the nonlinear plant model, is as follows:

$$X_{d.}(k+1) = G[\, X_d(k), U(k) \,]$$
$$Y_d(k) = F[\, X_d(k) \,]$$

Where: $\{Y_d(.), X_d(.), U(.)\}$ are output, state and input variables with dimensions l, n_d, m, respectively; F(.), G(.) are vector valued nonlinear functions with respective dimensions. Under the assumption of RTNN identifiability made, the application of the BP learning algorithm for A(.), B(.), C(.), in general matricial form, described by equation (6)-(13), and the learning rates η (k), α (k) (here they are considered as time-dependent and normalized with respect to the error) are derived using the following Lyapunov function:

$$L(k) = L_1(k)+L_2(k)$$

Where: $L_1(k)$ and $L_2(k)$ are given by:

$$L_1(k) = \tfrac{1}{2}e^2(k)$$

$$L_2(k) = tr\left(\tilde{W}_A(k)\tilde{W}_A^T(k)\right)+tr\left(\tilde{W}_B(k)\tilde{W}_B^T(k)\right)+tr\left(\tilde{W}_C(k)\tilde{W}_C^T(k)\right)$$

Where: $$\tilde{W}_A(k) = \hat{A}(k) - A^*, \tilde{W}_B(k) = \hat{B}(k) - B^*, \tilde{W}_C(k) = \hat{C}(k) - C^*$$

Are vectors of the estimation error and (A^*, B^*, C^*), $(\hat{A}(k), \hat{B}(k), \hat{C}(k))$ denote the ideal neural weight and the estimate of the neural weight at the k-th step, respectively, for each case. Then the identification error is bounded, i.e.:

$$L(k+1) = L_1(k+1)+L_2(k+1) < 0$$
$$\Delta L(k+1) = L(k+1) - L(k)$$

Where the condition for $L_1(k+1)<0$ is that:

$$\frac{\left(1-\frac{1}{\sqrt{2}}\right)}{\psi_{max}} < \eta_{max} < \frac{\left(1+\frac{1}{\sqrt{2}}\right)}{\psi_{max}}$$

And for $L_2(k+1)<0$ we have:

$$\Delta L_2(k+1) < -\eta_{max}\left|e(k+1)\right|^2 - \alpha_{max}\left|e(k)\right|^2 + d(k+1)$$

Note that η_{max} changes adaptively during the RTNN learning and:

$$\eta_{max} = \max_{i=1}^{3}\{\eta_i\}$$

Where all: the unmodelled dynamics, the approximation errors and the perturbations, are represented by the d-term. The Rate of Convergence Lemma used is given in [10]. The complete proof of that Theorem of stability is given in [9].

3 Direct Adaptive Neural Control

The Direct Adaptive Neural Control (DANC) using the RTNN as plant identifier and plant controller has been described in [9]-[13]. The block-diagram of the control system is given on Fig. 3. It contained a recurrent neural identifier RTNN 1, and two recurrent neural controllers (Feedback-FB and Feedforward-FF). Let us to write the following z-transfer- functions of the plant, the neural identifier and controllers:

$$W_p(z) = C_p\,(zI - A_p)^{-1}\,B_p \tag{14}$$

$$P_i(z) = (zI - A_i)^{-1}\,B_i \tag{15}$$

$$Q_1(z) = C_{cfb}\,(zI - A_{cfb})^{-1}\,B_{cfb} \tag{16}$$

$$Q_2(z) = C_{cff}\,(zI - A_{cff})^{-1}\,B_{cff} \tag{17}$$

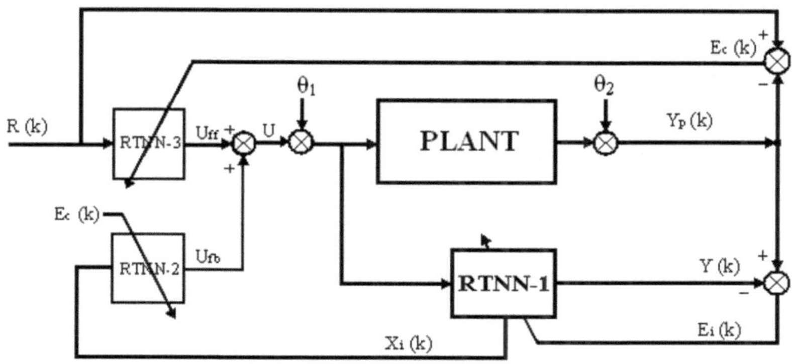

Fig. 3. Block diagram of the closed-loop system using RTNN identifier and two adaptive RTNN controllers

The z-transfer functions (14)-(17) are connected by the following equation, derived following the block-diagram of the Fig. 3, and given in z-operational form:

$$Y_p(z) = W_p(z)\,[I + Q_1(z)\,P_i(z)]^{-1}\,Q_2(z)\,R(z) + \theta(z) \tag{18}$$

$$\theta(z) = W_p(z)\,\theta_1(z) + \theta_2(z) \tag{19}$$

Where: θ (z) is a generalized noise term. The RTNN topology is controllable and observable. The BP algorithm of learning is convergent, [9]. Then the identification and control errors tend to zero.

$$E_i(k) = Y_p(k) - Y(k) \rightarrow 0;\; E_c(k) = R(k) - Y_p(k) \rightarrow 0;\; k \rightarrow \infty \tag{20}$$

This means that each transfer function given by equations (14)-(17) is stable with minimum phase. The closed-loop system is stable and the RTNN-2 feedback controller compensates the plant dynamics. The RTNN-3 feedforward controller dynamics is an inverse dynamics of the closed-loop system one, which assure a precise reference tracking in spite of the presence of process and measurement noises. If l=m, both RTNN-2, RTNN-3 controllers are learnt by the output control error applying the BP algorithm. The centralized DPS could be considered as a system with excessive measurements, where the DANC performed a data fusion so to elaborate the control action. If l>m, the needed input control error for BP learning is obtained from the output control error, using the C, B parameters, estimated by the RTNN-1:

$$E_u(k) = (CB)^+ E_c(k), (CB)^+ = [(CB)^T (CB)]^{-1} (CB)^T \quad (21)$$

4 Indirect Adaptive Control Using a RTNN Identifier

The indirect adaptive control using the RTNN as plant identifier has been described in [9]-[12] and applied for aerobic fermentation bioprocess plant control, [11], [12]. Here the same control approach will be represented as a Sliding Mode Control (SMC) and applied for wastewater anaerobic digestion bioprocess, modeled by a simplified rectangular ODE system, [14]. The block diagram of this control is shown on Fig. 4. The stable nonlinear plant is identified by a RTNN identifier with topology, given by equations (1)-(5) and learned by the stable BP-learning algorithm, given by equations (6)-(13), where the identification error tends to zero. Let us first to admit that l= m (square system). The identified local linear plant model described in both state-space and input-output form is given by the equations:

$$X(k+1) = A X(k) + B U(k) \quad (22)$$

$$Y(k) = C X(k) \quad (23)$$

$$Y(k+1) = C X(k+1) = C [AX(k) + BU(k)] \quad (24)$$

In [16], the sliding surface is defined with respect to the state variables and the SMC objective is to move the states from an arbitrary state-space position to the sliding surface in finite time. In [17], the sliding surface is also defined with respect to the states but the states of the SISO system are obtained from the plant outputs by differentiation. In [18], the sliding surface definition and the control objectives are the same. The equivalent control systems design is done with respect to the plant output, but the reachability of the stable output control depended on the plant structure. Here, the sliding surface is defined directly with respect to the plant outputs which facilitated the equivalent SMC systems design and decoupled the closed-loop system. Let us define the following sliding surface as an output tracking error function:

$$S(k+1)=E(k+1)+\sum_{i=1}^{p} \gamma_i E(k-i+1); \ |\gamma_i| < 1 \quad (25)$$

where S (.) is the Sliding Surface Error Function (SSEF) defined with respect to the plant output; E(.) is the systems output tracking error; γ_i are parameters of the desired stable SSEF; p is the order of the SSEF. The tracking error for k, k+1 is defined as:

$$E(k) = R(k) - Y(k); E(k+1) = R(k+1) - Y(k+1) \quad (26)$$

Here the l-dimensional vectors R (k), Y (k) are the system reference and the output of the local linear plant model. The objective of the sliding mode control

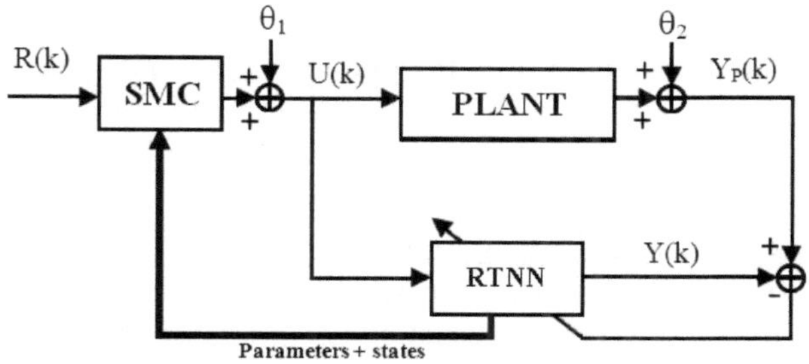

Fig. 4. Block-diagram of the closed-loop system using RTNN identifier and a SMC

systems design is to find a control action which maintained the system error on the sliding surface, assuring that the output tracking error reached zero in p steps, where p < n. So, the control objective is fulfilled if the SSEF is S (k+1) = 0. From (24)-(26), we obtained:

$$R(k+1) - CAX(k) - CBU(k) + \sum_{i=1}^{p} \gamma_i E(k-i+1) = 0 \qquad (27)$$

As the local approximation plant model (22), (23), is controllable, observable and stable, [9], the matrix A is diagonal, and l = m, then the matrix product (CB), representing the plant model static gain, is nonsingular, and the plant states X (k) are smooth non-increasing functions. Now, from (27) it is easy to obtain the equivalent control capable to lead the system to the sliding surface which yields:

$$U_{eq}(k) = (CB)^{-1} [- CAX(k) + R(k+1) + \sum_{i=1}^{p} \gamma_i E(k-i+1)] \qquad (28)$$

Following [16], the SMC avoiding chattering is taken using a saturation function instead of sign one. So the SMC took the form:

$$U^*(k) = \begin{cases} U_{eq}(k), & \text{if } \|U_{eq}(k)\| < Uo \\ -Uo\, U_{eq}(k)/\|U_{eq}(k)\|, & \text{if } \|U_{eq}(k)\| \geq Uo \end{cases} \qquad (29)$$

Here the saturation level Uo is chosen with respect to the load level perturbation. It is easy to see that the substitution of the equivalent control (28) in the input-output linear plant model (24) showed an exact complete plant dynamics compensation. The designed plant output sliding mode equivalent control substituted the multi-input multi-output coupled high order dynamics of the linearized plant with desired decoupled low order one. The centralized DPS could be considered as a system with excessive measurements, where the SMC performed a data fusion so to elaborate the control action. If l > m (rectangular system) the equivalent control law (28) is changed. The inverse of the (CB) became a pseudo-inverse $(CB)^+$ and a learnable threshold Of with dimension (mx1) is added to the right hand side of (28), so to obtain:

$$U_{eq}(k) = (CB)^{+} [- CAX(k) + R(k+1) + \sum_{i=1}^{p} \gamma_i E(k-i+1)] + Of \tag{30}$$

The threshold Of is trained using the input control error, applying the BP algorithm, given by (6). The input control error is obtained from the output control error backpropagating it through the RTNN adjoint model (see Fig.3). An approximate way to obtain the input error from the output error is to multiply it by the same (CB)-pseudo-inverse (see equation (21)). So, the threshold correction is obtained as:

$$\Delta Of\,(k) = (CB)^{+}E(k);\ (CB)^{+} = [(CB)^{T}(CB)]^{-1}\,(CB)^{T} \tag{31}$$

5 Analytical Model of the Anaerobic Digestion System

The anaerobic digestion systems block diagram is depicted on Fig.5. It is conformed by a fixed bed reactor and a recirculation tank. The physical meaning of all variables and constants (also its values), are summarized on Table 1. The complete analytical model of wastewater treatment anaerobic bioprocess, taken from [14], could be described by the following system of PDE and ODE (for the recirculation tank):

$$\frac{\partial X_1}{\partial t} = (\mu_1 - \varepsilon D) X_1, \quad \mu_1 = \mu_{1\max} \frac{S_1}{K_{s_1}^{'} X_1 + S_1}, \tag{32}$$

$$\frac{\partial X_2}{\partial t} = (\mu_2 - \varepsilon D) X_2, \quad \mu_2 = \mu_{2s} \frac{S_2}{K_{s_2}^{'} X_2 + S_2 + \frac{S_2^2}{K_{I_2}}}, \tag{33}$$

$$\frac{\partial S_1}{\partial t} = \frac{E_z}{H^2} \frac{\partial^2 S_1}{\partial z^2} - D \frac{\partial S_1}{\partial t} - k_1 \mu_1 X_1, \tag{34}$$

$$\frac{\partial S_2}{\partial t} = \frac{E_z}{H^2} \frac{\partial^2 S_2}{\partial z^2} - D \frac{\partial S_2}{\partial t} + k_2 \mu_1 X_1, \tag{35}$$

$$S_1(0,t) = \frac{S_{1,in}(t) + RS_{1T}}{R+1}, \quad S_2(0,t) = \frac{S_{2,in}(t) + RS_{2T}}{R+1}, \quad R = \frac{Q_T}{DV_{eff}}, \tag{36}$$

$$\frac{\partial S_1}{\partial z}(1,t) = 0, \quad \frac{\partial S_2}{\partial z}(1,t) = 0. \tag{37}$$

For practical purpose, the full PDE anaerobic digestion process model, [14], could be reduced to an ODE system using an early lumping technique and the

Table 1. Summary of the variables in the plant model

Variable	*Units*	*Name*	*Value*
z	$z \in [0,1]$	*Space variable*	
t	D	*Time variable*	
E_z	m^2/d	*Axial dispersion coefficient*	1
D	1/d	*Dilution rate*	0.55
H	m	*Fixed bed length*	3.5
X_1	g/L	*Concentration of acidogenic bacteria*	
X_2	g/L	*Concentration of methanogenic bacteria*	
S_1	g/L	*Chemical Oxygen Demand*	
S_2	mmol/L	*Volatile Fatty Acids*	
ε		*Bacteria fraction in the liquid phase*	0.5
k_1	g/g	*Yield coefficients*	42.14
k_2	mmol/g	*Yield coefficients*	250
k_3	mmol/g	*Yield coefficients*	134
μ_1	1/d	*Acidogenesis growth rate*	
μ_2	1/d	*Methanogenesis growth rate*	
μ_{1max}	1/d	*Maximum acidogenesis growth rate*	1.2
μ_{2s}	1/d	*Maximum methanogenesis growth rate*	0.74
K_{1s}'	g/g	*Kinetic parameter*	50.5
K_{2s}'	mmol/g	*Kinetic parameter*	16.6
K_{I2}'	mmol/g	*Kinetic parameter*	256
Q_T	m^3/d	*Recycle flow rate*	0.24
V_T	m^3	*Volume of the recirculation tank*	0.2
S_{1T}	g/L	*Concentration of Chemical Oxygen Demand in the recirculation tank*	
S_{2T}	mmol/L	*Concentration of Volatile Fatty Acids in the recirculation tank*	
Q_{in}	m^3/d	*Inlet flow rate*	0.31
V_B	m^3	*Volume of the fixed bed*	1
V_{eff}	m^3	*Effective volume tank*	0.95
$S_{1,in}$	g/l	*Inlet substrate concentration*	
$S_{2,in}$	mmol/L	*Inlet substrate concentration*	

Orthogonal Collocation Method (OCM), [14], in three points (0.25H, 0.5H, 0.75H) obtaining the following system of OD equations:

$$\frac{dX_{1,i}}{dt} = \left(\mu_{1,i} - \varepsilon D\right) X_{1,i}, \quad \frac{dX_{2,i}}{dt} = \left(\mu_{2,i} - \varepsilon D\right) X_{2,i}, \tag{38}$$

$$\frac{dS_{1,i}}{dx} = \frac{E_z}{H^2} \sum_{j=1}^{N+2} B_{i,j} S_{1,j} - D \sum_{j=1}^{N+2} A_{i,j} S_{1,j} - k_1 \mu_{1,i} X_{1,i}, \tag{39}$$

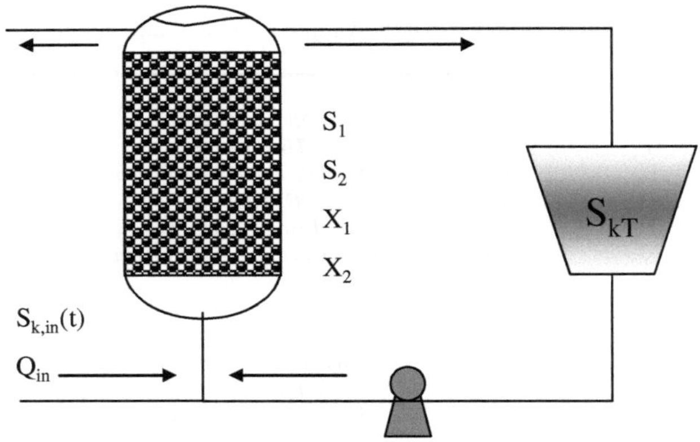

Fig. 5. Block-Diagram of Anaerobic Digestion Bioreactor

$$\frac{dS_{1T}}{dt}=\frac{Q_T}{V_T}\left(S_1(1,t)-S_{1T}\right),\quad \frac{dS_{2T}}{dt}=\frac{Q_T}{V_T}\left(S_2(1,t)-S_{2T}\right). \tag{40}$$

$$\frac{dS_{2,i}}{dx}=\frac{E_z}{H^2}\sum_{j=1}^{N+2}B_{i,j}S_{1,j}-D\sum_{j=1}^{N+2}A_{i,j}S_{2,j}+k_2\mu_{1,i}X_{2,i}-k_3\mu_{2,i}X_{2,i}\,, \tag{41}$$

$$\frac{dS_{1T}}{dt}=\frac{Q_T}{V_T}\left(S_{1,N+2}-S_{1T}\right),\quad \frac{dS_{2T}}{dt}=\frac{Q_T}{V_T}\left(S_{2,N+2}-S_{2T}\right), \tag{42}$$

$$S_{k,1}=\frac{1}{R+1}S_{k,in}(t)+\frac{R}{R+1}S_{kT}\,,\quad S_{k,N+2}=\frac{K_1}{R+1}S_{k,in}(t)$$
$$+\frac{K_1R}{R+1}S_{kT}+\sum_{i=2}^{N+1}K_iS_{k,i} \tag{43}$$

$$K_1=-\frac{A_{N+2,1}}{A_{N+2,N+2}}\,,\quad K_i=-\frac{A_{N+2,i}}{A_{N+2,N+2}}\,, \tag{44}$$

$$A=\Lambda\phi^{-1}\,,\quad \Lambda=\left[\varpi_{m,l}\right],\quad \varpi_{m,l}=(l-1)z_m^{l-2}\,, \tag{45}$$

$$B=\Gamma\phi^{-1}\,,\quad \Gamma=\left[\tau_{m,l}\right],\quad \tau_{m,l}=(l-1)(l-2)z_m^{l-3}\,,\phi_{m,l}=z_m^{l-1} \tag{46}$$

$$i=2,\dots,N+2,\quad m,l=1,\dots,N+2. \tag{47}$$

The reduced plant model (38)-(47) (here (40) represented the OD equations of the recirculation tank), could be used as unknown plant model which generate input/output data for the centralized Direct Adaptive Neural Control (DANC) system design.

6 Simulation Results of DPS Direct and Indirect Neural Control

The centralized DANC controlled 14 output plant variables, which are: 4 variables for each collocation point z=0.25H, z=0.5H, z=0.75H of the fixed bed as: X_1 (acidogenic bacteria), X_2 (methanogenic bacteria), S_1 (chemical oxygen demand) and S_2 (volatile fatty acids), and the following variables in the recirculation tank: S_{1T} (chemical oxygen demand) and S_{2T} (volatile fatty acids). The topologies of the RTNNs and the learning rate parameters are as follows: for the RTNN1 it is (2, 16, 14), the activation functions are tanh (.) for both layers, the learning rate parameters are α=0.001, η=0.1; for the RTNN2 the topology is (16, 18, 2); for the RTNN3 the topology is (14, 16, 2); the activation functions are tanh (.)-for the hidden layers, sigm (.) for the output layers of both control nets; the learning rate parameters for both control RTNNs are α=0.1, η=0.1. The simulation results of DANC are obtained on-line during 100 days with a step of 0.1 day (To=0.1 sec., Nt=1000 iterations). The Figs. 6-10 compared the plant outputs with the respective reference signals for X_1, X_2, S_1, S_2, in three collocation points of the bioreactor, and for both S_{1T}, S_{2T} of the recirculation tank. The Means Squared Error (MSE%) of the DANC plant variables for the fixed bed and the recirculation tank are shown in Table 2. For sake of comparison, in Table 3 are given the MSE% results, obtained with optimal control of the linearized plant model. The comparison of that MSE% results showed slight priority of the direct centralized adaptive neural control over the optimal control due to the adaptability of the first one.

Table 2. MSE of the DANC of all output plant variables

Collocation points of the fixed bed variables	X1	X2	S1	S2
z=0.25	2.8674E-10	1.0574E-9	6.1765E-8	1.7201E-8
z=0.5	3.4107E-11	6.2468E-11	8.6977E-9	1.2094E-12
z=0.75	6.5947E-13	3.3659E-12	1.3766E-10	1.6511E-10
Recirculation tank variables			2.6501E-9	2.4932E-9

Table 3. MSE of the proportional optimal control of all output plant variables

Collocation points of the fixed bed variables	X1	X2	S1	S2
z=0.25	5.3057E-8	1.7632E-7	1.1978E-5	2.1078E-5
z=0.5	6.6925E-9	4.2626E-8	1.4922E-6	4.4276E-6
z=0.75	3.0440E-10	2.0501E-9	6.8737E-8	2.0178E-7
Recirculation tank variables			2.7323E-7	6.0146E-7

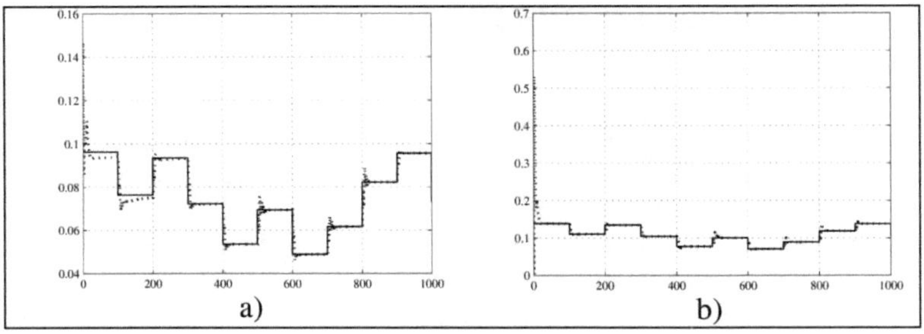

Fig. 6. a) DANC of S_{1T} (chemical oxygen demand in the recirculation tank) dotted line – plant output, continuous line –reference signal; b) SMC of S_{2T} (volatile fatty acids in the recirculation tank) dotted line- plant output, continuous line –reference signal

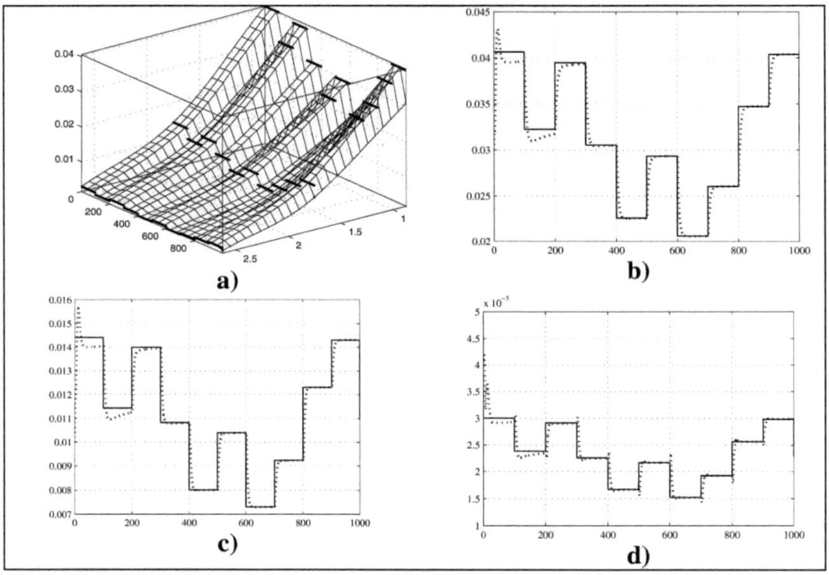

Fig. 7. DANC of X_1 (acidogenic bacteria in the fixed bed) dotted line -plant output, continuous line reference signal; a) 3d view of X_1; b) DANC of X_1 in z=0.25H; c) DANC of X_1 in z=0.5H; d) DANC of X_1 in z=0.75H

The graphical simulation results of SMC are obtained also on-line in real-time during 100 days with a step of 0.1 day (To = 0.1 sec., N_t = 1000 iterations). The Figs. 11-15 compared the respective plant outputs with the reference signals during N_t = 1000 iterations of RTNN learning and sliding mode bioprocess control. Fig. 12 b, c, d showed the variable X_1 (dotted line) and the respective reference (continuous line) in three measurement points of the fixed bed bioreactor (z=0.25H, z=0.5H, z=0.75H, H= 3.5 m). The Fig. 12 a show a 3-D view which gave a spatiotemporal representation of the output variable X_1 for the bioreactor

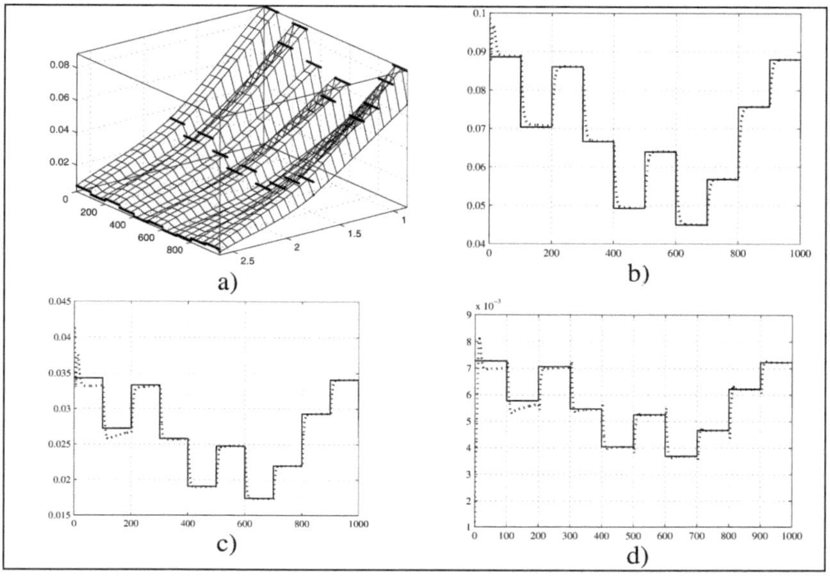

Fig. 8. DANC of X_2, (methanogenic bacteria in the fixed bed) dotted line – plant output, continuous line –reference signal; a) 3d view of X_2 b) DANC of X_2 in z=0.25H; c) DANC of X_2 in z=0.5H, d) DANC of X_2 in z=0.75H

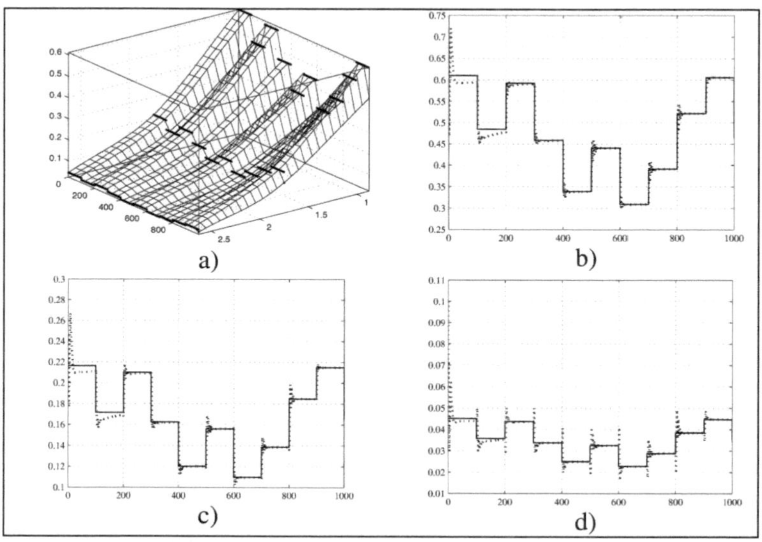

Fig. 9. DANC of S_1 (chemical oxygen demand in the fixed bed) dotted line-plant output, continuous line –reference signal; a) 3d view of S_1; b) DANC of S_1 in z=0.25H; c) DANC of S_1 in z=0.5H, d) DANC of S_1 in z=0.75H

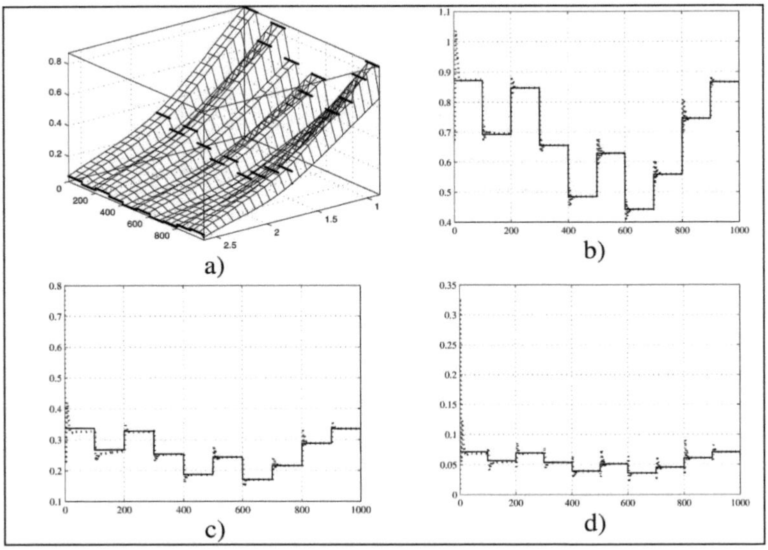

Fig. 10. DANC of DANC of S_2 (volatile fatty acids in the fixed bed) dotted line- plant output, continuous line –reference signal; a) 3d view of S_2 ; b) DANC of S_2 in z=0.25H; c) DANC of S_2 in z=0.5H; d) DANC of S_2 in z=0.75H

fixed bed length z and the time. The interpolation of the variable X_1 between the space measurement points (z=0.25H, z=0.5H, z=0.75H, H= 3.5 m) is linear. The graphical results of RTNN learning and SMC, given on Fig. 13 a, b, c, d, Fig. 14 a, b, c, d, Fig. 15 a, b, c, d showed similar results for the variables of the fixed bed X_2, S_1, S_2. The graphical results of RTNN learning and SMC, given on Fig. 11 a, b showed temporal results for the variables of the recirculation tank S_{1T}, S_{2T}. The 3-D views showed a decrease of the variables with the increase of the fixed-bed length z. The temporal graphics of RTNN learning and SM control showed some bigger discrepancies between the plant variables and the respective references at the beginning of the learning which decreased faster on time. In order to see the accuracy of the system identification and SM control, the values of the Mean Squared Error of all plant output variables for the fixed bed and the recirculation tank at the end of the 1000 iterations of learning are shown on Table 4.

Table 4. MSE of the SMC of all output plant variables

Collocation points of the fixed bed variables	X1	X2	S1	S2
z=0.25H	1.4173E-10	1.1231E-9	3.2284E-8	1.4994E-8
z=0.5H	1.6442E-11	8.0341E-12	4.7359E-9	1.8789E-9
z=0.75H	1.9244E-13	3.1672E-13	4.3831E-11	2.9076E-12
Recirculation tank variables			2.6501E-9	2.4932E-9

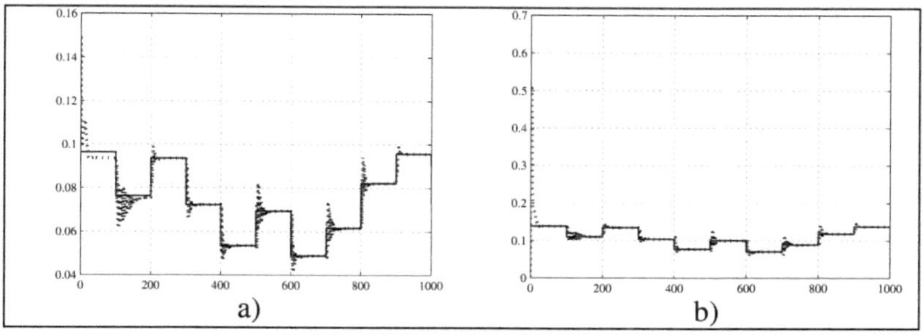

Fig. 11. SMC of the recirculation tank variables; a) SMC of S_{1T} (chemical oxygen demand in the recirculation tank) -dotted line –plant output, continuous line –reference signal; b) SMC of S_{2T} (volatile fatty acids in the recirculation tank) dotted line- plant output, continuous line –reference signal

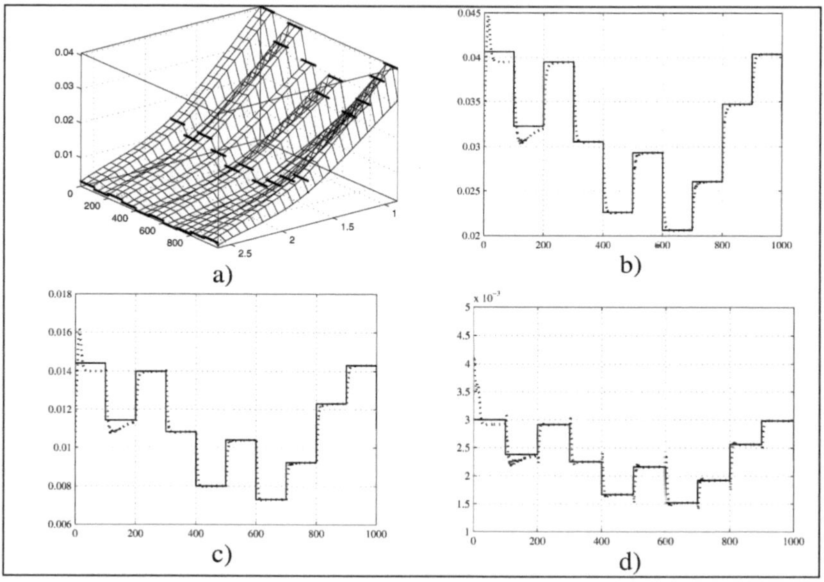

Fig. 12. SMC of X_1 (acidogenic bacteria in the fixed bed) (*dotted line -plant output, continuous line- reference signal*); a) 3-D view of X_1; b) SMC of X_1 in z=0.25H; c) SMC of X_1 in z=0.5H; d) SMC of X_1 in z=0.75H.

The value of the MSE% obtained for the worse case (S_1 for z=0.25H) is 3.2284. 10^{-8}. The obtained in Table 4 results of MSE% are compared with the results of MSE% for proportional linear optimal control, given on Table 3. The comparison of that MSE% results showed slight priority of the indirect centralized adaptive neural control over the linearized optimal control due to the adaptability of the first one.

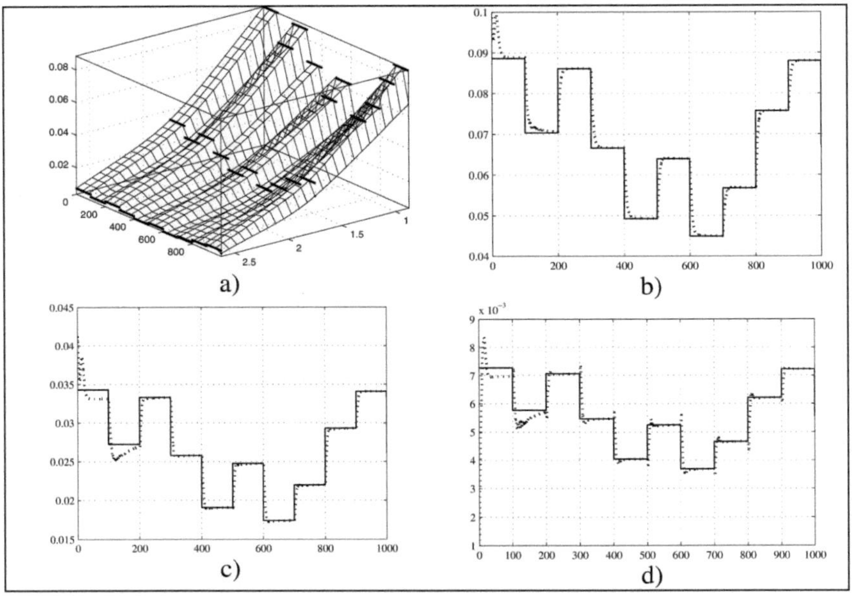

Fig. 13. SMC of X_2, (methanogenic bacteria in the fixed bed) (*dotted line – plant output, continuous line –reference signal)*; a) 3d view of X_2 b) SMC of X_2 in z=0.25H; c) SMC of X_2 in z=0.5H, d) SMC of X_2 in z=0.75H.

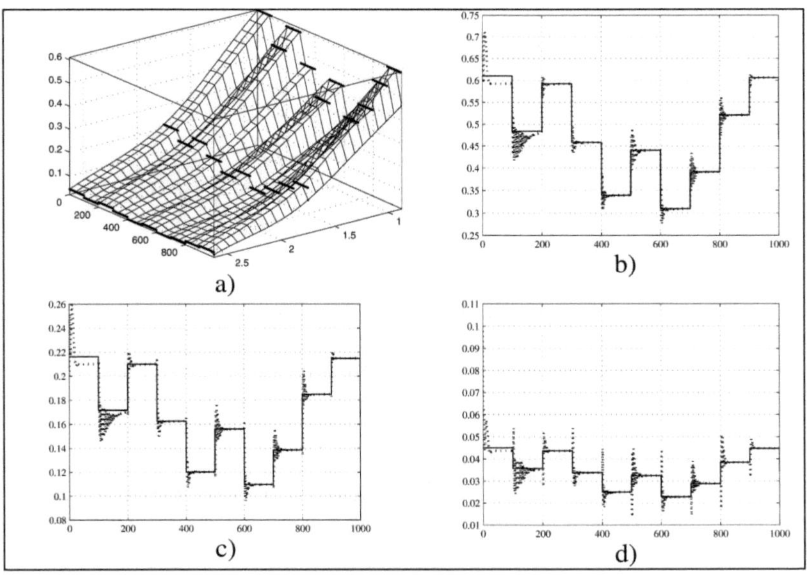

Fig. 14. SMC of S_1 (chemical oxygen demand in the fixed bed) (dotted line-plant output, continuous line –reference signal); a) 3d view of S_1; b) SMC of S_1 in z=0.25H; c) SMC of S_1 in z=0.5H; d) SMC of S_1 in z=0.75H

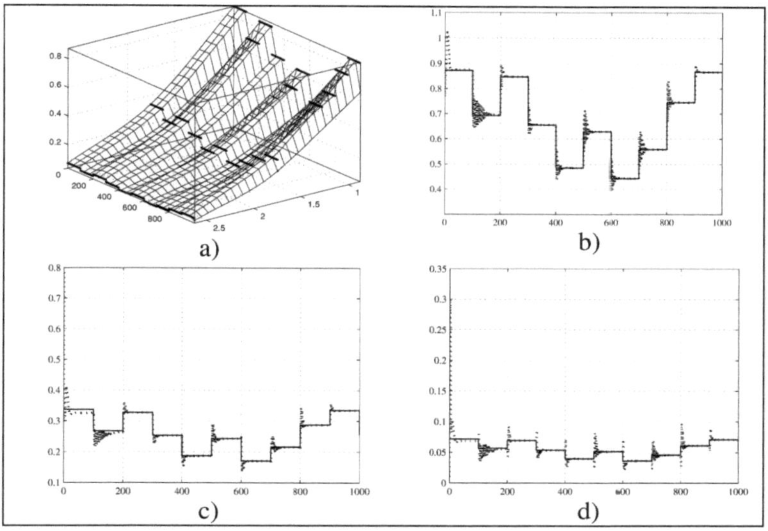

Fig. 15. SMC of S_2 (volatile fatty acids in the fixed bed) (dotted line- plant output, continuous line –reference signal); a) 3d view of S_2 ; b) SMC of S_2 in z=0.25H; c) SMC of S_2 in z=0.5H; d) SMC of S_2 in z=0.75H

In total, the MSE% results of bioprocess identification and control showed a very good RTNN convergence and precise reference tracking with slight priority of the SMC over the linear optimal control and the DANC due to the better plant dynamics compensation of the SMC.

7 Conclusions

The paper proposed to use a Recurrent Neural Network for centralized identification, direct and indirect control of Distributed Parameter Systems. An anaerobic digestion bioprocess plant model described by partial differential equations is used as an example of such system. The simplification of the DPS described by PDE equations is realized using the orthogonal collocation method in three collocation points, converting the PDE plant description in ODE one. The ODE system has excessive measurements and it is identified using a centralized RTNN model. The state and parameters estimated by the RTNN identifier are used for direct and indirect adaptive neural control, performing data fusion. The applied neural approach to complex systems identification and control exhibits a good RTNN convergence and precise plant outputs tracking. The MSE% of plant outputs tracking for the SMC and DANC are of order of E-008 in the worse case, outperforming the optimal control (E-005 in the worse case) due to its adaptability. The comparison of the MSE% of reference tracking gave slight priority of the SMC over DANC due to its better compensation ability.

Acknowledgements. The Ph.D. student Rosalba Galvan-Guerra is thankful to CONACYT for the scholarship received during her studies at the Department of Automatic Control, CINVESTAV-IPN, Mexico City, MEXICO.

References

1. Haykin, S.: Neural Networks, a Comprehensive Foundation, 2nd edn., Section 2.13, 84–89; Section 4.13, 208–213. Prentice-Hall, Upper Saddle River (1999)
2. Bulsari, A., Palosaari, S.: Application of Neural Networks for System Identification of an Adsorption Column. Neural Computing and Applications 1, 160–165 (1993)
3. Deng, H., Li, H.X.: Hybrid Intelligence Based Modeling for Nonlinear Distributed Parameter Process with Applications to the Curing Process. IEEE Transactions on Systems, Man and Cybernetics 4, 3506–3511 (2003)
4. Deng, H., Li, H.X.: Spectral-Approximation-Based Intelligent Modeling for Distributed Thermal Processes. IEEE Transactions on Control Systems Technology 13, 686–700 (2005)
5. Gonzalez-Garcia, R., Rico-Martinez, R., Kevrekidis, I.: Identification of Distributed Parameter Systems: A Neural Net Based Approach. Computers and Chemical Engineering 22(4-suppl. 1), 965–968 (1998)
6. Padhi, R., Balakrishnan, S., Randolph, T.: Adaptive Critic based Optimal Neuro-Control Synthesis for Distributed Parameter Systems. Automatica 37, 1223–1234 (2001)
7. Padhi, R., Balakrishnan, S.: Proper Orthogonal Decomposition Based Optimal Neurocontrol Synthesis of a Chemical Reactor Process Using Approximate Dynamic Programming. Neural Networks 16, 719–728 (2003)
8. Pietil, S., Koivo, H.N.: Centralized and Decentralized Neural Network Models for Distributed Parameter Systems. In: Proc. of the Symposium on Control, Optimization and Supervision, CESA 1996. IMACS Multiconference on Computational Engineering in Systems Applications, Lille, France, pp. 1043–1048 (1996)
9. Baruch, I.S., Mariaca-Gaspar, C.R., Barrera-Cortes, J.: Recurrent Neural Network Identification and Adaptive Neural Control of Hydrocarbon Biodegradation Processes. In: Hu, X., Balasubramaniam, P. (eds.) Recurrent Neural Networks, ch. 4, pp. 61–88. I-Tech Education and Publishing KG, Vienna (2008)
10. Baruch, I.S., Flores, J.M., Nava, F., Ramirez, I.R., Nenkova, B.: An Advanced Neural Network Topology and Learning, Applied for Identification and Control of a D.C. Motor. In: Proc. 1st International IEEE Symposium on Intelligent Systems, IS 2002, Varna, Bulgaria, vol. 1, pp. 289–295 (2002)
11. Baruch, I.S., Georgieva, P., Barrera-Cortes, J., Feyo de Azevedo, S.: Adaptive Recurrent Neural Network Control of Biological Wastewater Treatment (Special Issue on Soft Computing for Modeling, Simulation and Control of Nonlinear Dynamical Systems, Guest Eds: Castillo, O., Melin, P.) 20(2), 173–194 (2005)
12. Baruch, I.S., Flores, J.M., Thomas, F., Garrido, R.: Adaptive Neural Control of Nonlinear Systems. In: Dorffner, G., Bischof, H., Hornik, K. (eds.) ICANN 2001. LNCS, vol. 2130, pp. 930–936. Springer, Heidelberg (2001)
13. Baruch, I.S., Barrera-Cortes, J., Hernandez, L.A.: A Fed-Batch Fermentation Process Identification and Direct Adaptive Neural Control with Integral Term. In: Monroy, R., Arroyo-Figueroa, G., Sucar, L.E., Sossa, H. (eds.) MICAI 2004. LNCS (LNAI), vol. 2972, pp. 764–773. Springer, Heidelberg (2004)

14. Aguilar-Garnica, E., Alcaraz-Gonzalez, V., Gonzalez-Alvarez, V.: Interval Observer Design for an Anaerobic Digestion Process Described by a Distributed Parameter Model. In: Proc. of the Second International Meeting on Environmental Biotechnology and Engineering (2IMEBE), Mexico City, Mexico, 26-29 September, paper 117, pp. 1–16 (2006)
15. Wan, E., Beaufays, F.: Diagrammatic Method for Deriving and Relating Temporal Neural Networks Algorithms. Neural Computations 8, 182–201 (1996)
16. Young, K.D., Utkin, V.I., Ozguner, U.: A Control Engineer's Guide to Sliding Mode Control. IEEE Transactions on Control Systems Technology 7(3), 328–342 (1999)
17. Levent, A.: Higher Order Sliding Modes, Differentiation and Output Feedback Control. International Journal of Control, Special Issue Dedicated to Vadim Utkin on the Occasion of his 65th Birthday (Guest editor: Leonid M. Fridman) 76(9/10), 924–941 (2003)
18. Eduards, C., Spurgeon, S.K., Hebden, R.G.: On the Design of Sliding Mode Output Feedback Controllers International Journal of Control. Special Issue Dedicated to Vadim Utkin on the Occasion of his 65th Birthday (Guest editor: Leonid M. Fridman) 76(9/10), 893–905 (2003)

Part II
Pattern Recognition

An Ensemble Neural Network Architecture with Fuzzy Response Integration for Complex Time Series Prediction

Martha Pulido, Alejandra Mancilla, and Patricia Melin

Tijuana Institute of Technology, Tijuana, México
marthapulido_84@hotmail.com, alejandra.mancilla@gmail.com, epmelin@hafsamx.org

Abstract. In this paper we describe the application of an architecture for an ensemble neural network for Complex Time Series Prediction. The times series we are considering are: the Mackey-Glass, Dow Jones and Mexican Stock Exchange and we show the results of a set of trainings with the ensemble neural network, and its integration with the methods of average, weighted average and Fuzzy Integration. Simulation results show very good prediction of the ensemble neural network with fuzzy logic integration.

1 Introduction

Time series predictions are very important because we can analyze past events to know the possible behavior of futures events and thus can take preventive or corrective decisions to help avoid unwanted circumstances.

The choice and implementation of an appropriate method of prediction has always been a major issue for enterprises that seek to ensure the profitability and survival of business. The predictions give the company the ability to make decisions in the medium and long term, and due to the accuracy or inaccuracy of data could mean predicted growth or profits and financial losses.

It is very important for companies to know the behavior that will be the future development of their business, and thus be able to make decisions that can improve the company's activities, and avoid unwanted situations which in some cases can lead to the company's failure.

2 Time Series and Prediction

A time-series is defined as a sequence of observations on a set of values that takes a variable (quantitative) at different points in time. The time series are widely used today because organizations need to know the future behavior of certain phenomena in order to plan, prevent, and so on, their actions. That is, to predict what will happen with a variable in the future from the behavior of that variable in the

O. Castillo et al. (Eds.): Evolutionary Design of Intelligent Systems, SCI 257, pp. 85–110.
springerlink.com

past [1]. The data can behave in different ways over time, this may be a trend, which is the component that represents a long-term growth or decline in value over a period of time high. You can also have a cycle, which refers to the wave motion that occurs around the trend, or may not have a defined or random manner; there are seasonal variations (annual, biannual, etc.). , which is a behavior pattern that is repeated year after year at a particular time. [2].

The word "prediction" comes from the Latin prognosticum, which means I know in advance. Prediction is to issue a statement about what is likely to happen in the future, based on analysis and considerations of experiments. Making a forecast is to obtain knowledge about uncertain events that are important in decision-making [3]. Time series prediction tries to predict the future based on past data, it take a series of real data $x_t - n, ... x_t - 2, x_t - 1, x_t$ and then obtains the prediction of data $x_t + 1, x_t + 2 ... x_t + n$. The goal of time series prediction or a model is to observe the series of real data, so that future data may be accurately predicted. [4]

3 Neural Networks

Neural networks are composed of many elements (Artificial Neurons), grouped into layers and are highly interconnected (with the synapses), this structure has several inputs and outputs, which are trained to react (or give values) in a way you want to input stimuli (R values). These systems emulate in some way, the human brain. Neural networks are required to learn to behave (Learning) and someone should be responsible for the teaching or training (Training), based on prior knowledge of the environment problem [5].

Artificial neural networks are inspired by the architecture of the biological nervous system, which consists of a large number of relatively simple neurons that work in parallel to facilitate rapid decision-making [6].

A neural network is a system of parallel processors connected as a directed graph. Schematically each processing element (neuron) of the network is represented as a node. These connections establish a hierarchical structure that is trying to emulate the physiology of the brain as it looks for new ways of processing to solve real world problems. What is important in developing the techniques of NN is if its useful to learn behavior, recognize and apply relationships between objects and plots of real-world objects themselves. In this sense, artificial neural networks have been applied to many problems of considerable complexity. Its most important advantage is in solving problems that are too complex for conventional technologies, problems that have no solution or that the algorithm of the solution is very difficult to find [5].

4 Methods of Integration

There exists a diversity of methods of integration or aggregation of information, and we mention some of these methods below:

Integration by average: this method is used in the ensembles of networks. This integration method is the simplest and most straightforward, consists in the sum of the results generated by each module divided by the number of modules, and the disadvantage is that there are cases in which the prognosis is not good.

Integration of Weighted Average: this method is an extension of the integration by average, with the main difference that the weighted average assigns importance weights to each of the modules. These weights are assigned to a particular module based on several factors; the most important is the knowledge product of experience. This integration method belongs to the well known aggregation operators.

Fuzzy logic was proposed for the first time in the mid-sixties at the University of California Berkeley by the brilliant engineer Lotfi A. Zadeh. Who proposed what it's called the principle of incompatibility: "As the complexity of a system increases, our ability to give precise instructions and build on their behavior decreases to the threshold beyond which the accuracy and meaning are mutually exclusive characteristics." Then introduced the concept of a fuzzy set (Fuzzy Set), under which lies the idea that the elements on which to build human thinking are not numbers but linguistic labels. Fuzzy logic can represent the common knowledge that natural of language is mostly qualitative and not necessarily quantitative in a mathematical language by means of fuzzy set theory and function characteristics associated with them. [7].

Fuzzy Set: Let X be a space of objects and x be generic element of X. A classical set A, $A \subseteq X$, is defined as a collection of elements or objects $x \in X$, such that each x can either belong or not belong to the set A. By defining a characteristic function for each element x in X, we can represent a classical set A by a set of ordered pairs $(x, 0)$ or $(x, 1)$, which indicates $x \notin A$ or $x \in A$, respectively. Unlike the aforementioned conventional set, a fuzzy set expresses the degree to which an element belongs to a set. Hence the characteristic function of a fuzzy set is allowed to have values between 0 and 1, which denotes the degree of membership of an element in a given set. The basic structure of a fuzzy inference system consists of three conceptual components: a rule base, which contains a selection of fuzzy rules; (or dictionary), which defines the membership functions used in the fuzzy rules; and a reasoning mechanism, which performs the inference produce (usually the fuzzy reasoning). [8].

5 Genetic Algorithms

Genetic algorithms were introduced by the first time by a professor of the University of Michigan named John Holland [9]. A genetic algorithm, is a mathematical highly parallel algorithm that transforms a set of mathematical individual objects with regard to the time using operations based on evolution. The Darwinian laws of reproduction and survival of the fittest can be used, and after having appeared of natural form a series of genetic operations between(among the object) that stands out the sexual recombination [10,11]. Each of these mathematical objects is in the habit of being a chain of characters (letters or numbers) of fixed length that adjusts to the model of the chains of chromosomes, and one associates to them with a certain mathematical function that reflects the fitness.

6 Problem Statement and Proposed Method

This paper is concerned with the study of a fuzzy integration method that can be applied to ensemble neural networks with applications to complex time series, in addition to developing alternative methods for the integration of ensemble network, such as the average and weighted average. Figure 1 shows the general architecture used in this work.

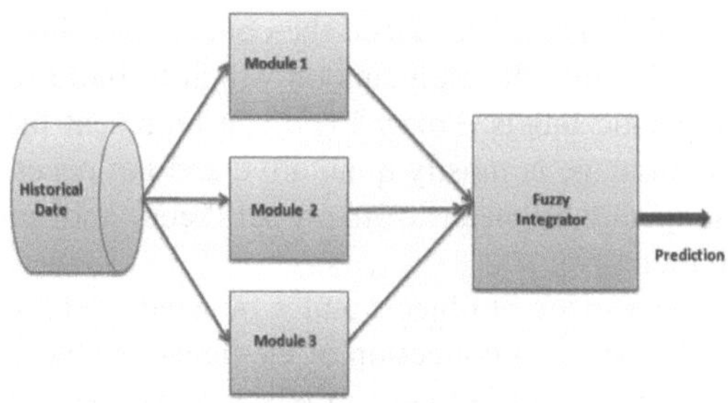

Fig. 1. General Architecture of the ensemble neural network.

Historical data of the Mackey-Glass time series was used for the ensemble neural network trainings, where each module was fed with the same information, to find a suitable architecture for a module of the ensemble will be same or very similar to the other modules, unlike the modular networks, where each module is fed with different data, which leads to architectures that are not uniform. Integration by the average method was very easy to implement, just joining the results of each module and the result was divided by the number of elements, the main problem of this method is that if one of the modules produces an unfortunate result, it can greatly affect the result the integration. Integration by weighted average includes assigning values from 0 to 1, where the module that has the best prediction is the one that will have a greater weight. The network consists of three modules, the allocation of weights we used is 0.50 for the module that produces better results, 0.30 for second best and 0.20 for the worst module, we do this only if the three modules meet the above conditions. Fuzzy integration for the Mackey-Glass time series was implemented in a system of traditional Mamdani fuzzy inference, which consists of three input variables (the results of each module of our ensemble network) and one output variable (the result of integration), is shown in Figure 2.

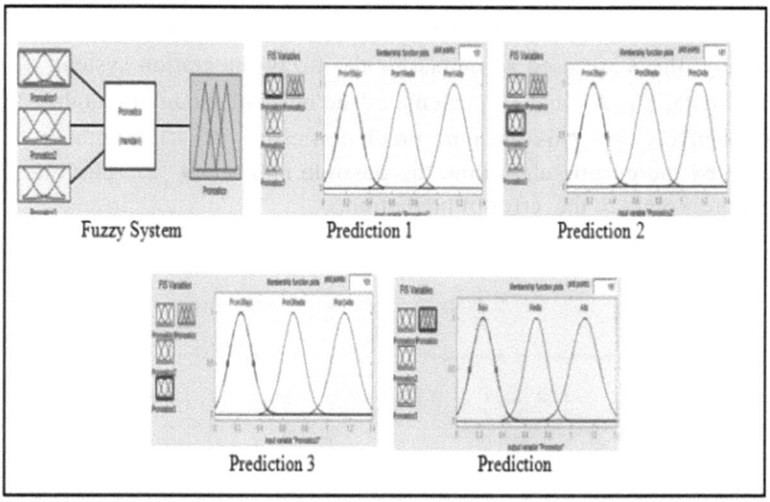

Fig. 2. Fuzzy Inference System.

To Optimize the Fuzzy System for Integration of the Mackey-Glass, Dow Jones and the Mexican Stock Exchange time series we implemented a genetic algorithm to optimize the membership functions and rules of the fuzzy system. Where the objective function is defined to minimize the prediction error:

$$f_1 = \left(\sum_{i=1}^{297} |a_i - x_i| \right) / 297$$

$$f_2 = \left(\sum_{i=1}^{297} |b_i - x_i| \right) / 297 \tag{1}$$

$$f_3 = \left(\sum_{i=1}^{297} |c_i - x_i| \right) / 297$$

$$f = \left(\left((f_1 + f_2 + f_3) / 3 \right) + \left((R/27) / 100 \right) \right)$$

Where $a, b, and\ c,$ correspond to prediction1, prediction2 and prediction3, respectively; these are used as inputs for the fuzzy integration system, X represents real data, f_1, f_2 and f_3 calculates the error of module1, module2, module3 respectively and f is the total prediction error, R is the number of rules generated by the genetic algorithm, the possible number of **27** rules and **100** to assign more weight to the error of prediction.

The corresponding chromosome structure is shown in Figure 3.

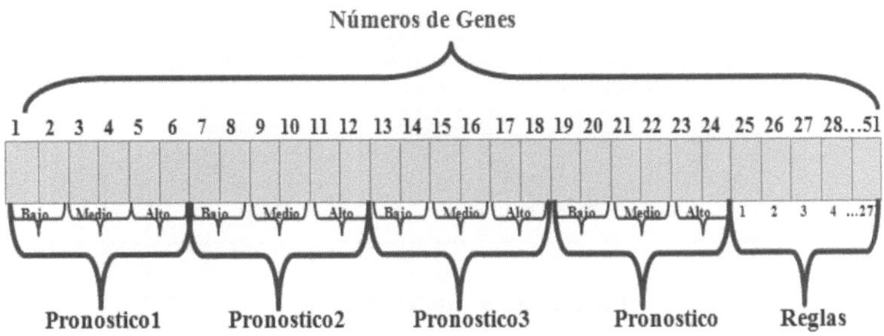

Fig. 3. Chromosome Structure

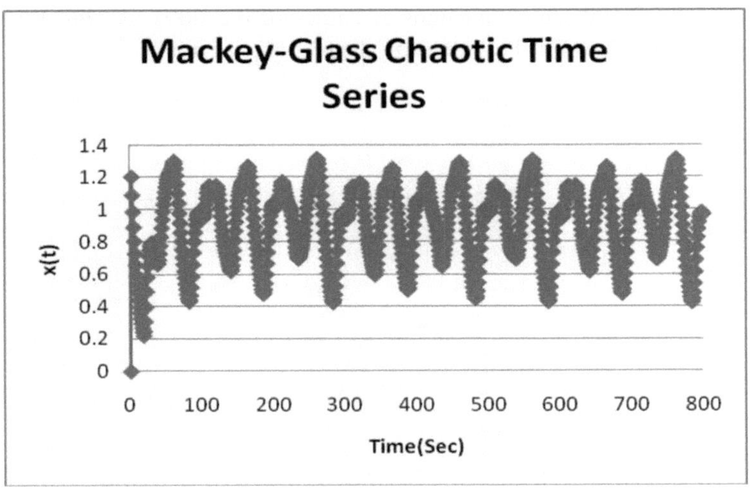

Fig. 4. Series data Mackey-Glass.

Data of the Mackey-Glass time series was generated using equation (1). We are using 800 points. We use 70% of the data for the ensemble neural network trainings and 30% to test the network.

The Mackey-Glass Equation is defined as follows:

$$\dot{x}(t) = \frac{0.2x(t-\tau)}{1+x^{10}(t-\tau)} - 0.1x(t) \tag{2}$$

Where it is assumed x (0) = 1.2, t = 17, and x (t) = 0 for t <0. Figure 4 shows a plot of the time series for these parameter values.

This time series is chaotic, and there is no clearly defined period over time. The series does not converge or diverge, and the trajectory is extremely sensitive to the initial conditions. The time series is measured in number of points, and we apply the fourth order Runge-Kutta method to find the numerical solution of the equation [12].

Data of the Dow Jones time series: We are using 800 points that correspond from 11/03/05 to 01/08/09. We used 70% of the data for the ensemble neural network trainings and 30% to test the network [13].

Figure 5 shows a plot of the time series for these parameter values.

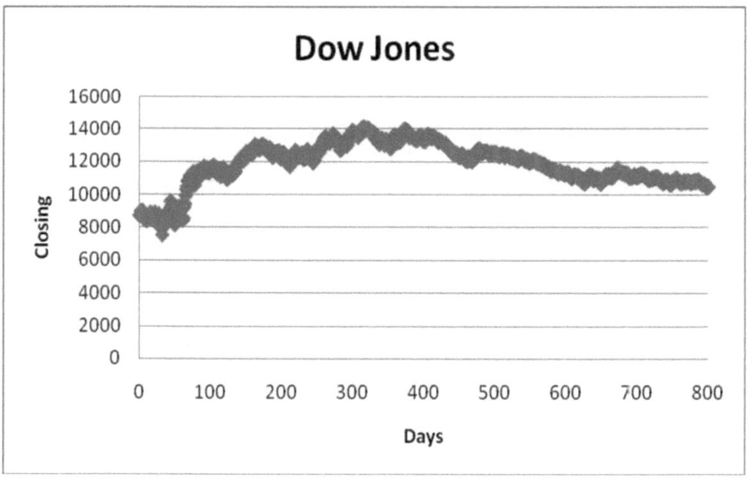

Fig. 5. Series data Dow Jones.

Data of the Mexican Stock Exchange time series: We are using 800 points that correspond from 11/09/05 to 01/15/09. We used 70% of the data for the ensemble neural network trainings and 30% to test the network [14]. We show in Figure 6 the plot of this time series.

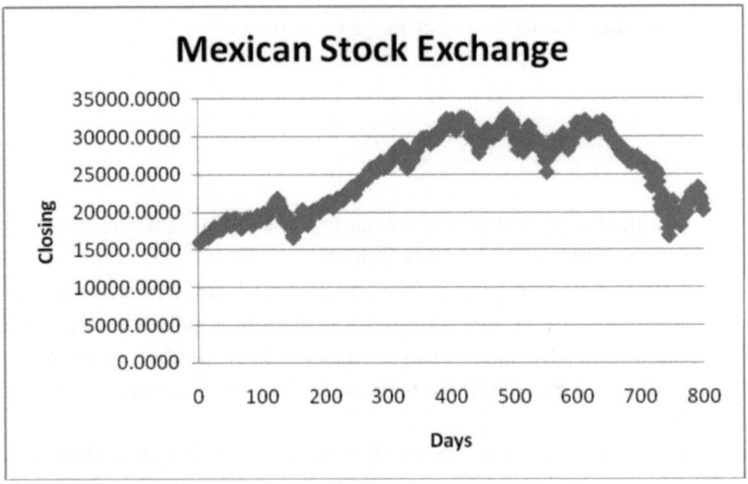

Fig. 6. Mexican Stock Exchange.

7 Simulation Results

In this section we show results for the three time series using the Ensemble NN with fuzzy Integration.

7.1 Simulation Results for the Mackey-Glass Time Series

The architecture of the ensemble network that produced the best results is shown in Figure 7 for the Mackey-Glass Time Series [14].

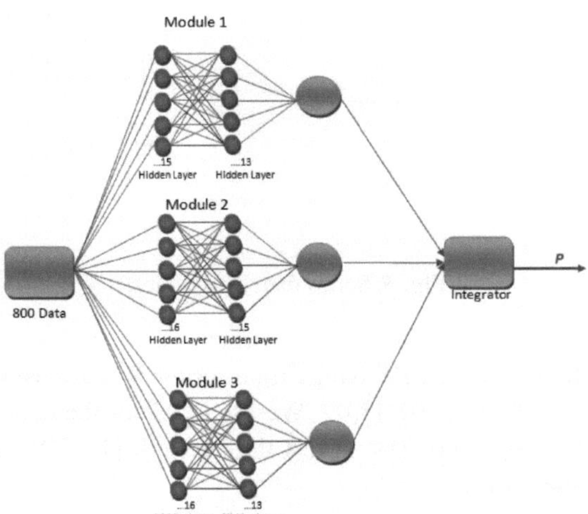

Fig. 7. The best Network Architecture for Mackey-Glass.

In this architecture we used two layers in each module. In module 1, in the first layer we used 15 neurons and 13 neurons in second layer. In module 2 we used 15 neurons in the first layer and 13 neurons in the second, and in module 3 we used 16 neurons in the first layer and 13 neurons in the second. The training method used was the Levenberg-Marquardt (LM); we also applied 3 delays to the network.

The best training was the one shown in row number 3, using 1 layer in each of the modules of the ensemble network, implementing 3 delays in the network, the training method used was Levenberg-Marquardt (LM), (as shown in Table 1) where the total error of this training was: 0.1296.

Table 1. Results of Training using 1 layer for Mackey-Glass.

	Number of Layers	Number of Neurons	Number of Delays	Target Error	Number Max Epoch	Epoch	Time	Error by Module	Error Total
Training1	1	37 35 35	3	0.00000001 0.00001 0.00001	16000 20000 20000	4 1 1	1 Seg 1 Seg 1 Seg	0.0144 0.0617 0.0417	0.2703
Training2	1	37 35 35	3	0.00000001 0.00001 0.00001	16000 20000 20000	4 2 1	1 Seg 1 Seg 1 Seg	0.0983 0.074 0.024	0.1803
Training3	1	29 30 31	3	0.000001 0.00001 0.00001	6000 2000 5000	3 1 1	1 Seg 1 Seg 1 Seg	0.0777 0.0347 0.0517	0.1296
Training4	1	25 28 29	3	0.0000001 0.0000001 0.00000001	6000 3000 5000	2 2 4	1 Seg 1 Seg 1 Seg	0.1072 0.0356 0.0973	0.1752
Training5	1	27 30 28	3	0.0000001 0.0000001 0.00000001	6000 3000 5000	2 2 7	1 Seg 1 Seg 1 Seg	0.0894 0.0577 0.058	0.1664

The best training was the one shown in row number 5 using 2 layers in each of the modules of the ensemble network, implementing 3 delays in the network, the training method used was Levenberg-Marquardt (LM), (as shown in Table 2) where the total error of this training was: 0.01396.

Table 2. Results of Training using 2 layers for Mackey-Glass.

	Number of Layers	Number of Neurons	Number of Delays	Target Error	Number Max Epoch	Epoch	Time	Error by Module	Total Error
Training1	2	29,10 18,18 22,12	3	0.00000001 0.00000001 0.000001	11000 20000 10000	7 8 5	1 Seg 1 Seg 1 Seg	0.076 0.0425 0.1157	0.1571
Training2	2	29,12 18,16 20,17	3	0.00000001 0.00000001 0.000001	11000 20000 10000	7 9 8	1 Seg 1 Seg 1 Seg	0.0533 0.0479 0.0795	0.1277
Training3	2	18,12 18,17 21,19	3	0.0000001 0.000000001 0.000001	7000 1000 9000	9 11 7	1 Seg 1 Seg 1 Seg	0.0924 0.0543 0.1672	0.2025
Training4	2	18,13 15,18 15,18	3	0.0000001 0.000000001 0.000000001	7000 1000 1000	10 26 9	1 Seg 1 Seg 1 Seg	0.0579 0.0798 0.073	0.162
Training5	2	15,13 16,15 11,13	3	0.00000001 0.00000001 0.0000001	6000 3000 10000	11 8 10	1 Seg 1 Seg 1 Seg	0.0196 0.0116 0.0098	0.0136

The best training was the one shown in row number 5 varying between 1 and 2 layers in each of the modules of the ensemble network, implementing 3 delays in the network, the training method used was Levenberg-Marquardt (LM), (as shown in Table 3) where the total error of this training was: 0.01396.

Table 3. Results of Training using 1and 2 layers for Mackey-Glass.

	Number of Layers	Number of Neurons	Number of Delays	**Target Error**	Number Max Epoch	Epoch	Time	**Error by** Module	Total Error
Training1	1 y 2	10,17 30 18,11	3	0.0000001 0.0000001 0.00000001	6000 3000 5000	6 1 10	1 Seg 1 Seg 1 Seg	0.0622 0.0295 0.0255	0.1002
Training2	1 y 2	13,12 30 10,17	3	0.00000001 0.00000001 0.00000001	6000 2000 5000	8 11 9	1 Seg 1 Seg 1 Seg	0.027 0.1237 0.0613	0.1712
Training3	1 y2	13,15 31 15,10	3	0.00000001 0.00000001 0.0000001	6000 2000 3000	11 4 11	1 Seg 1 Seg 1 Seg	0.1134 0.1016 0.0407	0.2285
Training4	1 y 2	31 31 15,10	3	0.00000001 0.00000001 0.00000001	2000 2000 5000	55 4 9	1 Seg 1 Seg 1 Seg	0.0848 0.1058 0.0926	0.2215
Training5	1 y 2	23 25 10,14	3	0.000000001 0.0000001 0.0000001	6000 1000 3000	44 1 8	1 Seg 1 Seg 1 Seg	0.0144 0.0617 0.417	0.0899

The result is training number 3 (as shown in Table 4), where the integration method used was average integration; and we obtained an error of 0.025, with weighted average integration we obtained an error of 0.0461, with fuzzy integration we obtained an error of 0.0789 and with optimized fuzzy integration we obtained an error of 0.038131.

The best method of integration for this architecture was the average integration with an error of 0.0205.

Table 4. Results of the Integration Methods with 1 layer for Mackey-Glass.

	Average Integration	Weighted Average Integration	Fuzzy Integration	Optimized Fuzzy Integration
Training1	0.0417	0.0558	0.1151	0.05127
Training2	0.317	0.0436	0.0715	0.037832
Training3	0.025	0.0461	0.0789	0.038131
Training4	0.0205	0.0245	0.0694	0.028319
Training5	0.0261	0.0326	0.0787	0.036614

The best result is training 5 (as shown in Table 5), where the integration method used was average integration; we obtained an error of 0.0106, with weighted average integration we obtained an error of 0.0126, with fuzzy integration we

Table 5. Results of the Integration Methods with 2 layers for Mackey-Glass.

	Average Integration	Weighted Average Integration	Fuzzy Integration	Optimized Fuzzy Integration
Training1	0.0396	0.0737	0.0813	0.040998
Training2	0.0388	0.0439	0.0815	0.033412
Training3	0.0583	0.075	0.0826	0.052223
Training4	0.0272	0.0355	0.0899	0.04853
Training5	0.0106	0.0126	0.0708	0.023317

obtained an error of 0.0708 and with optimized fuzzy integration we obtained an error of 0.023317.

The best result is training 5 (as shown in Table 6), where the integration method used was average integration; we obtained an error of 0.0241, with weighted average integration we obtained an error of 0.0348, with fuzzy integration we obtained an error of 0.0751 and with optimized fuzzy integration we obtained an error of 0.023317.

Where the best method for this architecture was the average integration with an error of 0.0241.

Table 6. Results of the Integration Methods with 1 and layers for Mackey-Glass

	Average Integration	Weighted Average Integration	Fuzzy Integration	Optimized Fuzzy Integration
Training1	0.0264	0.037	0.0681	0.040998
Training2	0.0258	0.053	0.0728	0.033412
Training3	0.03	0.0307	0.0738	0.052223
Training4	0.0312	0.0297	0.0963	0.048869
Training5	0.0241	0.0348	0.0751	0.026237

The best result obtained for the Mackey-Glass time series was number 5 using 2 layers in each of the modules of the ensemble network, implementing 3 delays, the training method used was the Levenberg-Marquardt (LM), (as shown in Table 2 and in Figure 8).

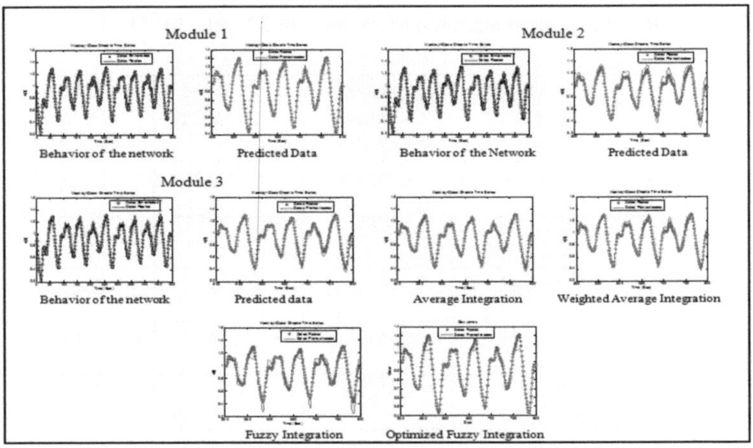

Fig. 8. Simulation results of training 5 for Mackey-Glass.

Table 7. Genetic algorithm results for training 5 for Mackey-Glass.

Num.	Ind.	Gen.	GGP	selection	Mutation	Pm	Crossover	Pc	Pm Rules	Núm. Rules	Duration	GA Error	Prediction Error
1	100	100	0.85	rws	mutbga	0.07	xovdprs	0.5	0.002	22	01:56:46	0.031767	0.023619
2	200	100	0.85	rws	mutbga	0.03	xovdprs	0.01	0.002	22	03:10:08	0.030951	0.022803
3	250	100	0.85	rws	mutbga	0.04	xovdprs	0.02	0.002	22	04:17:40	0.033902	0.025754
4	150	100	0.85	rws	mutbga	0.06	xovdprs	0.8	0.002	27	02:36:06	0.035352	0.025352
5	100	100	0.85	rws	mutbga	0.1	xovdprs	1	0.002	27	01:36:32	0.03835	0.02835
6	200	400	0.85	rws	mutbga	0.05	xovdprs	0.75	0.002	20	13:13:53	0.030779	0.0233716
7	100	200	0.85	rws	mutbga	0.05	xovdprs	0.9	0.002	27	03:42:17	0.060882	0.050882
8	100	200	0.85	rws	mutbga	0.06	xovdprs	1	0.002	27	03:36:33	0.045956	0.035956
9	100	150	0.85	rws	mutbga	0.08	xovdprs	0.05	0.002	27	02:19:40	0.046778	0.035778
10	100	150	0.85	rws	mutbga	0.5	xovdprs	0.7	0.002	27	02:23:40	0.051033	0.041033
11	100	150	0.85	rws	mutbga	0.06	xovdprs	0.09	0.002	27	02:23:18	0.057214	0.047214
12	100	150	0.85	rws	mutbga	0.06	xovdprs	0.5	0.002	27	02:17:30	0.049478	0.039478
13	100	150	0.85	rws	mutbga	0.05	xovdprs	0.65	0.002	27	02:22:37	0.043053	0.033053

Table7 shows the genetic algorithm results for training 5 (as shown in Table 2 and in Figure 9). Where the prediction error is of 0.023372.

The fuzzy integration system generated by the genetic algorithm is represented in figure 10:

Comparing Figure 2, which belongs to the base fuzzy integrator (not optimized) with Figure 9, which belongs to the fuzzy integrator generated by the genetic algorithm, it can be seen that the parameters of membership functions are different, the number of rules obtained are 20 and this helps to improve the prediction error.

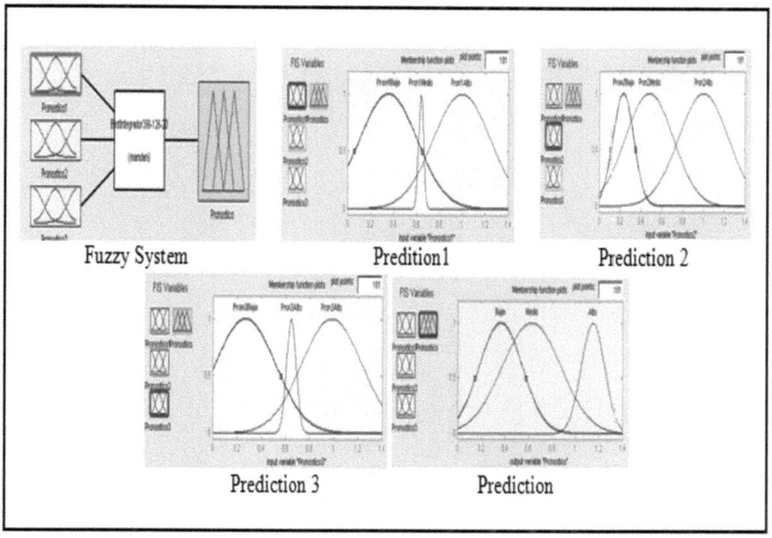

Fig. 9. Fuzzy system generated by genetic algorithm for Mackey-Glass.

1. If (pronostico1 is Pron1Bajo) and (Pronostico2 is Pron2Bajo) and (Pronostico3 is Pron3Alto) then (Pronostico is Bajo)
2. If (Pronostico1 is Pron1Bajo) and (Pronostico2 is Pron2Medio) and (Pronostico3 is Pron3Bajo) then (Pronostico is Bajo)
3. If (Pronostico1 is Pron1Bajo) and (Pronostico2 is Pron2Medio) and (Pronostico3 is Pron3Alto) then (Pronostico is Bajo)
4. If (Pronostico1 is Pron1Bajo) and (Pronostico2 is Pron2Medio) and (Pronostico3 is Pron3Alto) then (Pronostico is Medio)
5. If (Pronostico1 is Pron1Bajo) and (Pronostico2 is Pron2Alto) and (Pronostico3 is Pron3Medio) then (Pronostico is Bajo)
6. If (Pronostico1 is Pron1Bajo) and (Pronostico2 is Pron2Alto) and (Pronostico3 is Pron3Alto) then (Pronostico is Medio)
7. If (Pronostico1 is Pron1Medio) and (Pronostico2 is Pron2Bajo) and (Pronostico3 is Pron3Bajo) then (Pronostico is Alto)
8. If (Pronostico1 is Pron1Medio) and (Pronostico2 is Pron2Bajo) and (Pronostico3 is Pron3Alto) then (Pronostico is Medio)
9. If (Pronostico1 is Pron1Medio) and (Pronostico2 is Pron2Medio) and (Pronostico3 is Pron3Bajo) then (Pronostico is Alto)
10. If (Pronostico1 is Pron1Medio) and (Pronostico2 is Pron2Medio) and (Pronostico3 is Pron3Medio) then (Pronostico is Medio)
11. If (Pronostico1 is Pron1Medio) and (Pronostico2 is Pron2Medio) and (Pronostico3 is Pron3Alto) then (Pronostico is Medio)
12. If (Pronostico1 is Pron1Medio) and (Pronostico2 is Pron2Alto) and (Pronostico3 is Pron3Medio) then (Pronostico is Bajo)
13. If (Pronostico1 is Pron1Alto) and (Pronostico2 is Pron2Bajo) and (Pronostico3 is Pron3Bajo) then (Pronostico is Alto)
14. If (Pronostico1 is Pron1Alto) and (Pronostico2 is Pron2Bajo) and (Pronostico3 is Pron3Alto then (Pronostico is Medio)
15. If (Pronostico1 is Pron1Alto) and (Pronostico2 is Pron2Medio) and (Pronostico3 is Pron3Bajo) then (Pronostico is Alto)
16. If (Pronostico1 is Pron1Alto) and (Pronostico2 is Pron2Medio) and (Pronostico3 is Pron3Alto) then (Pronostico is Medio)
17. If (Pronostico1 is Pron1Alto) and (Pronostico2 is Pron2Alto) and (Pronostico3 is Pron3Bajo) then (Pronostico is Alto)
18. If (Pronostico1 is Pron1Alto) and (Pronostico2 is Pron2Alto) and (Pronostico3 is Pron3Medio) then (Pronostico is Alto)
19. If (Pronostico1 is Pron1Alto) and (Pronostico2 is Pron2Alto) and (Pronostico3 is Pron3Alto) then (Pronostico is Alto)
20. If (Pronostico1 is Pron1Bajo) and (Pronostico2 is Pron2Bajo) and (Pronostico3 is Pron3Bajo) then (Pronostico is Alto)

Fig. 10. Rules Fuzzy system generated by genetic algorithm for Mackey-Glass.

7.2 Simulation Results for the Dow Jones Time Series

The architecture of the ensemble network that produced the best results for Dow Jones Time Series is shown in Figure 11.

In this architecture we used one layer in each module. In module 1, in the first layer we used 37 neurons in module 2 we used 38 neurons in the first layer and in module 3 we used 38 neurons in first layer. The training method used was the Levenberg-Marquardt (LM); we also applied 3 delays to the network.

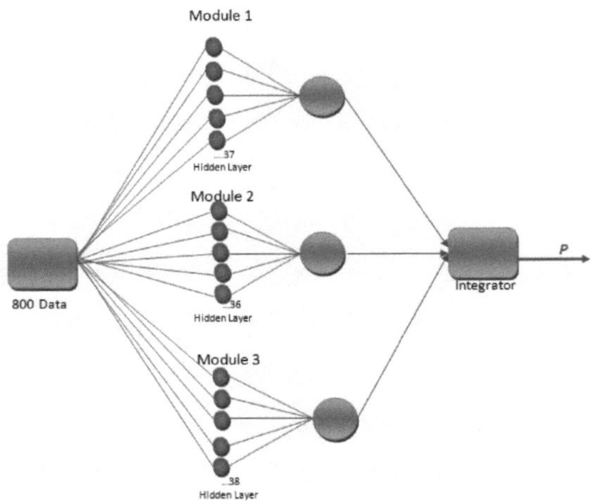

Fig. 11. The Best Network Architecture for Dow Jones.

The best training was the one shown in row number 4, using 1 layer in each of the modules of the ensemble network, implementing 3 delays in the network, the training method used was Levenberg-Marquardt (LM), (as shown in Table 8) where the total error of this training was: 0.00000886.

Table 8. Results of Training using 1 layer for Dow Jones.

	Number of Layers	Number of Neurons	Number of Delays	Target Error	Number Max Epoch	Epoch	Time	Error by Module	Error Total
Training1	1	25	3	0.1	22000	12	00:00:01	0.0005738	
		22		0.1	30000	4	00:00:01	0.00004885	0.00016782
		21		0.1	3000	12	00:00:01	0.00039722	
Training2	1	26	3	0.1	2000	8	00:00:01	0.00003269	
		25		0.1	3000	7	00:00:01	0.00027969	0.00024572
		24		0.1	3500	10	00:00:01	0.00042478	
Training3	1	27	3	0.1	2000	5	00:00:01	0.00024953	
		26		0.1	3000	6	00:00:01	0.00003122	0.00013306
		27		0.1	3500	4	00:00:01	0.00011843	
Training4	1	37	3	0.1	2000	6	00:00:01	0.00000242	
		36		0.1	3000	7	00:00:01	0.00002334	0.00000886
		38		0.1	3500	5	00:00:01	0.00000084	
Training5	1	25	3	0.1	2000	8	00:00:01	0.00001953	
		31		0.1	3000	7	00:00:01	0.00009723	0.00009188
		33		0.1	3500	6	00:00:01	0.00016338	

The best training was the one shown in row number 1 using 2 layers in each of the modules of the ensemble network, implementing 3 delays in the network, the training method used was Levenberg-Marquardt (LM), (as shown in Table 9) where the total error of this training was: 0.00018046.

Table 9. Results of Training using 2 layers for Dow Jones.

	Number of Layers	Number of Neurons	Number of Delays	Target Error	Number Max Epoch	Epoch	Time	Error by Module	Error Total
Training1	2	19,17	3	0.1	5000	8	00:00:02	0.00001458	
		19,17		0.01	5000	7	00:00:02	0.00013627	0.00018046
		15,19		0.1	3500	7	00:00:01	0.00039054	
Training2	2	20,18	3	0.1	5000	7	00:00:01	0.00019284	
		19,18		0.1	5000	6	00:00:01	0.00026851	0.00030629
		18,19		0.1	3500	6	00:00:01	0.00045751	
Training3	2	20,22	3	0.1	2000	6	00:00:03	0.00016737	
		21,22		0.1	5000	9	00:00:03	0.00072676	0.00032187
		21,23		0.1	3500	8	00:00:03	0.00007148	
Training4	2	12,14	3	0.1	2000	8	00:00:02	0.00028912	
		15,13		0.1	5000	7	00:00:01	0.00053855	0.00036097
		16,15		0.1	3500	4	00:00:01	0.00025523	
Training5	2	13,14	3	0.1	2000	6	00:00:01	0.00020369	
		15,16		0.1	5000	5	00:00:01	0.00047477	0.00028602
		16,18		0.1	3500	6	00:00:01	0.00017961	

The best training was the one shown in row number 4 varying between 1 and 2 layers in each of the modules of the ensemble network, implementing 3 delays in the network, the training method used was Levenberg-Marquardt (LM), (as shown in Table 10) where the total error of this training was: 0.00008543.

Table 10. Results of Training using 1 y 2 layers for Dow Jones.

	Number of Layers	Number of Neurons	Number of Delays	Target Error	Number Max Epoch	Epoch	Time	Error by Module	Error Total
Training1	1 y 2	21,21	3	0.1	2000	12	00:00:12	0.000155	
		38		0.1	3500	10	00:00:01	0.000412	0.000357
		21,21		0.1	3500	16	00:00:16	0.000504	
Training2	1 y 2	17,15	3	0.1	2000	7	00:00:03	0.0001679	
		32		0.1	3500	4	00:00:01	0.0000573	0.0001957
		15,16		0.1	3500	6	00:00:02	0.0001334	
Training3	1 y 2	17,19	3	0.1	2000	6	00:00:02	0.00019984	
		34		0.1	3500	5	00:00:00	0.00007986	0.00010505
		17		0.1	3500	5	00:00:01	0.00003545	
Training4	1 y 2	30	3	0.1	2000	7	00:00:01	0.00007017	
		19,17		0.1	3500	7	00:00:01	0.00013854	0.00008543
		32		0.1	3500	6	00:00:00	0.00004757	
Training5	1 y 2	34	3	0.1	2000	5	00:00:01	0.00017922	
		18,16		0.1	3500	7	00:00:01	0.00009751	0.00009281
		34		0.1	3500	7	00:00:00	0.0000017	

The result is training number 4 (as shown in Table 11), where the integration method used was average integration; and we obtained an error of 0.0053, with weighted average integration we obtained an error of 0.0060, with fuzzy integration we obtained an error of 0.0194 and with optimized fuzzy integration we obtained an error of 0.006863.

The best method of integration for this architecture was the average integration with an error of 0.0053.

Table 11. Results of the Integration Methods with 1 layer for Dow Jones.

	Average Integration	Weighted Average Integration	Fuzzy Integration	Optimized Fuzzy Integration
Training1	0.0215	0.0290	0.0207	0.007044
Training2	0.0408	0.0466	0.0225	0.007108
Training3	0.0271	0.0336	0.0190	0.007176
Training4	0.0053	0.0060	0.0194	0.006863
Training5	0.0085	0.0117	0.0257	0.007879

The best result is training 1 (as shown in Table 12), where the integration method used was average integration; we obtained an error of 0.0208, with weighted average integration we obtained an error of 0.0406, with fuzzy integration we obtained an error of 0.0315 and with optimized fuzzy integration we obtained an error of 0.008987.

The best method of integration for this architecture was the optimized fuzzy Integration with an error of 0.008987.

Table 12. Results of the Integration Methods with 2 layers for Dow Jones.

	Average Integration	Weighted Average Integration	Fuzzy Integration	Optimized Fuzzy Integration
Training1	0.0208	0.0406	0.0226	0.008987
Training2	0.0205	0.0338	0.0315	0.006189
Training3	0.0588	0.0656	0.0282	0.009745
Training4	0.0419	0.0537	0.0356	0.006242
Training5	0.0214	0.0460	0.0378	0.012783

The best result is training 4 (as shown in Table 13), where the integration method used was average integration; we obtained an error of 0.0069, with weighted average integration we obtained an error of 0.0098, with fuzzy integration we obtained an error of 0.0455 and with optimized fuzzy integration we obtained an error of 0.008387.

The best method of integration for this architecture was the average integration with an error of 0.008387.

Table 13. Results of the Integration Methods with 1and 2 layers for Mackey-Glass.

	Average Integration	Weighted Average Integration	Fuzzy Integration	Optimized Fuzzy Integration
Training1	0.0095	0.0146	0.0305	0.008766
Training2	0.0124	0.0157	0.0312	0.007788
Training3	0.0097	0.0174	0.0241	0.010623
Training4	0.0069	0.0098	0.0455	0.008387
Training5	0.0102	0.0132	0.0365	0.009729

The best result obtained for the Dow Jones time series was number 4 using 1 layer in each of the modules of the ensemble network, implementing 3 delays, the training method used was Levenberg-Marquardt (LM), (as shown in Table 13 and in Figure 12).

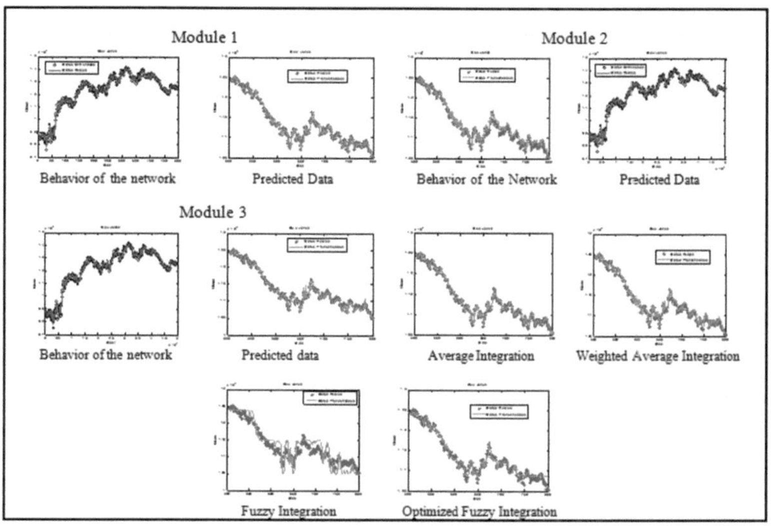

Fig. 12. Simulation results of training 4 for Dow Jones time series.

Table 14. Genetic algoritm results for training 4 for the Dow Jones time series.

Num.	Ind.	Gen.	GGP	Selection	Mutation	Pm	Crossover	Pc	Pm Rules	Núm. Rules	Duration	GA Error	Prediction Error
1	100	100	0.85	rws	mutbga	0.8	xovdprs	1	0.002	27	02:06:04	0.018556	0.017556
2	100	100	0.85	rws	mutbga	0.07	xovdprs	0.5	0.002	22	02:07:44	0.015011	0.006863
3	100	100	0.85	rws	mutbga	0.6	xovdprs	1	0.002	22	02:05:34	0.016055	0.007907
4	100	100	0.85	rws	mutbga	0.06	xovdprs	0.8	0.002	20	02:06:33	0.01828	0.01051
5	100	100	0.85	rws	mutbga	0.5	xovdprs	1	0.002	22	02:06:08	0.018319	0.010171
6	100	100	0.85	rws	mutbga	0.05	xovdprs	0.8	0.002	21	02:07:49	0.015718	0.007941
7	100	100	0.85	rws	mutbga	0.06	xovdprs	0.9	0.002	20	02:07:49	0.016283	0.008876
8	100	100	0.85	rws	mutbga	0.06	xovdprs	0.5	0.002	22	02:09:36	0.016539	0.008391
9	100	100	0.85	rws	mutbga	0.7	xovdprs	0.5	0.002	22	02:34:40	0.021852	0.013704
10	100	100	0.85	rws	mutbga	0.09	xovdprs	0.5	0.002	22	02:39:27	0.020939	0.012791

Table14 shows the genetic algorithm results for training 4 (as shown in Table 13 and in Figure 12) where the prediction error is of 0.006863.

The fuzzy integration system generated by the genetic algorithm is represented in Figure 13 and the fuzzy rules in Figure 14.

Comparing Figure 2, which belongs to the base fuzzy integrator (not optimized) with Figure 13, which belongs to the fuzzy integrator generated by the genetic algorithm, it can be seen that the parameters of membership functions are different, the number of rules obtained are 22 and this helps to improve the error prediction.

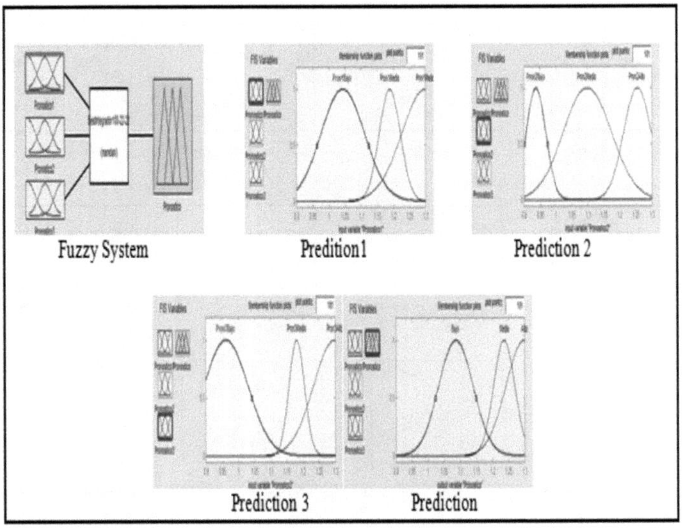

Fig. 13. Fuzzy system generated by the genetic algorithm for the Dow Jones time series.

1. If (pronostico1 is Pron1Bajo) and (Pronostico2 is Pron2Bajo) and (Pronostico3 is Pron3Medio) then (Pronostico is Bajo)
2. If (Pronostico1 is Pron1Bajo) and (Pronostico2 is Pron2Bajo) and (Pronostico3 is Pron3Alto) then (Pronostico is Bajo)
3. If (Pronostico1 is Pron1Bajo) and (Pronostico2 is Pron2Medio) and (Pronostico3 is Pron3Bajo) then (Pronostico is Bajo)
4. If (Pronostico1 is Pron1Bajo) and (Pronostico2 is Pron2Medio) and (Pronostico3 is Pron3Alto) then (Pronostico is Medio)
5. If (Pronostico1 is Pron1Bajo) and (Pronostico2 is Pron2Medio) and (Pronostico3 is Pron3Bajo) then (Pronostico is Bajo)
6. If (Pronostico1 is Pron1Bajo) and (Pronostico2 is Pron2Alto) and (Pronostico3 is Pron3Bajo) then (Pronostico is Alto)
7. If (Pronostico1 is Pron1Bajo) and (Pronostico2 is Pron2Alto) and (Pronostico3 is Pron3Medio) then (Pronostico is Bajo)
8. If (Pronostico1 is Pron1Bajo) and (Pronostico2 is Pron2Alto) and (Pronostico3 is Pron3Alto) then (Pronostico is Medio)
9. If (Pronostico1 is Pron1Medio) and (Pronostico2 is Pron2Bajo) and (Pronostico3 is Pron3Bajo) then (Pronostico is Alto)
10. If (Pronostico1 is Pron1Medio) and (Pronostico2 is Pron2Bajo) and (Pronostico3 is Pron3Medio) then (Pronostico is Alto)
11. If (Pronostico1 is Pron1Medio) and (Pronostico2 is Pron2Bajo) and (Pronostico3 is Pron3Alto) then (Pronostico is Medio)
12. If (Pronostico1 is Pron1Medio) and (Pronostico2 is Pron2Medio) and (Pronostico3 is Pron3Bajo) then (Pronostico is Alto)
13. If (Pronostico1 is Pron1Medio) and (Pronostico2 is Pron2Medio) and (Pronostico3 is Pron3Medio) then (Pronostico is Medio)
14. If (Pronostico1 is Pron1Medio) and (Pronostico2 is Pron2Alto) and (Pronostico3 is Pron3Bajo) then (Pronostico is Medio)
15. If (Pronostico1 is Pron1Medio) and (Pronostico2 is Pron2Bajo) and (Pronostico3 is Pron3Bajo) then (Pronostico is Alto)
16. If (Pronostico1 is Pron1Medio) and (Pronostico2 is Pron2Bajo) and (Pronostico3 is Pron3Medio) then (Pronostico is Bajo)
17. If (Pronostico1 is Pron1Medio) and (Pronostico2 is Pron2Bajo) and (Pronostico3 is Pron3Alto) then (Pronostico is Medio)
18. If (Pronostico1 is Pron1Medio) and (Pronostico2 is Pron2Medio) and (Pronostico3 is Pron3Alto) then (Pronostico is Medio)
19. If (Pronostico1 is Pron1Medio) and (Pronostico2 is Pron2Alto) and (Pronostico3 is Pron3Bajo) then (Pronostico is Alto)
20. If (Pronostico1 is Pron1Medio) and (Pronostico2 is Pron2Alto) and (Pronostico3 is Pron3Medio) then (Pronostico is Alto)
21. If (Pronostico1 is Pron1Medio) and (Pronostico2 is Pron2Alto) and (Pronostico3 is Pron3Alto) then (Pronostico is Alto)
22. If (Pronostico1 is Pron1Bajo) and (Pronostico2 is Pron2Bajo) and (Pronostico3 is Pron3Bajo) then (Pronostico is Alto)

Fig. 14. Rules of the Fuzzy system generated by genetic algorithm for Mexican Stock Exchange.

7.3 Simulation Results for the Mexican Stock Exchange Time Series

The architecture of the ensemble network that produced the best results for Mexican Stock Exchange Time Series is shown in Figure 15.

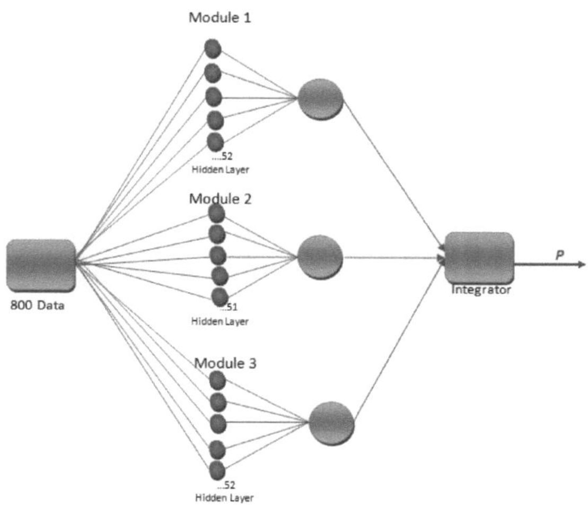

Fig. 15. The Best Network Architecture for Mexican Stock Exchange.

In this architecture we used one layer in each module. In module 1, in the first layer we used 52 neurons in module 2 we used 51 neurons in the first layer and in module 3 we used 52 neurons in first layer. The training method used was the Levenberg-Marquardt (LM); we also applied 3 delays to the network.

The best training was the one shown in row number 4, using 1 layer in each of the modules of the ensemble network, implementing 3 delays in the network, the training method used was Levenberg-Marquardt (LM), (as shown in Table 15) where the total error of this training was: 0.00015.

Table 15. Results of Training using 1 layer for Mexican Stock Exchange.

	Number of Layers	Number of Neurons	Number of Delays	Target Error	Number Max Epoch	Epoch	Time	Error by Module	Error Total
Training1	1	40	3	0.1	22000	82	00:00:09	0.00019	0.00029
		65		0.1	30000	17	00:00:03	0.00046	
		70		0.1	30000	33	00:00:06	0.00022	
Training2	1	42	3	0.1	22000	31	00:00:03	0.00007	
		67		0.1	30000	16	00:00:02	0.00047	0.00021
		74		0.1	30000	80	00:00:11	0.00009	
Training3	1	50	3	0.1	22000	27	00:00:03	0.00022	0.00021
		50		0.1	30000	69	00:00:06	0.00009	
		50		0.1	30000	10	00:00:01	0.00033	
Training4	1	52	3	0.1	22000	23	00:00:03	0.00022	0.00015
		51		0.1	30000	38	00:00:03	0.00012	
		52		0.1	30000	7	00:00:01	0.00011	
Training5	1	53	3	0.1	2000	6	00:00:02	0.00067	0.00031
		52		0.1	3000	18	00:00:02	0.00006	
		54		0.1	3000	57	00:00:05	0.00022	

The best training was the one shown in row number 1 using 2 layers in each of the modules of the ensemble network, implementing 3 delays in the network, the training method used was Levenberg-Marquardt (LM), (as shown in Table 16) where the total error of this training was: 0.000195.

Table 16. Results of Training using 2 layers for Mexican Stock Exchange.

	Number of Layers	Number of Neurons	Number of Delays	Target Error	Number Max Epoch	Epoch	Time	Error by Module	Error Total
Training1	2	26,26	3	0.1	2000	17	00:00:30	0.000407	0.000195
		32,32		0.1	5000	16	00:00:16	0.000086	
		22,22		0.1	3500	9	00:00:09	0.000093	
Training2	2	27,27		0.1	2000	15	00:00:30	0.000479	
		25,25	3	0.1	5000	12	00:00:16	0.000427	0.000389
		27,27		0.1	3500	15	00:00:09	0.000270	
Training3	2	28,28	3	0.1	2000	10	00:00:08	0.0000045	
		29,29		0.1	5000	14	00:00:09	0.0006578	0.000646
		28,27		0.1	3500	12	00:00:07	0.0013	
Training4	2	27,28	3	0.1	2000	7	00:00:06	0.000348	
		26,27		0.1	5000	9	00:00:07	0.000214	0.000220
		30,30		0.1	3500	832	00:08:55	0.000997	
Training5	2	29,29	3	0.1	2000	10	00:00:09	0.000613	
		28,28		0.1	5000	12	00:00:10	0.000310	0.000552
		30,30		0.1	3500	9	00:00:09	0.000732	

The best training was the one shown in row number 5 varying between 1 and 2 layers in each of the modules of the ensemble network, implementing 3 delays in the network, the training method used was Levenberg-Marquardt (LM), (as shown in Table 17) where the total error of this training was: 0.00016.

Table 17. Results of Training using 1and 2 layers for Mexican Stock Exchange.

	Number of Layers	Number of Neurons	Number of Delays	Target Error	Number Max Epoch	Epoch	Time	Error by Module	Error Total
Training1	1 y 2	30,30	3	0.1	2000	12	00:00:12	0.0020	0.0009
		50		0.1	5000	10	00:00:01	0.0002	
		52,52		0.1	3500	16	00:02:37	0.0006	
Training2	1 y 2	54	3	0.1	2000	6	00:00:01	0.0003	
		52		0.1	5000	9	00:00:01	0.0006	0.00052
		30,28		0.1	3500	15	00:00:13	0.0006	
Training3	1 y 2	54	3	0.1	2000	11	00:00:11	0.00034	
		32,32		0.1	5000	11	00:00:11	0.00046	0.00045
		55		0.1	3500	9	00:00:09	0.00056	
Training4	1 y 2	54	3	0.1	2000	10	00:00:02	0.00016	
		22,21		0.1	5000	10	00:00:03	0.00004	0.00016
		55		0.1	3500	8	00:00:02	0.00029	
Training5	1 y 2	55	3	0.1	2000	8	00:00:02	0.00015	
		19,19		0.1	5000	160	0:00:28	0.00089	0.00038
		54		0.1	3500	8	00:00:01	0.00012	

The result is training number 4 (as shown in Table 18), where the integration method used was average integration; and we obtained an error of 0.0479, with weighted average integration we obtained an error of 0.0519, with fuzzy integration we obtained an error of 0.1094 and with optimized fuzzy integration we obtained an error of 0.045848.

The best method of integration for this architecture was the optimized fuzzy integration with an error of 0.045848.

Table 18. Results of the Integration Methods with 1 layer for Mexican Stock Exchange

	Average Integration	Weighted Average Integration	Fuzzy Integration	Optimized Fuzzy Integration
Training1	0.0410	0.0619	0.1224	0.043856
Training2	0.0919	0.1017	0.1575	0.046531
Training3	0.0468	0.0500	0.1223	0.046123
Training4	0.0479	0.0519	0.1094	0.045848
Training5	0.0570	0.0701	0.1062	0.044292

The best result is training 1 (as shown in Table 19), where the integration method used was average integration; we obtained an error of 0.0815, with weighted average integration we obtained an error of 0.0926, with fuzzy integration we obtained an error of 0.1286 and with optimized fuzzy integration we obtained an error of 0.05337.

The best method of integration for this architecture was the optimized fuzzy integration with an error of 0.05337.

Table 19. Results of the Integration Methods with 2 layers for Mexican Stock Exchange

	Average Integration	Weighted Average Integration	Fuzzy Integration	Optimized Fuzzy Integration
Training1	0.0815	0.0926	0.1286	0.05337
Training2	0.0647	0.0831	0.1341	0.06387
Training3	0.0612	0.0830	0.1386	0.065554
Training4	0.0536	0.0693	0.1237	0.049567
Training5	0.0504	0.0696	0.1336	0.050638

The best result is training 4 (as shown in Table 20), where the integration method used was average integration; we obtained an error of 0.0458, with weighted average integration we obtained an error of 0.0529, with fuzzy integration we obtained an error of 0.1023 and with optimized fuzzy integration we obtained an error of 0.046131.

The best method for this architecture was the average integration with an error of 0.0458.

Table 20. Results of the Integration Methods with 1y 2 layers for Mexican Stock Exchange

	Average Integration	Weighted Average Integration	Fuzzy Integration	Optimized Fuzzy Integration
Training1	0.0883	0.1187	0.1491	0.050886
Training2	0.0474	0.0804	0.1250	0.052601
Training3	0.0561	0.0774	0.1094	0.051523
Training4	0.0458	0.0529	0.1023	0.046131
Training5	0.0731	0.0843	0.1270	0.051713

The best result obtained for the Mexican Stock Exchange Time Series. was number 4 using 1 layers in each of the modules of the ensemble network, implementing 3 delays, the training method used was Levenberg-Marquardt (LM), (as shown in Table 20 and in Figure16).

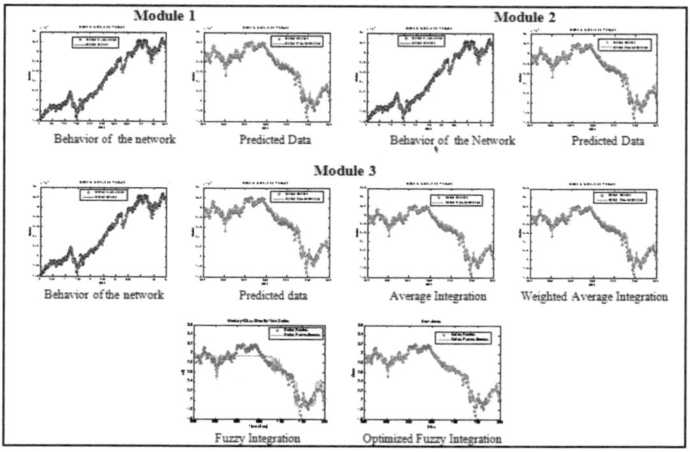

Fig. 16. Simulation results of training 4 for the Mexican Stock Exchange.

Table 21. Genetic algoritm results for training 4 for Mexican Stock Exchange.

Ind.	Gen.	GGP	Selection	Mutation	Pm	Crossover	Pc	Pm Rules	Núm. Rules	Duration	GA Error	Prediction Error
100	100	0.85	rws	mutbga	0.8	xovdprs	1	0.002	27	03:25:59	0.069886	0.059886
100	100	0.85	rws	mutbga	0.07	xovdprs	0.5	0.002	20	03:39:59	0.053679	0.046272
100	100	0.85	rws	mutbga	0.6	xovdprs	1	0.002	22	03:51:53	0.067592	0.059444
100	100	0.85	rws	mutbga	0.09	xovdprs	0.5	0.002	27	04:09:37	0.063615	0.053615
100	100	0.85	rws	mutbga	0.06	xovdprs	0.8	0.002	22	02:58:50	0.065992	0.057884
100	100	0.85	rws	mutbga	0.5	xovdprs	1	0.002	27	03:08:20	0.071177	0.061177
100	100	0.85	rws	mutbga	0.05	xovdprs	0.8	0.002	20	03:14:33	0.069961	0.062554
100	100	0.85	rws	mutbga	0.06	xovdprs	0.9	0.002	22	03:30:04	0.053996	0.045848
100	100	0.85	rws	mutbga	0.6	xovdprs	0.05	0.002	22	03:23:30	0.058665	0.050517
100	100	0.85	rws	mutbga	0.7	xovdprs	0.5	0.002	27	03:28:07	0.065534	0.055534

Table 21 shows the genetic algorithm results for training 4 (as shown in Table 20 and in Figure 16) where the prediction error is of 0.045848.

The fuzzy integration system generated by the genetic algorithm is represented in Figure 17 and the fuzzy rules in Figure 18.

Comparing Figure 2, which belongs to the base fuzzy integrator (not optimized) with Figure 17, which belongs to the fuzzy integrator generated by the genetic algorithm, it can be seen that the parameters of the membership functions are different , the number of rules obtained are 22 and this helps to improve the prediction error.

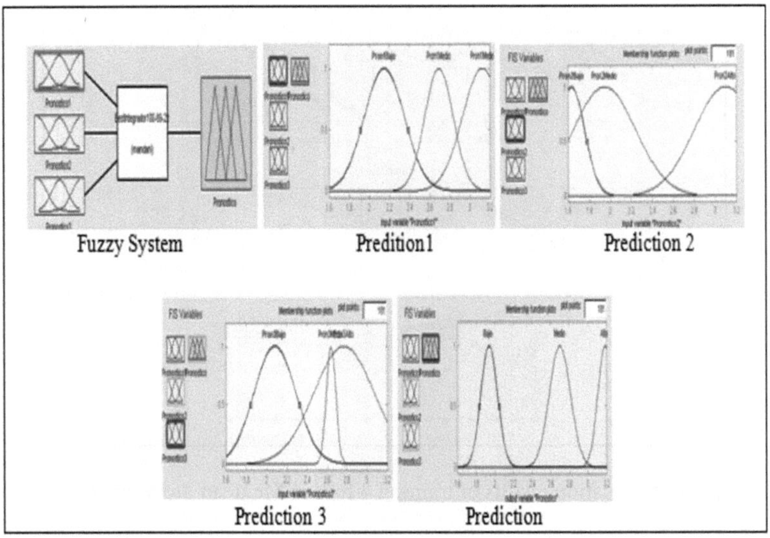

Fig. 17. Fuzzy system generated by genetic algorithm for Mexican Stock Exchange.

1. If (pronostico1 is Pron1Bajo) and (Pronostico2 is Pron2Bajo) and (Pronostico3 is Pron3Medio) then (Pronostico is Bajo)
2. If (Pronostico1 is Pron1Bajo) and (Pronostico2 is Pron2Bajo) and (Pronostico3 is Pron3Alto) then (Pronostico is Bajo)
3. If (Pronostico1 is Pron1Bajo) and (Pronostico2 is Pron2Medio) and (Pronostico3 is Pron3Bajo) then (Pronostico is Bajo)
4. If (Pronostico1 is Pron1Bajo) and (Pronostico2 is Pron2Medio) and (Pronostico3 is Pron3Medio) then (Pronostico is Bajo)
5. If (Pronostico1 is Pron1Bajo) and (Pronostico2 is Pron2Medio) and (Pronostico3 is Pron3Alto) then (Pronostico is Medio)
6. If (Pronostico1 is Pron1Bajo) and (Pronostico2 is Pron2Alto) and (Pronostico3 is Pron3Bajo) then (Pronostico is Alto)
7. If (Pronostico1 is Pron1Bajo) and (Pronostico2 is Pron2Alto) and (Pronostico3 is Pron3Medio) then (Pronostico is Bajo)
8. If (Pronostico1 is Pron1Bajo) and (Pronostico2 is Pron2Alto) and (Pronostico3 is Pron3Alto) then (Pronostico is Medio)
9. If (Pronostico1 is Pron1Medio) and (Pronostico2 is Pron2Bajo) and (Pronostico3 is Pron3Bajo) then (Pronostico is Alto)
10. If (Pronostico1 is Pron1Medio) and (Pronostico2 is Pron2Bajo) and (Pronostico3 is Pron3Medio) then (Pronostico is Bajo)
11. If (Pronostico1 is Pron1Medio) and (Pronostico2 is Pron2Bajo) and (Pronostico3 is Pron3Alto) then (Pronostico is Medio)
12. If (Pronostico1 is Pron1Medio) and (Pronostico2 is Pron2Medio) and (Pronostico3 is Pron3Medio) then (Pronostico is Medio)
13. If (Pronostico1 is Pron1Medio) and (Pronostico2 is Pron2Alto) and (Pronostico3 is Pron3Bajo then (Pronostico is Medio)
14. If (Pronostico1 is Pron1Medio) and (Pronostico2 is Pron2Alto) and (Pronostico3 is Pron3Alto) then (Pronostico is Medio)
15. If (Pronostico1 is Pron1Medio) and (Pronostico2 is Pron2Bajo) and (Pronostico3 is Pron3Bajo) then (Pronostico is Alto)
16. If (Pronostico1 is Pron1Medio) and (Pronostico2 is Pron2Bajo) and (Pronostico3 is Pron3Medio) then (Pronostico is Bajo)
17. If (Pronostico1 is Pron1Medio) and (Pronostico2 is Pron2Bajo) and (Pronostico3 is Pron3Alto) then (Pronostico is Medio)
18. If (Pronostico1 is Pron1Medio) and (Pronostico2 is Pron2Medio) and (Pronostico3 is Pron3Medio) then (Pronostico is Medio)
19. If (Pronostico1 is Pron1Medio) and (Pronostico2 is Pron2Medio) and (Pronostico3 is Pron3Alto) then (Pronostico is Medio)
20. If (Pronostico1 is Pron1Medio) and (Pronostico2 is Pron2Alto) and (Pronostico3 is Pron3Medio) then (Pronostico is Alto)
21. If (Pronostico1 is Pron1Medio) and (Pronostico2 is Pron2Alto) and (Pronostico3 is Pron3Alto) then (Pronostico is Alto)
22. If (Pronostico1 is Pron1Bajo) and (Pronostico2 is Pron2Alto) and (Pronostico3 is Pron3Alto) then (Pronostico is Alto)

Fig. 18. Rules of the Fuzzy system generated by genetic algorithm for Mexican Stock Exchange.

8 Conclusions

The best result obtained for the Mackey-Glass series training was using the ensemble neural network architecture with 2 layers using 3 delays. The error obtained by the average integration was 0.0106, the error obtained by the average weighted integration was 0.0126 and the error obtained by the fuzzy integration was 0.0708. The best result obtained for the Dow Jones series training was using the ensemble neural network architecture with 1 layer using 3 delays. The error obtained by the average integration was 0.0053, the error obtained by the average weighted integration was 0.0060 and the error obtained by the fuzzy integration was 0.0194. The best result for the Mexican Stock Exchange series training was obtained using an ensemble neural network architecture with 1 layer using 3 delays. The error obtained by the average integration was 0.0479, by the average weighted integration was 0.0519 and by the fuzzy integration was 0.1094, respectively.

Since the results were not satisfactory with the fuzzy integration system, it was decided to implement an evolutionary approach to optimize the membership functions and rules of this system. The method chosen to optimize this integration system was a genetic algorithm. After applying the genetic algorithm we were able to obtain an error of 0.023317 for the Mackey-Glass Series, an error of 0.006863 for the Dow Jones time series and an error of 0.045848 for the Mexican Stock Exchange time series, with this improving the results obtained with respect to the non optimized system.

Acknowledgment

We would like to express our gratitude to the CONACYT, Tijuana Institute of Technology for the facilities and resources granted for the development of this research.

References

[1] Vásquez, G., Muñoz, Tapia, J.: "Series de Tiempo" Catholic University of the Holy Concepción, Chile (2001), `http://zip.rincondelvago.com/?00033622` (December 05, 2008)

[2] Arellano, M.: Introduction to time series analysis (2001), `http://ciberconta.unizar.es/LECCION/seriest/100.HTM` (September 06, 2008)

[3] Davey, N., Hunt, S., Frank, R.: Time Series Prediction and Neural Networks, University of Hertfordshire, Hatfield, UK (1999) (December 04, 2008)

[4] Plummer, E.A.: Time series forecasting with feed-forward neural Networks: uidelines and limitations. University of Wyoming (July 2000), `http://www.karlbranting.net/papers/plummer/Paper_7_12_00.htm` (January 20, 2009)

[5] Neural Network Tutorial, `http://www.answermath.com/neural-networks/tutorial-esp-2-concepto.htm` (December 05, 2008)

[6] Multaba, I.M., Hussain, M.A.: Application of Neural Networks and Other Learning. Technologies in Process Engineering. Imperial Collage Press (2001) (December 03, 2008)

[7] Basic Concepts of Fuzzy Logic, `http://www.tdx.cbuc.es/TESIS_UPC/AVAILABLE/TDX0207105105056//04Rpp04de11.pdf` (September 10, 2008)

[8] Jang, J.S.R., Sun, C.T., Mizutani, E.: Neuro-Fuzzy and Soft Computing. Prentice Hall, New Jersey
[9] Holland, J.: Adaptation in natural and artificial systems. University of Michigan Press (1975)
[10] Golberg, D.: Genetic Algorithms in search, optimization and machine learning. Addison Wesley, Reading (1989)
[11] Yen, J., Langari, R.: Fuzzy Logic: intelligence, control and Information. Prentice Hall, New Jersey (1999)
[12] MATLAB, Historical Mackey Glass data (2007)
[13] Dow Jones Indexes, http://www.djindexes.com (September 5, 2008)
[14] Mexico Bank, http://www.banxico.org.mx (December 10, 2008)

Optimization of Fuzzy Response Integrators in Modular Neural Networks with Hierarchical Genetic Algorithms: The Case of Face, Fingerprint and Voice Recognition

Ricardo Muñoz, Oscar Castillo, and Patricia Melin

Tijuana Institute of Technology, Tijuana México
ocastillo@hafsamx.org

Abstract. In this paper we describe the development of Fuzzy Response Integrators of Modular Neural Networks (MNN) for Face, Fingerprint and Voice Recognition, and their Optimization with a Hierarchical Genetic Algorithm (HGA). The optimization of the integrators consists of optimizing their membership functions, fuzzy rules, type of model (Mamdani or Sugeno), type of fuzzy logic (type-1 or type-2). The MNN architecture consists of three modules; face, fingerprint and voice. Each of the modules is divided again into three sub modules. The same information is used as input to train the sub modules. Once we have trained and tested the MNN modules, we proceed to integrate these modules with an optimized fuzzy integrator. In this paper we show that using a HGA as an optimization technique for the fuzzy integrators is a good option to solve MNN integration problems.

1 Introduction

Person identification has become a very important activity in different areas of application and with different purposes; in business applications based on punctuality and attendance biometric systems are implemented using the fingerprint or the full hand of the employees; in judicial systems biometric pattern recognition has become a routine tool in the police force during the criminal investigation, allowing the arrest of criminals worldwide, but also other specific applications are known, such as controlling access to any type of transaction or access to data [1].

Neural networks provide a methodology to work in the field of pattern recognition. Modular neural networks are often used to simplify the problem and obtain good results. Because of this, there arises the need to develop methods to integrate the responses provided by the modules of a modular neural network and thus provide a good result. This paper develops a fuzzy integrator to combine the responses provided by the modular neural network and achieve a good recognition of persons.

The work described in this paper is divided into two main areas, the first one is about the Modular Neural Network (MNN) and the second one is about the Fuzzy

O. Castillo et al. (Eds.): Evolutionary Design of Intelligent Systems, SCI 257, pp. 111–129.
springerlink.com

Integrators. The main goal of the first part is to create and test different MNNs architectures for face, fingerprint and voice recognition. The main goal of the second part is to develop fuzzy integrators to integrate the outputs given by the modules of the modular neural network.

This paper is organized as follows: in sections 2, 3 and 4 we present the basic concepts of neural networks, fuzzy logic theory and genetic algorithms respectively; in sections 5, 6 and 7 we describe the problem and mention how to solve it making use of a MNN with fuzzy integration and the results obtained. In section 8 we describe the HGA to optimize the fuzzy integrator structure and the optimization results; in section 9 we make a comparison with other work. Finally, section 10 presents the conclusions.

2 Neural Networks

Artificial neural networks (NN) are an abstract simulation of a real nervous system that consists of a set of neural units connected to each other via axon connections. These connections are very similar to the dendrites and axons in biological nervous systems. Figure 1 shows the abstract simulation of a real nervous system [2].

Models of artificial neural networks can be classified as:

- Biological models: Networks that try to simulate the biological neural systems, as well as the functions of hearing or some basic functions of vision.
- Models for applications: Those models are less dependent on models of biological systems. These are models in which their architectures are strongly linked to the application requirements.

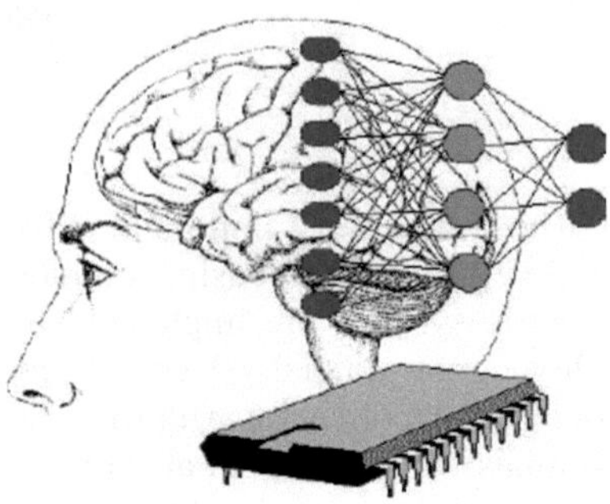

Fig. 1. Simulation of an abstract real nervous system.

One of the missions of a neural network is to simulate the properties observed in biological neural systems through mathematical models recreated through artificial mechanisms (such as an integrated circuit, a computer or a set of valves). The aim is that the machines give similar responses to those the brain is able to give, which are characterized by their robustness and generalization [3].

Artificial neural networks are models that attempt to reproduce the behavior of the brain. As such model, a simplification is made, identifying the relevant elements of the system, either because the amount of information available is excessive or because it is redundant. An appropriate choice of features and an appropriate structure is the conventional procedure used to construct networks capable of performing certain tasks [1].

Modularity is defined as the ability of a system being studied, seen or understood as the union of several parts interacting with each other and working towards a common goal, each one performing a task necessary to achieve that objective. Each of these parts in which the system is divided is called a module. Ideally, a module must be able to work as a black box, i.e. be independent of other modules and communicate with them (all or only a part) through well-defined inputs and outputs [4].

It is said that a neural network is modular if the computations performed by the network can be decomposed into two or more modules (subsystems) that operate on different inputs with no communication between them. The outputs of the modules are measured by an integration unit, which is not allowed to feed information to the modules. In particular, the unit decides: (1) how the modules are combined to form the final output of the system, and (2) modules that must learn that patterns of training [5].

As mentioned previously, modular neural networks require an integration unit, which allows some form of joining or combining the responses provided by each module. Listed below are some methods that can be used to perform this task:

- Average operators
- Gating Network
- *Fuzzy Integrators*
- Voting mechanism using the Softmax function
- Etc.

3 Fuzzy Logic

Fuzzy logic has gained a great reputation as a good methodology for a variety of applications, ranging from control of complex industrial processes to the design of artificial devices for automatic deduction through the construction of electronic devices for home use and entertainment, as well as diagnostic systems. It has been considered generally that the concept of fuzzy logic appeared in 1965 at the University of California at Berkeley, introduced by Lotfi A. Zadeh.

The basic structure of a fuzzy inference system consists of three conceptual components: a rule base, which contains a selection of fuzzy rules, a database (or dictionary) which defines the membership functions used in the rules, and a reasoning mechanism that performs the inference procedure [6].

The basic fuzzy inference system can take either fuzzy or traditional inputs, but the outputs it produces are always fuzzy sets. Sometimes you need a traditional output, especially when a fuzzy inference system is used as a controller. So we

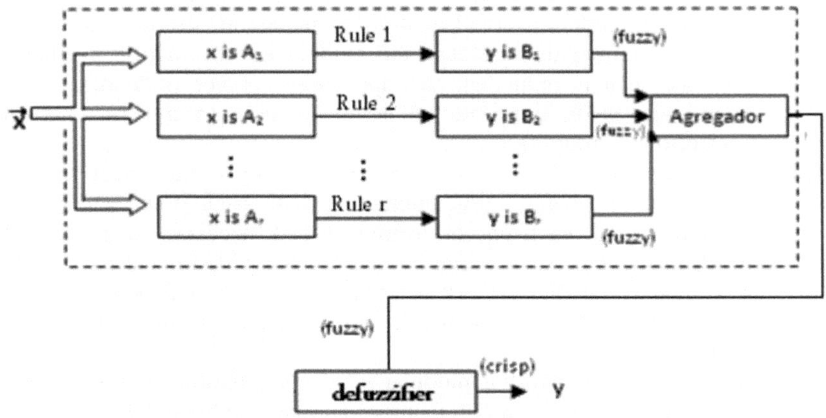

Fig. 2. Structure of a Fuzzy Inference System.

need a method of "Defuzzification" to extract the numerical value of the output (as shown in Figure 2).

4 Genetic Algorithms

Genetic algorithms or evolutionary algorithms are stochastic search techniques guided or inspired by the mechanisms of natural selection, genetics and evolution. Early research on the subject was developed by John Holland at the University of Michigan in the 70's with two objectives: to abstract and rigorously explain the adaptive processes of natural systems and the design of software systems that determine the retention of these important mechanisms. [7]

This technique is based on the selection mechanisms used by nature [8], according to which the fittest individuals of a population are those who survive, which are those who adapt more easily to changes in their environment. We now know that these changes are made in the genes (the basic encoding unit of each of the attributes of a living being) of an individual, and that the most desirable attributes (for example, that allow an individual to better adapt to its environment) of the individuals are transmitted to their descendants, if reproduced sexually.

Imitating the mechanics of biological evolution in nature, genetic algorithms operate on a population of possible solutions to the problem. Each element of the population is called "chromosome". A chromosome is the representative within the genetic algorithm, of a possible solution to the problem. The process begins by selecting a number of chromosomes to form the initial population. Then we evaluate the role of adaptation for these individuals. The fitness function measures the chromosome's fitness to survive in its environment. This function must be defined so that the chromosomes that represent better solutions have higher values of fitness. The fittest individuals are selected in pairs to breed. Reproduction generates new chromosomes that combine characteristics of both parents. These new chromosomes replace individuals with lower values of adaptation. After this, some chromosomes are randomly selected to be mutated. The mutation consists of applying a random change in its structure. Then, the new chromosomes must be inserted into the population; these chromosomes should replace existing chromosomes. There are different criteria that can be used to select chromosomes to be replaced. The cycle of selection, reproduction and mutation is repeated until it meets the criterion of termination of the algorithm, at which the chromosome that is better suited as a solution is returned.

5 Modular Neural Network with Fuzzy Integration for Face, Fingerprint and Voice Recognition

The proposed MNN architecture defined in this paper consists of three modules: face, fingerprint and voice. Each of the mentioned modules will be divided into three sub modules. The same information is used as input to train the sub modules. After the MNN trainings are done, we proceed to integrate the three modules (face, fingerprint ad voice) with a fuzzy integrator.

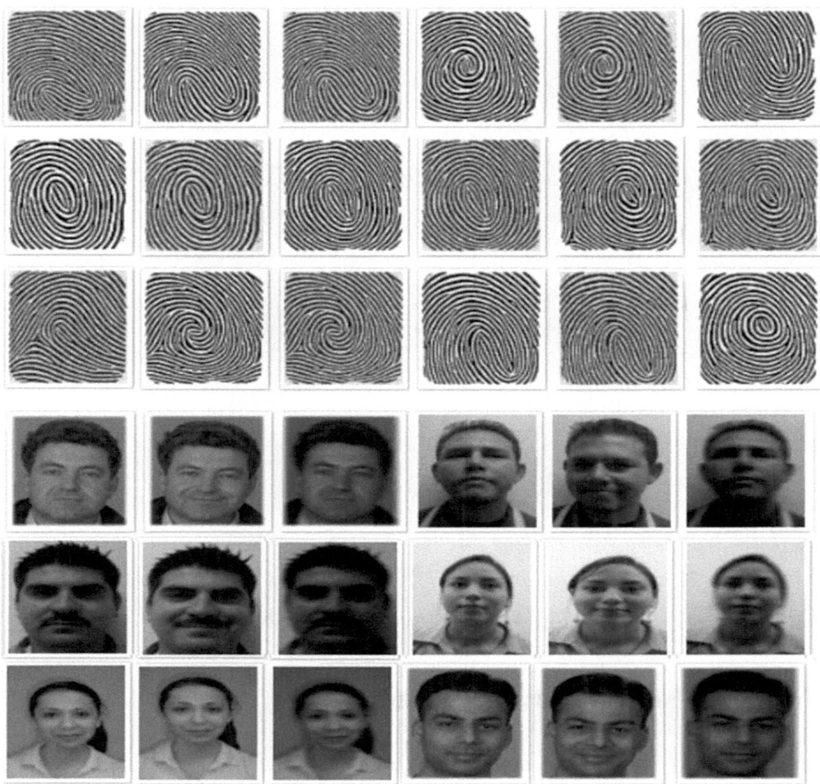

Fig. 3. Fingerprint and face samples.

The data used for training the MNN correspond to 30 persons taken from the ORL database and from students of a master degree in computer science from Tijuana Institute of Technology, Mexico. The details are as follows:

- Face: 90 images taken from [9]. The size of these images is of 268x338 pixels with bmp extension and were preprocessed with the Wavelet transform. Two images are used for training (with different gestures) and one with Gaussian noise to test the recognition.
- Fingerprints taken from [9]: 60 fingerprint images. The size of these images is of 268x338 pixels with bmp extension and were preprocessed with the Wavelet function. One fingerprint per person for training and one with Gaussian noise to test the recognition.
- Voices taken from [10]: There are 3 word files per person; the words are "hola", "accesar" and "presentacion" in Spanish. The network was trained with the 90 words preprocessed with Mel-cepstral Coefficients and the network is tested with any of them but with noise.

Examples of face and fingerprint are shown in Figure 3 and a general Scheme of the architecture is shown in Figure 4.

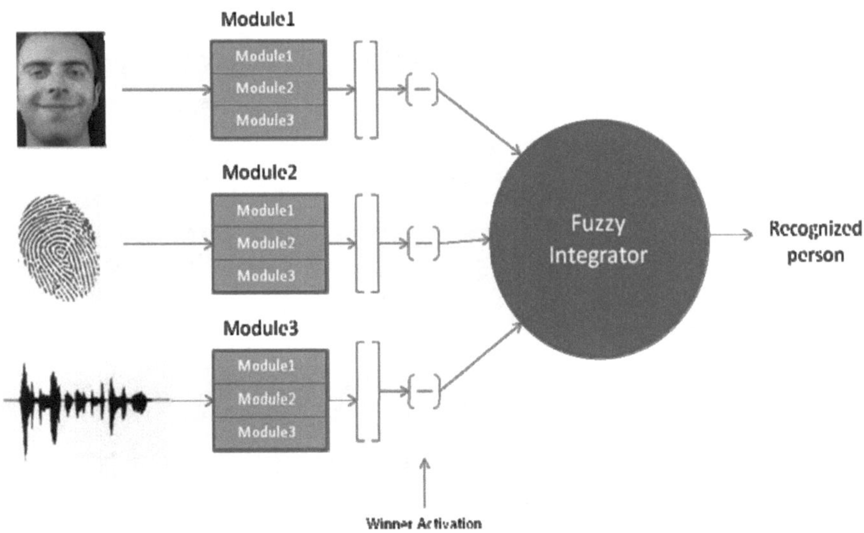

Fig. 4. Architecture of the MNN.

6 Modular Neural Network Results

Several trainings were made separately to each of the modules until we accomplish good results. Tables 1, 2, 3 and 4 show the results of the MNN trainings for face, fingerprint and voice, respectively. Figure 5 shows the results from table 4.

Table 1. Face training results.

Training	Method	Mod	Architecture	Error	Duration	Rec.	%
		1	350,300				
ER1	trainscg	2	350,300	0.01	16:10	19/30	63.33
		3	350,300				
		1	300,300				
ER2	trainscg	2	300,300	0.01	20:34	21/30	70
		3	300,300				
		1	300,240				
ER3	trainscg	2	300,250	0.001	49:39	28/30	93.33
		3	300,260				
		1	350,300				
ER4	trainscg	2	350,300	0.001	40:58	28/30	93.33
		3	350,300				
		1	300,150				
ER5	trainscg	2	300,150	0.001	01:12:30	28/30	93.33
		3	350,150				
		1	290,250				
ER6	trainscg	2	305,210	0.001	50:50	29/30	96.66
		3	320,200				
		1	300,300				
ER7	trainscg	2	300,300	0.001	46:07	30/30	100
		3	300,300				

Table 2. Fingerprint training results.

Training	Method	Mod	Architecture	Error	Duration	Rec.	%
		1	340,175				
EH1	trainscg	2	280,105	0.01	12:51	30/30	100
		3	295,138				
		1	200,100				
EH2	trainscg	2	200,100	0.01	15:56	30/30	100
		3	200,100				
		1	150,80				
EH3	trainscg	2	150,80	0.01	59:34	30/30	100
		3	150,80				
		1	150,100				
EH4	trainscg	2	150,90	0.01	18:09	30/30	100
		3	150,110				
		1	100,50				
EH5	trainscg	2	100,60	0.01	01:39:13	30/30	100
		3	100,55				
		1	100,50				
EH6	trainscg	2	100,60	0.02	16:37	26/30	86.66
		3	100,55				
		1	150,70				
EH7	trainscg	2	117,90	0.02	08:07	26/30	86.66
		3	157,87				

Table 3. Voice training configuration.

Training	Method	Architecture	Error	Duration
EV1	trainscg	450,240 420,190 410,225	0.001	01:44
EV2	trainscg	150,80 170,90 170,75	0.001	12:27
EV3	trainscg	150,100 150,100 150,100	0.001	02:03
EV4	trainscg	350,140 320,90 310,125	0.001	02:17
EV5	trainscg	180,80 180,80 180,80	0.001	29:01
EV6	trainscg	180,120 180,120 180,120	0.001	01:41
EV7	trainscg	300,100 300,100 300,100	0.001	02:24

Table 4. Voice training results.

	Recognized persons with noise									
Training	0.1	0.2	0.3	0.4	0.5	0.6	0.7	0.8	0.9	1.0
EV1	90	90	90	90	90	90	90	86	81	74
EV2	90	90	90	90	89	86	85	74	62	62
EV3	90	90	90	90	90	88	88	75	50	49
EV4	90	90	90	90	90	90	90	83	78	68
EV5	90	90	90	90	90	89	89	79	64	65
EV6	90	90	90	90	90	89	89	72	64	56
EV7	90	90	90	90	90	89	89	88	81	68

We can observe in table 1 that the best result obtained for face module recognition is for training ER7 with 100 percent of recognition.

We can observe in table 2 that in several trainings (EH1, EH2, EH3, EH4 and EH5) we obtained good results, having 100 percent of recognition for the fingerprint module.

We can observe in tables 3 and 4 that the best result obtained for the voice module recognition is for training EV1, because it has better performance when different noise levels are applied to the input when testing the module. It is important to know that the noise level consists of adding to each voice signal, a data array of normally distributed random numbers with a standard deviation of 0.1 to 1 (where noise level 0.1 represents a standard deviation of 0.1, 0.2 a standard deviation of 0.2…1.0 a standard deviation of 1). When the recognized persons equals 90 we have a 100 percent of recognition.

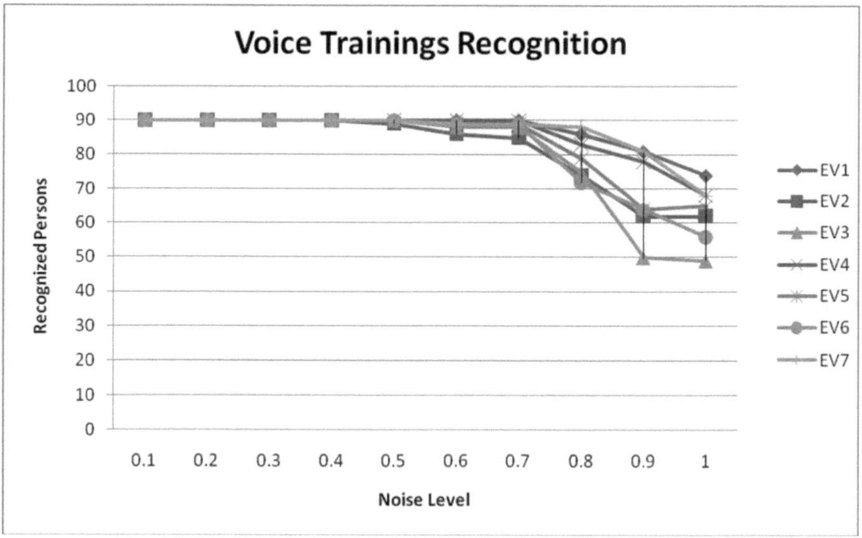

Fig. 5. Results of voice trainings with different noise level.

7 Fuzzy Integration Results

A fuzzy integrator was created using the MATLAB fuzzy logic toolbox, so that we can use it to integrate the outputs of the MNN modules and to generate a final result. This fuzzy integrator consists of 27 rules; three inputs that are provided by the MNN: face activation, fingerprint activation and voice activation; one output: winner activation that will indicate which person is recognized. It is important to notice that each input and output has three Gaussian membership functions. Figure 6 shows the architecture of the fuzzy integrator.

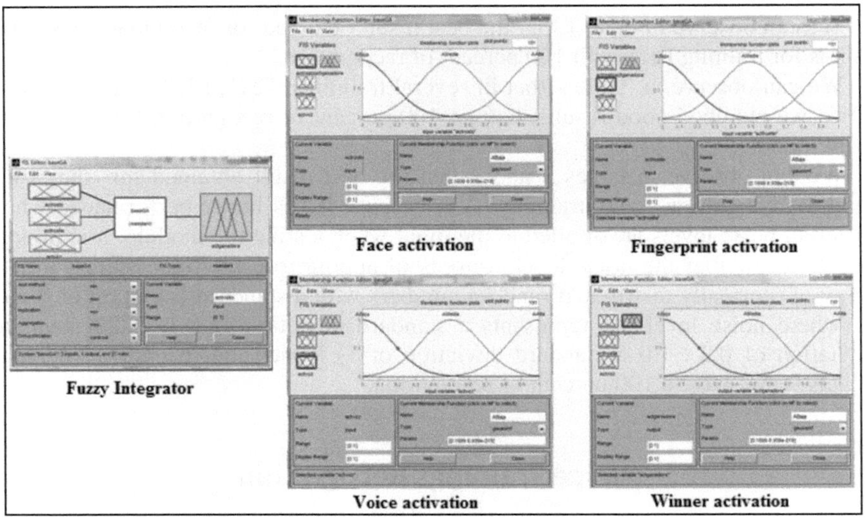

Fig. 6. Architecture of the fuzzy integrator.

Table 5 shows the results from the fuzzy integration of the MNN. These results were obtained by making some combinations of the MNN modules. We can observe that when all three modules have 100% of recognition, fuzzy integration is perfect (i.e. No. 8). Even in some cases where one or two of the modules do not give good recognition, fuzzy integration is perfect too (i.e. no.4, 6 and 7). But, when all three modules have poor recognition, the fuzzy integration is not good (i.e. no. 1).

Table 5. Fuzzy integration results.

No.	Face	Rec.	Fingerprint	Rec.	Voice-noise		Rec.	Integration
1	ER1	63.33%	EH6	86.66%	EV3	1.0	49%	71/90(78.89%)
2	ER1	63.33%	EH1	100%	EV3	1.0	49%	90/90 (100%)
3	ER7	100%	EH6	86.66%	EV3	1.0	49%	80/90(88.89%)
4	ER7	100%	EH1	100%	EV3	1.0	49%	90/90 (100%)
5	ER1	63.33%	EH6	86.66%	EV1	0.5	100%	73/90(81.10%)
6	ER1	63.33%	EH1	100%	EV1	0.5	100%	90/90(100%)
7	ER7	100%	EH6	86.66%	EV1	0.5	100%	81/90(90%)
8	ER7	100%	EH1	100%	EV1	0.5	100%	90/90(100%)

8 Hierarchical Genetic Algorithm

Analyzing the results shown in Table 5. We can observe that the integration results for the cases 1, 3, 5 and 7 are not the most desirable, so we decided to develop a genetic algorithm to optimize the structure of the fuzzy integrator and thus increase the percentage of recognition. Figure 7 shows the chromosome structure we designed to optimize the fuzzy integrator.

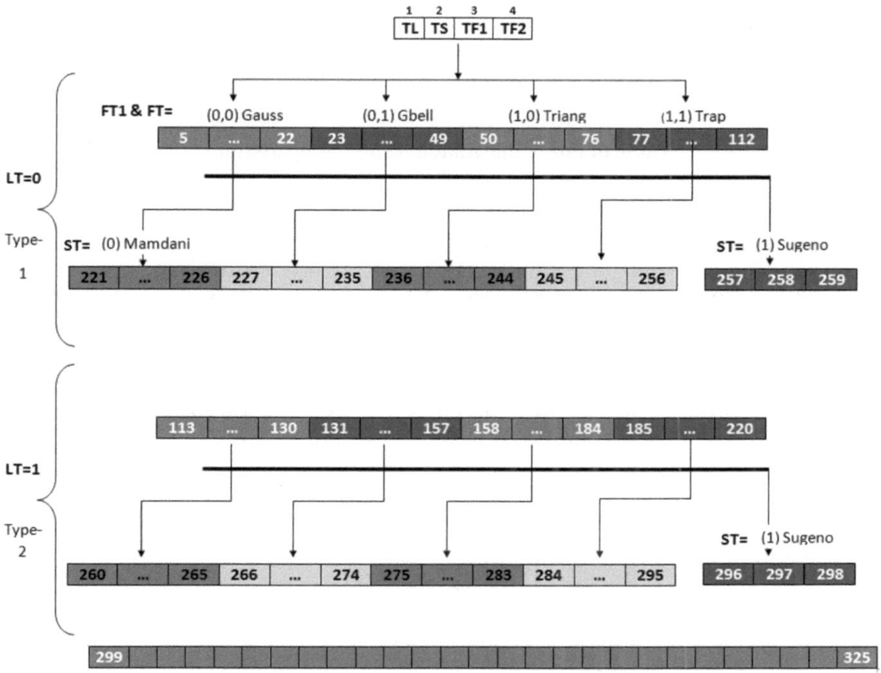

Fig. 7. HGA Chromosome structure.

The chromosome shown in figure 7 consists of 325 gens that will help us optimize the fuzzy integrator structure. Gens 1-4 (Binary) are activation gens that define the structure of the fuzzy integrator such as the logic type (FL type-1 or type-2), the system's type (Mamdani or Sugeno) and membership function type (Gaussian, Gbell, Triangular or Trapezoidal). Gens 5-298 (Real Numbers) allow us to manage the MFs parameters for the input and output (depending of FL type, system type and MFs type we use part of the gens); gens 299-325 (Binary) allows us to reduce the set of possible fuzzy rules activating or deactivating them.

The optimization of the fuzzy integrator is proposed to be made based directly on the performance of the neural network, because we want to: (1) reduce the number of recognition errors and (2) reduce the number of fuzzy rules. The objective Function (equation 1) is the one we want to minimize, so that we can achieve our two objectives.

$$Obj = e_r + (n_r / m_r) \tag{1}$$

Where:

e_r is the error of recognition given when simulating the MNN.
n_r is the number of fuzzy rules of the current individual.
m_r is the maximum number of fuzzy rules.

We executed the HGA 20 times with different parameters (generation, cross-over, mutation) for each of the cases we wanted to optimize (in table 5: cases 1, 3, 5 and 7). We show the results in table 6.

Table 6. Fuzzy integration results comparisson.

No.	Face	Rec.	Fingerprint	Rec.	Voice-noise		Rec.	Non Optimized Integration	Optimized Integration
1	ER1	63.33%	EH6	86.66%	EV3	1.0	49%	71/90(78.89%)	87/90 (96.67%)
2	ER1	63.33%	EH1	100%	EV3	1.0	49%	90/90 (100%)	NO NEED
3	ER7	100%	EH6	86.66%	EV3	1.0	49%	80/90(88.89%)	90/90 (100%)
4	ER7	100%	EH1	100%	EV3	1.0	49%	90/90 (100%)	NO NEED
5	ER1	63.33%	EH6	86.66%	EV1	0.5	100%	73/90(81.10%)	85/90 (94.44%)
6	ER1	63.33%	EH1	100%	EV1	0.5	100%	90/90(100%)	NO NEED
7	ER7	100%	EH6	86.66%	EV1	0.5	100%	81/90(90%)	90/90 (100%)
8	ER7	100%	EH1	100%	EV1	0.5	100%	90/90(100%)	NO NEED

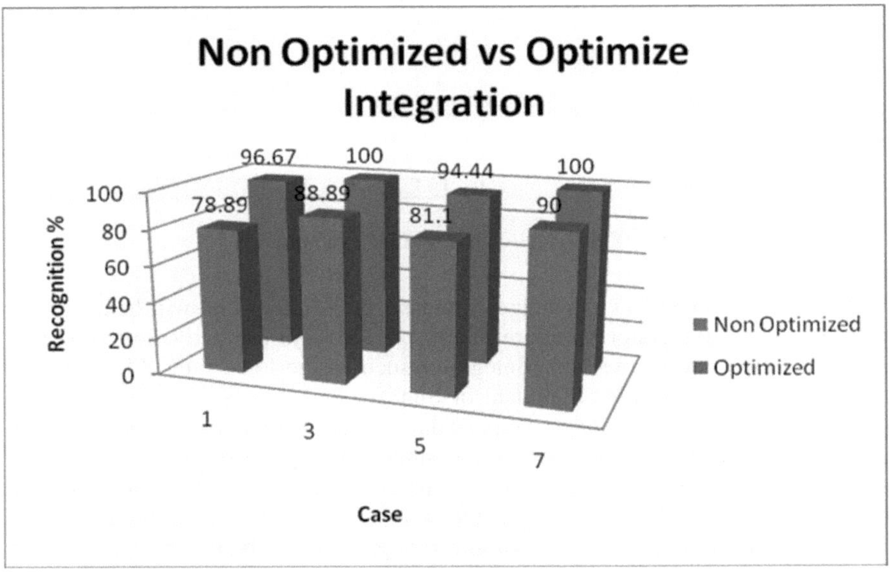

Fig. 8. Integration Comparison.

We can observe in figure 8 that the proposed HGA help us increase the percentage of recognition in all cases. Despite these good results we decided to manually change some rules consequent in the best fuzzy integrator obtained with the HGA for case 5 because we believed that we could improve that result. When we tested the manually changed fuzzy integrator we obtained a 100% of recognition for case 5, so we decided to make some changes to our HGA chromosome structure so that it would be able not only to activate or deactivate fuzzy rules but also to change the rules' consequents. Figure 9 shows the new HGA chromosome

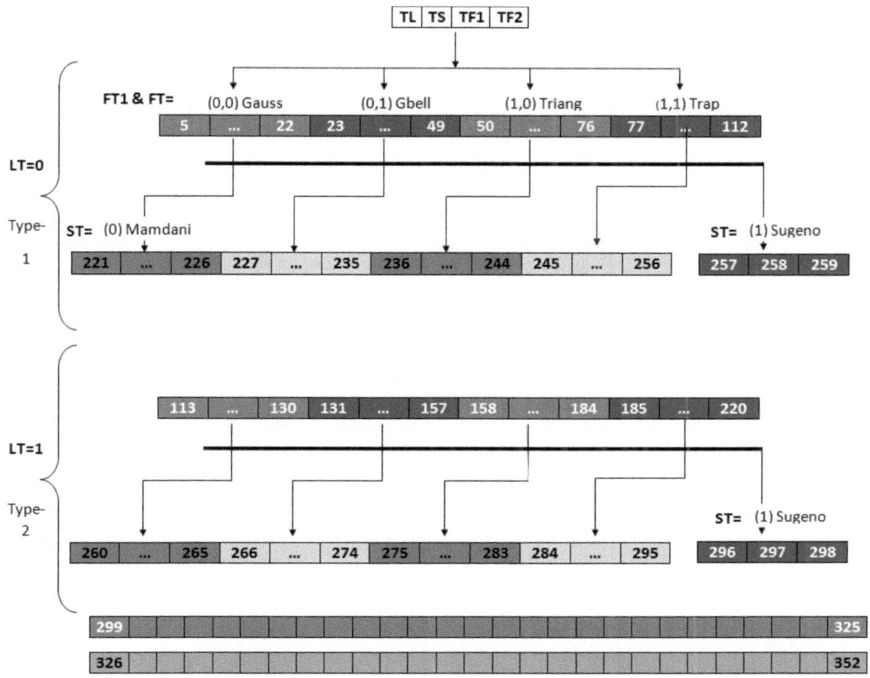

Fig. 9. New HGA Chromosome structure.

structure; the change we made to it, in comparison with the previous is that we added 27 gens more that will allow the HGA to manage the consequents of each of the 27 rules.

We also decided to change the objective function for this new HGA as we can appreciate in equation 2.

$$Obj = \alpha(e_r) + \beta(n_r) \quad (2)$$

Where:

e_r is the error of recognition given when simulating the MNN.
n_r is the number of fuzzy rules of the current individual.
α is the weight assigned to the error recognition objective.
β is the weight assigned to the fuzzy rules objective.

We executed the HGA 20 times with different parameters (generation, crossover, mutation) and for each case we needed to optimize. We show the results in table 8.

Table 8 shows a comparison of all the fuzzy integrations that were made. We can observe in the last column that we achieve the best fuzzy integration result for No. 1 with α=0.9 β=0.1.

Table 7. Trainings information for fuzzy integration.

No.	Face	Rec.	Fingerprint	Rec.	Voice-noise		Rec.
1	ER1	63.33%	EH6	86.66%	EV3	1.0	49%
2	ER1	63.33%	EH1	100%	EV3	1.0	49%
3	ER7	100%	EH6	86.66%	EV3	1.0	49%
4	ER7	100%	EH1	100%	EV3	1.0	49%
5	ER1	63.33%	EH6	86.66%	EV1	0.5	100%
6	ER1	63.33%	EH1	100%	EV1	0.5	100%
7	ER7	100%	EH6	86.66%	EV1	0.5	100%
8	ER7	100%	EH1	100%	EV1	0.5	100%

Table 8. Fuzzy integration results comparison.

No.	Non-Optimized Fuzzy Integration	Optimized Integration HGA 1	Optimized Integration HGA 2 **(α=0.5 β=0.5)**	Optimized Integration HGA 2 **(α=0.7 β=0.3)**	Optimized Integration HGA 2 **(α=0.9 β=0.1)**
1	71/90 (78.89%)	87/90 (96.67%)	84/90 (93.33%)	86/90 (95.56%)	88/90 (97.78%)
2	90/90 (100%)	NO NEED	NO NEED	NO NEED	NO NEED
3	80/90 (88.89%)	90/90 (100%)	NO NEED	NO NEED	NO NEED
4	90/90 (100%)	NO NEED	NO NEED	NO NEED	NO NEED
5	73/90(81.10%)	85/90 (94.44%)	90/90 (100%)	NO NEED	NO NEED
6	90/90(100%)	NO NEED	NO NEED	NO NEED	NO NEED
7	81/90(90%)	90/90 (100%)	NO NEED	NO NEED	NO NEED
8	90/90(100%)	NO NEED	NO NEED	NO NEED	NO NEED

Because the fuzzy integration of case 1 was the most difficult to optimize, figure 10 shows the behavior of the HGA evolutions with different weights (for case 1).

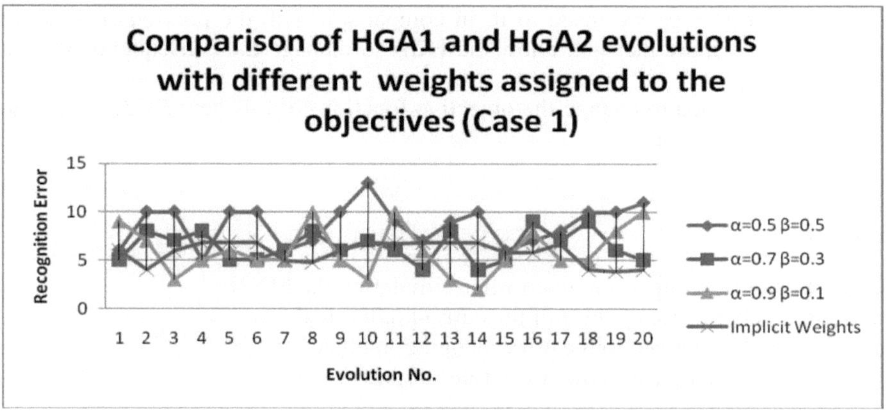

Fig. 10. HGA evolutions for case 1.

We can observe in figure 11 that the best weights configuration of the HGA for case 1 are α=0.9 and β=0.1, this means that assigning a very high weight to the error recognition objective and a smaller weight to the fuzzy rules objective allowed us to increase the recognition percentage in case 1.

Figure 10 shows the final comparison of non-optimized integration vs. optimized integration.

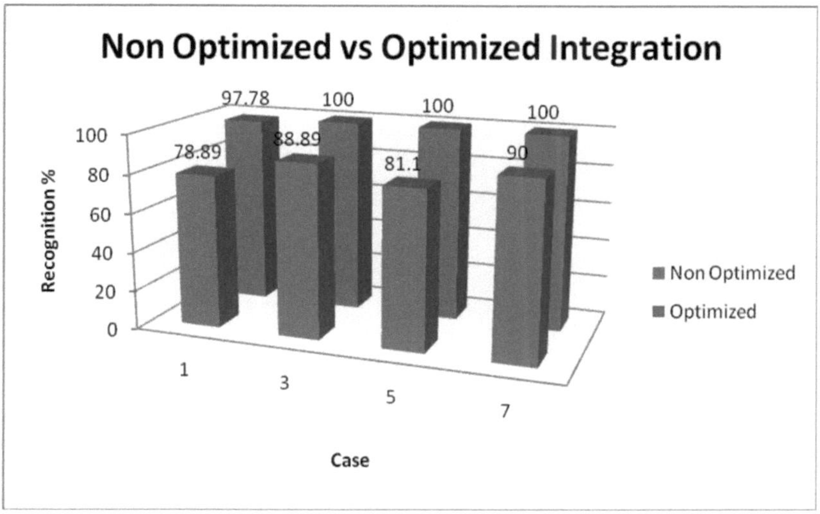

Fig. 11. New Fuzzy integration comparison.

9 Comparison with Other Works

Many research works have been made in this area. This paper is based directly on the work by Hidalgo et al. [11]. In this work a comparison of type-1 and type-2 fuzzy integration method was made for a MNN for face, fingerprint and voice recognition and optimized these systems with a GA (only MFs parameters). The work described in this paper is different from Hidalgo's work in the following form: the way of training the MNN (we trained each module separately), Optimization (we used a HGA to optimize the structure of a FLS not only MFs parameters). We used t-student statistic to make the comparison of results. Figures 12, 13, 14 and 15 show the results of the hypothesis testing (mean of two independent samples) using the Statdisct 9.5.2 software.

We can observe from figure 14 and 15 that there is statistical evidence to say that there is significant difference from the results of our paper with respect to the ones in Hidalgo's work (with almost 90% of confidence). In other words, the HGA provided significant improvement in the recognition rate by improving the responses of the MNN.

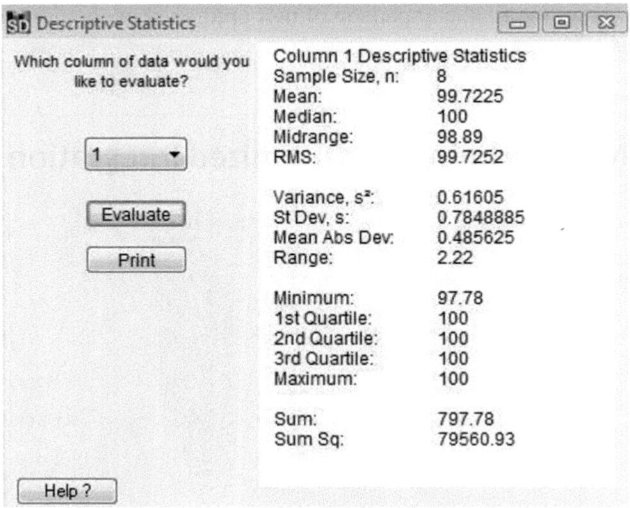

Fig. 12. Descriptive statistics for the method proposed in this paper.

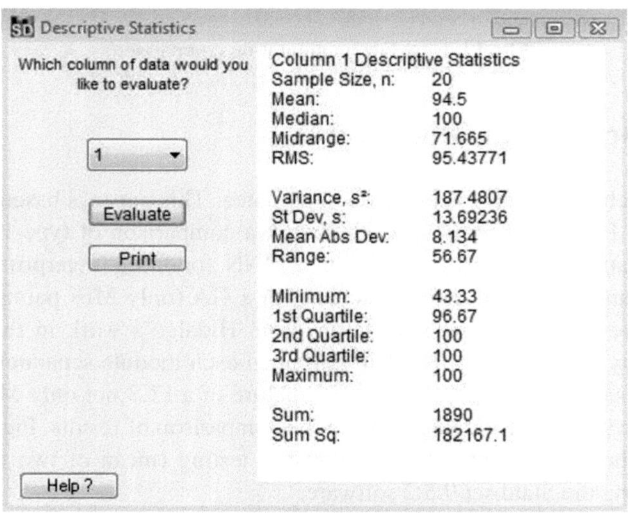

Fig. 13. Descriptive statistics for Hidalgo's work.

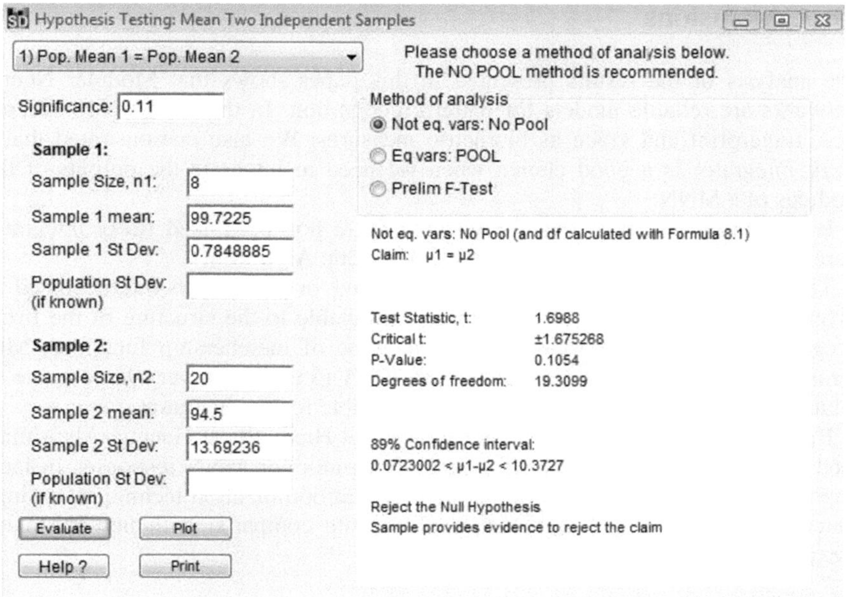

Fig. 14. Hypothesis testing.

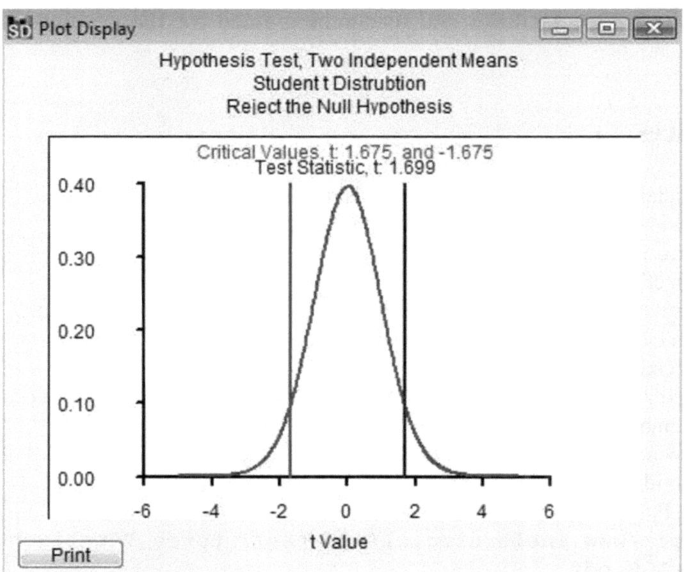

Fig. 15. Hipothesis testing plot.

10 Conclusions

The analysis of the results presented in this paper shows that Modular Neural Networks are reliable models for pattern recognition. In this case we considered face, fingerprint and voice as biometric measures. We also demonstrated that a fuzzy integrator is a good choice when we need to integrate the outputs of the modules of a MNN.

It is important to note that the results with a non optimized fuzzy integrator were improved using a HGA (Hierarchical Genetic Algorithm).

The hierarchical genetic algorithm presented in this paper was designed to cover a considerable number of features attributable to the structure of the fuzzy integrator (type of system, type of logic, type of membership functions, MFs parameters and fuzzy rules), allowing the HGA to explore a very large space of solutions and thus obtain the most optimal possible fuzzy integrator.

Based on the results obtained we can say that Hierarchical Genetic Algorithms used to optimize the structure of a fuzzy integrator for MNN responses in face, fingerprint and voice recognition are an efficient optimization technique, through which we can achieve a higher recognition rate compared with non-optimized fuzzy integrators as shown in this paper.

Acknowledgment

We would like to express our gratitude to the CONACYT and Tijuana Institute of Technology for the facilities and resources granted for the development of this research.

References

[1] Artificial neural networks fundamentals, models and applications, `http://www.monografias.com/trabajos12/redneur/redneur.shtml` (December 2008)
[2] Connectionist systems, `http://carpanta.dc.fi.udc.es/~cipenedo/cursos/scx/archivospdf/Tema1-0.pdf` (December 2008)
[3] Artificial neural networks, `http://es.wikipedia.org/wiki/Red_neuronal_artificial` (December 2008)
[4] Baldwin, C., Clark, K.: Design Rules, The Power of Modularity, vol. 1. MIT Press, Cambridge (2000)
[5] Kuri, F.: Neural Networks and Genetic Algorithms. PDF document, `http://www.inele.ufro.cl/apuntes/Tutores_Inteligentes/RNSyAGS.pdf`
[6] Jang, J.S.R., Sun, C.T., Mizutani, E.: Neuro-Fuzzy and Soft Computing. Prentice Hall, New Jersey (1997)

[7] Rodriguez, M.: Optimization of DNA genotyping as a problem of selection of features, thesis, Americas University of Puebla, Puebla Mexico (2005)
[8] Golberg, D.: Genetic Algorithms in search, optimization and machine learning. Addison Wesley, USA (1989)
[9] Alvarado, J.M.: Recognition of the person through his face and fingerprint using modular neural networks and wavelet transform, thesis, Tijuana Institute of Technology, Tijuana Mexico (2006)
[10] Ramos, J.: Neural networks applied to speaker identification by voice using feature extraction, thesis, Tijuana Institute of Technology, Tijuana México (2006)
[11] Hidalgo, D., Castillo, O., Melin, P.: Type-1 and type-2 fuzzy inference systems as integration methods in modular neural networks for multimodal biometry and its optimization with genetic algorithms, thesis, Tijuana Institute of Technology, Tijuana México (2009)
[12] Ripley, B.D.: Pattern Recognition and Neural Networks. Cambridge University Press, Oxford (1996)
[13] Morales, G.: Introduction to Fuzzy Logic, Research center and Advanced studies of the IPN, `http://delta.cs.cinvestav.mx/~gmorales/ldifll/node1.html`
[14] Castro, J.R.: Tutorial Type-2 Fuzzy Logic: Theory and Applications. UABC University and Tijuana Institute of Technology, Tijuana México (2006)
[15] Fuzzy Logic: Introduction and basic concepts, `http://members.tripod.com/jesus_alfonso_lopez/FuzzyIntro2.html` (January 2009)
[16] Cristea, M., Dinu, A., McCormick, M., Ghee, J.: Neural and Fuzzy Logic Control of Drives and Power Systems. Elsevier, Oxford (2002)
[17] Genetic Algorithms, `http://es.wikipedia.org/wiki/Algoritmo_gen%C3%A9tico` (December 2008)
[18] Langari, R.: A Framework for analysis and synthesis of fuzzy linguistic control systems, Ph.D. thesis, University of California, Berkeley (1990)
[19] Alvarado, J.M.: Recognition of the person through his face and fingerprint using modular neural networks and wavelet transform, thesis, Tijuana Institute of Technology, Tijuana Mexico (2006)

Modular Neural Network with Fuzzy Integration of Responses for Face Recognition

Erika Ayala, Miguel Lopez, and Patricia Melin

Tijuana Institute of Technology, Tijuana México
epmelin@hafsamx.org

Abstract. This paper presents a modular neural network with fuzzy integration of responses for face recognition. We describe its architecture and simulation results using the ORL database. We show results from different integrators, such as the Gating Network, fuzzy Sugeno integrals and type-1 fuzzy systems. We also show that the results with type-1 fuzzy systems are good, but we decided to optimize this fuzzy system with genetic algorithms (MFs parameters and fuzzy rules) to improve the results.

1 Introduction

Pattern recognition is not only a field of science but also a fundamental process that is found in almost all human actions. Over time we have developed systems capable of imitating the behavior of the human brain, many of these applied to the recognition of the individual according to their biometric measures, such systems are used for security or verification of identity for various purposes.

Face recognition is a problem that has been considered since the early stages of computer vision. This problem has been studied more thoroughly in recent years thanks to advances in computational power that have allowed to implement more complex algorithms using different techniques.

There are different techniques and methods that can be used for feature extraction, and it is now easier to recognize a person through biometric methods that exist, for example recognition by their irises, fingerprints, face, recognized by the voice, signature, hand geometry, ears, vein structure, retina, facial thermography, and others that exist.

This paper describes a modular neural network architecture with fuzzy response integration applied to face recognition. The proposed modular architecture was tested with the benchmark ORL database with good results.

Until now biometric methods have been implemented using different devices for the creation of patterns and generating the code that identifies the individual biometric. This is why, in this work we consider one of the most used biometric methods throughout history, which is the face recognition.

O. Castillo et al. (Eds.): Evolutionary Design of Intelligent Systems, SCI 257, pp. 131–158.
springerlink.com

2 Neural Networks

Neural networks are just another way to emulate certain characteristics of humans, such as the ability to associate and memorize facts. If you look carefully at the problems that cannot be expressed through an algorithm, it appears that they all have one thing in common: experience. Man is able to solve these situations by resorting to experience. Thus, it seems clear that one way to approach the problem is to build systems that are able to reproduce this human characteristic [1].

In short, neural networks are nothing more than an artificial and simplified model of the human brain, which is the most perfect example we have of a system that is capable of acquiring knowledge through experience. A neural network is "a new system for processing information, whose basic unit of processing is based on the fundamental cell of the human nervous system: the neuron."

Therefore, Neural Networks:

- consist of processing units that exchange data or information.
- They are used to recognize patterns, including images, manuscripts and sequences of time, financial trends.
- They have the ability to learn and improve their functionality.

Artificial neural networks are made up of interconnecting artificial neurons (programming constructs that mimic the properties of biological neurons). Artificial neural networks may either be used to gain an understanding of biological neural networks, or for solving artificial intelligence problems without necessarily creating a model of a real biological system.

To summarize these ideas we have the following similarities:

Biological Neurons - Artificial -Neurons
synaptic connections- Weighted Connections
Effectiveness synapse - connections of Weight
Combined effect of the synapse-spread function or network
Activation -> firing rate function activation -> Output [2].

2.1 Structure of an Artificial Neural System

The artificial neural system is composed of several components that are necessary to structure the system. In Figure 1 we show the structure of an artificial neuron, which is a very simplified representation of a real neuron.

2.2 Modular Neural Networks

A neural network can be considered modular if it can be decomposed into two or more subsystems in which each individual subsystem evaluates various entries without having communication with other subsystems. The output of the system depends on a modular unit, which receives the outputs of individual subsystems and combines them into a predefined fashion to produce the output of the system [3].

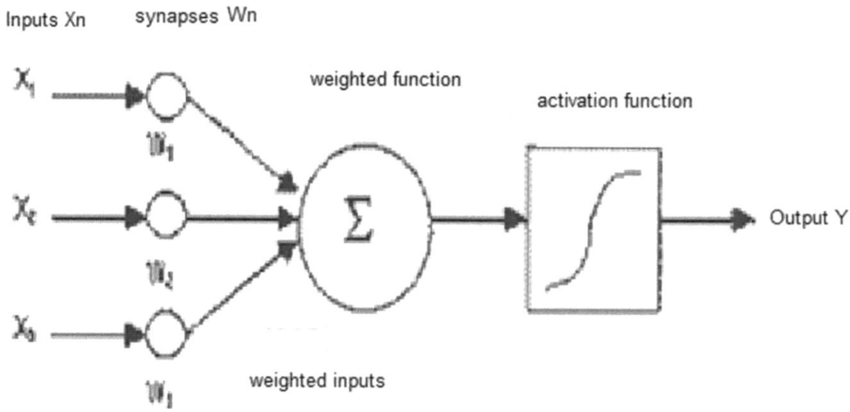

Fig. 1. Artificial Neuron

Based on biological evidence, the first modular complex systems were described by Jacobs and Jordan [4], which involved two distinct types of learning:

- Supervised learning, during which a teacher provides for each external stimulus input the correct output. However, this teacher does not specify that the module is to teach the corresponding pair (stimulus input, desired output).
- Unsupervised learning, which basically consists of a competitive learning, in which the various modules compete to learn the example presented.

The advantage is that if the model supports naturally a breakdown into more simple functions, the application of a modular network translates into faster learning. Each module can be built differently, in a way that meets the requirements of each subtask. In Figure 2 there are n modules, each one is an expert on a specific task.

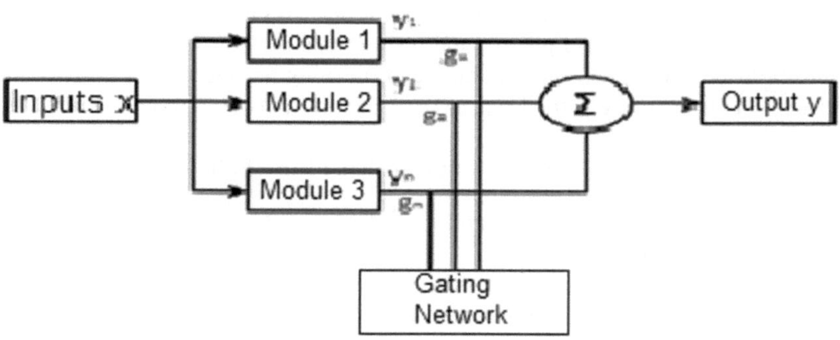

Fig. 2. Schematic of a modular neural network

The basic architecture consists of two main components: local experts and an integration unit* [5]. The outputs of a number of local experts (Oi) are measured by an integration unit. The unit outputs the board using estimated combination weights (gi). And together the outputs of the modular network are given by:

$$Yi=\sum_{i=1}^{k} giOi \tag{1}$$

2.3 Reason for Using Modular Neural Networks

The following points highlight some of the important reasons that make the design of a modular neural network more attractive than the design of a conventional monolithic neural network.

- Model complexity reduced
- Robustness
- Scalability
- Learning
- Computational efficiency
- Learning ability
- Learning economy
- Integration of knowledge
- Insight into the neural network model
- Biological analogy

3 Fuzzy Logic

Fuzzy logic was investigated for the first time in the mid-sixties at the University of California Berkeley by the brilliant Iranian engineer Lotfi A. Zadeh, when he realized what it is called the principle of incompatibility: "As the complexity of a system increases, our ability to build precise and instructions on their behavior decreases to the threshold beyond which the accuracy and meaning are mutually exclusive characteristics." This principle is the basic to the fuzzy systems used today.

3.1 The Structure of a Fuzzy System Consists of the Following Elements

1. The rule base that contains linguistic rules provided by the expert, or can be extracted from numerical data.
2. The fuzzifier, which assigns to the numerical entries in their corresponding membership functions. This is needed to activate rules, which are specified in terms of linguistic variables; fuzzifier takes the input values and determines the degree to which they belong to each of the fuzzy sets via membership functions.

3. The fuzzy inference engine defines the allocation of input fuzzy sets to the corresponding values of the output fuzzy sets. This determines the degree to which each part of the antecedent is satisfied for each rule.
4. The defuzzifier in the case of Mamdani-assigned sets of outputs in a number that is a fact not fuzzy. Given a fuzzy set which manages a range of output values, the defuzzifier returns a number within a set of fuzzy numbers.

Many methods are used for defuzzification, including the centroid, the maximum, among others. The most popular is the centroid, which calculates and produces the center of gravity of an aggregate fuzzy set [7] [8].

4 Genetic Algorithms

In 1975, John Holland proposed the now called "Genetic Algorithms". They are so called because they are inspired by biological evolution and its genetic and molecular basis.

These algorithms do evolve a population of individuals subjected to random actions similar to those involved in biological evolution (mutation and genetic recombination), as well as a selection according to some criteria, according to which individuals are decided to survive, and which are less suitable, are discarded.

We also have what is known as evolutionary algorithms, and includes the development of evolutionary strategies, programming and genetic programming.

A genetic algorithm is a programming technique that mimics biological evolution as a strategy to solve problems. Given a specific problem to resolve, the input AG is a set of potential solutions to that problem, encoded in some way, and a metric called fitness function that allows quantitative assessment of each candidate.

AG then evaluates each candidate according to the fitness function. In a pool of candidates generated randomly, of course, most will not work at all, and will be eliminated. However, by pure chance, a few may be able to show promising activity, albeit imperfect and weak activity towards solving the problem. [9].

4.1 Operation of a Basic Genetic Algorithm

A genetic algorithm can present several variations, depending on how they apply genetic operators (crossing, mutation) also on how the selection is done and how it will determine the replacement of individuals to form the new population. In general, the pseudo code consists of the following steps: i: initialization, f(x): evaluation, completion status, Se: selection, Cr: mating, Mu: mutation, Re: substitution, X: better solution. See Figure 3.

4.2 Selection Methods

A genetic algorithm can use many different techniques to select the individuals to be copied to the next generation, but below are some of the most common. Some of these methods are mutually exclusive, but others may be used in combination, something that is often done.

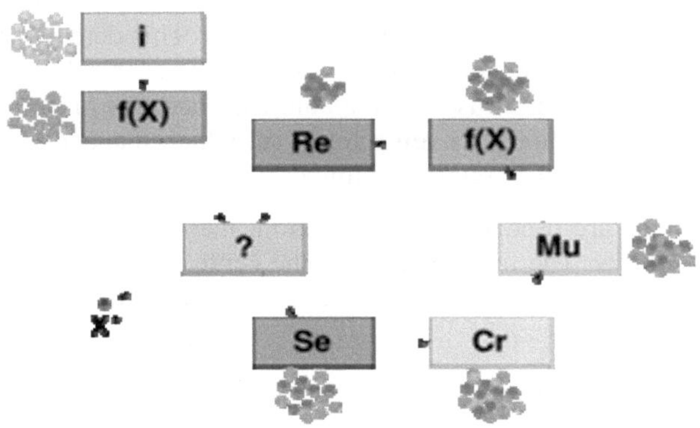

Fig. 3. Basic genetic algorithm

- Elitist selection
- Proportional to fitness selection
- Roulette wheel selection
- Climbing selection
- Tournament selection
- Ranking selection
- Generational selection
- Browse by state
- Hierarchical Selection [10].

4.3 Crossover Methods

Once the selection has been chosen to fit the individual, they must be altered at random with the hope of improving its fitness for the next generation. There are two basic strategies to do this. The first and simplest is called a mutation. Like a mutation in the gene changes a life on the other hand, a mutation in a genetic algorithm also causes small changes in specific points of the code of an individual.

The second method is called mating, and involves selecting two individuals to exchange segments of their code, resulting in an `` artificial" descendants of individuals who are combinations of their parents.

5 ORL Data Base

The Database of Faces, (formerly 'The ORL Database of Faces'), contains a set of face images taken between April 1992 and April 1994 at the lab. The database was used in the context of a face recognition project carried out in collaboration with the Speech, Vision and Robotics Group of the Cambridge University Engineering Department.

There are ten different images of each of 40 distinct subjects. For some subjects, the images were taken at different times, varying the lighting, facial expressions (open / closed eyes, smiling / not smiling) and facial details (glasses / no glasses). All the images were taken against a dark homogeneous background with the subjects in an upright, frontal position (with tolerance for some side movement) [11].

The files are in BPM format (bitmap), the size of each image is 92x112 pixels, with 256 grey levels per pixel. The images are organized in 40 directories (one for each subject), which have names of the form sX, where X indicates the subject number (between 1 and 40). In each of these directories, there are ten different images of that subject, which have names of the form Y.bpm, where Y is the image number for that subject (between 1 and 10), giving a total of 400 images. The following is a part of the ORL database, see Figure 4.

Fig. 4. ORL Database example of faces.

6 Characteristics of a Biometric Measure

A biometric measure is a feature that can be used to make an identification. Whatever the measure, it must meet the following requirements:

1. Universality: means that anyone should have that characteristic.
2. Uniqueness: the existence of two people with identical characteristics has a very small probability.
3. Permanent Characteristics: the characteristic does not change over time
4. Quantification: the characteristic can be measured in a quantitative form.

6.1 Architecture of a Biometric System for Person Identification

A biometric system has three basic components: The first is responsible for the acquisition of any analog or digital biometric feature of a person, such as the acquisition of a fingerprint image using a scanner. The second handles the compression, processing, storage and comparison of acquired data with the stored data. The third component provides an interface to applications on the same or another system. In figure 5 we show the phases of a biometrical identification system.

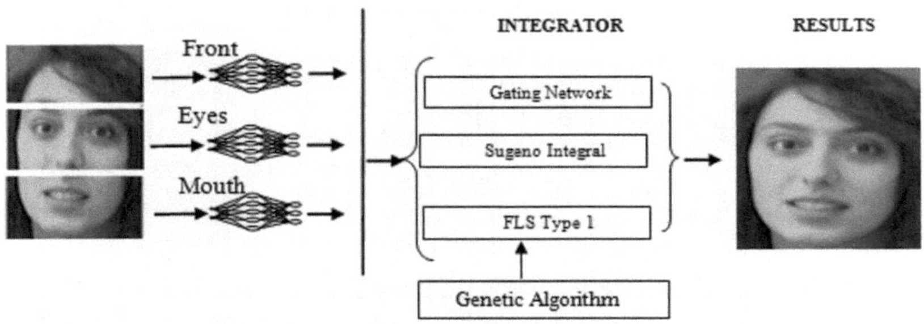

Fig. 5. Architecture of the modular neural network

6.2 Design of the Neural Network Structure for Face Recognition

The number of neurons is allocated using an empirical expression created by Renato Salinas [17] and can be explained as follows:

- Input neurons: 2 * (k+m), the activation function is a Tangent sigmoid.
- Hidden neurons: (k+m), the activation function is a Tangent sigmoid.
- Output neurons: the activation function is a Tangent logarithm sigmoid.

Where k is the number of individuals to train and m is the number of samples to train for each individual. In this case, these were 40 individuals for the database of ORL, and 30 for the database of the institute and 7 samples therefore, $k = 40$, $m = 7$ and $k = 30$, $m = 7$.

Taking into account the previous data to train the modules. The architecture is shown in Figure 6.

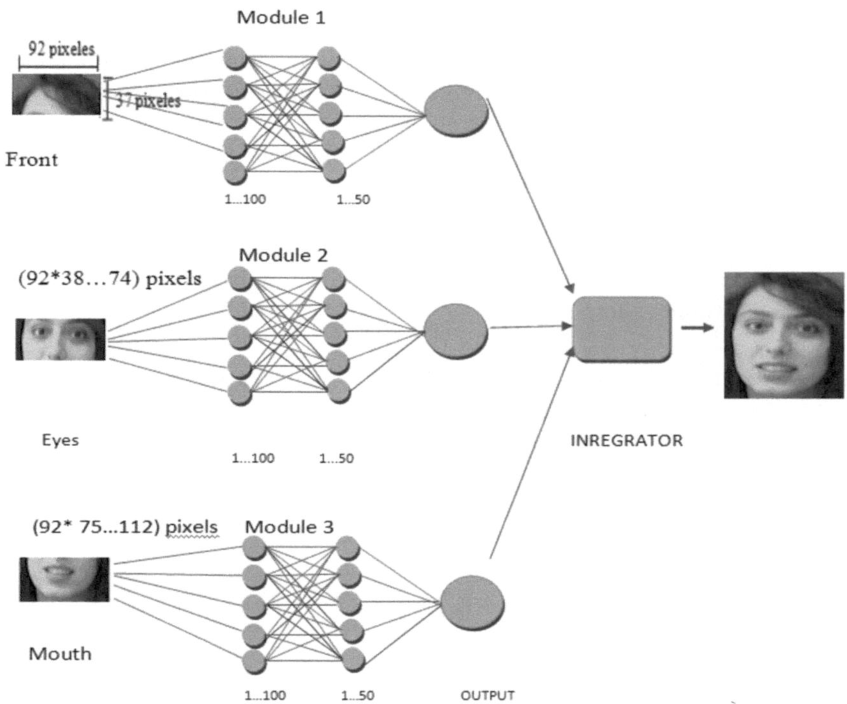

Fig. 6. Architecture of the neural network for face recognition.

7 Simulation Results

In this section, we only show the results for the biometric measures, testing with 3 methods of training and selecting the best results. Table 1 shows the different training methods used in the databases.

Table 1. Training methods for the neural networks.

Abbreviation	Training method
TRAINGDX	Gradient descendent with momentum and adaptive learning rate backpropagation
TRAINSCG	Scaled conjugate gradient backpropagation
TRAINCGB	Conjugate gradient backpropagation with Powell-Beales restarts

We used 400 images for training without pre-processing, Figure 4 shows that the first 7 images of each individual were used for training the neural network and the last 3 are used for identification, both the training images the identification of images without noise are from a photo editor. The difficult part of the problem is to identify the last 3 images that were not trained.

Table 2 shows the results of the trainings done with the ORL database. We show the results of the integration methods: Sugeno measures and Gating Network.

Several training was done using different methods, changing the learning rate. We can observe in table 2 that the greater identification has been found with the gating network integrator.

Table 2. Results with gradient Conjugate Escalade (trainscg), Conjugate gradient with Powell (traincgb), taking better performance the trainscg, both in training the neural network as in the identification results.

TRAINING METHOD	LEARNING RATE	ERROR GOAL	Trained images 70%	Images to Identify 30%	GATING NETWORK		Fuzzy and Sugeno integral	
					RECOGNITION	IDENTIFY	RECOGNITION	IDENTIFY
TRAINSCG 1	0.0001	0.001	280	120	280/280 100%	110/120 92%	280/280 100%	107/120 89%
	0.0001	0.001						
	0.0001	0.001						
TRAINCGB 2	0.000001	0.001	280	120	280/280 100%	113/120 94%	280/280 100%	114/120 91%
	0.0001	0.001						
	0.00001	0.001						
TRAINSCG	0.00001	0.001	280	120	280/280 100%	117/120 97%	280/280 100%	114/120 95%
	0.001	0.001						
	0.0001	0.001						

After having obtained the above mentioned results, three Cross-validations (permutations) were made to the original database, making 20 different trainings with each permutation. Using others methods of training, like the gradient descent with momentum and adaptive learning rate back propagation (TRAINGDX), and the above-mentioned gradient Conjugate Escalade (Trainscg), Conjugate gradient with Powell (Traincgb).

We show an example of how the 3 cross validations were made. We used the first 7 images for training and the last 3 to identify (ORL database), for the first crossover validation the last 3 images (that were for identification) were moved to the beginning and now were used for training, and now the last 3 were used for identification, and same was done for the other 2 crossover validations. See the Figure 7.

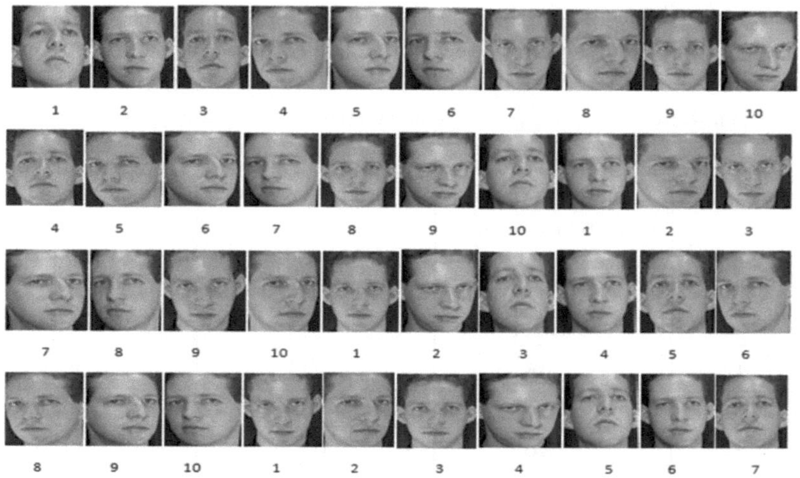

Fig. 7. Demonstration of the Cross Validations.

We achieved more results with integrators: Gating network, Integrals Sugeno and measures and with fuzzy type-1 integrator developed in this work, showing the best training with the different cross-validations and underscoring our best training, training with different characteristics.

The following tables show the results of the first cross-validation indicating the best training for each integrator.

Next we show the table containing data of the modular neural networks, with an overall error of 0001 Training 1-6 and 0.0001 de1 7-10, a learning rate of 0.00001 for the first module, 0.001 for the second and 0.0001 for the third module, and the training method is different for each module in different trainings, ranging from Trainscg, Traincgb and Traingdx. To verify the data see Table 3.

Table 3. Content of the modular neural networks

TRAINED	ERROR GOAL	LEARNING RATE	TRAINING METHOD	TIME
1	0.001	0.00001, 0.001, 0.0001	Traingdx, Traingdx, Traingdx	00:04:50,00:05:25, 00:04:33
2	0.001	0.00001, 0.001, 0.0001	Traincgb, Traingdx, Trainscg	00:04:13, 00:05:08, 00:07:09
3	0.001	0.00001, 0.001, 0.0001	Trainscg, Trainscg, Trainscg	00:02:25, 00:03:49, 00:02:34
4	0.001	0.00001, 0.001, 0.0001	Trainscg, Traincgb, Trainscg	00:03:53, 00:03:29, 00:03:14
5	0.001	0.00001, 0.001, 0.0001	Traincgb, Trainscg, Traincgb	00:03:49, 00:04:34, 00:05:53
6	0.001	0.00001, 0.001, 0.0001	Trainscg, Traingdx, Trainscg,	00:04:27, 00:03:50, 00:02:56
7	0.0001	0.00001, 0.001, 0.0001	Trainscg, Traincgb, Trainscg	00:04:63, 00:05:01, 00:07:54
8	0.0001	0.00001, 0.001, 0.0001	Traincgb, Trainscg, Traincgb	00:03:19, 00:04:34, 00:04:44
9	0.0001	0.00001, 0.001, 0.0001	Traincgb, Traincgb, Traincgb	00.04:35, 00:04:35, 00.06:30
10	0.0001	0.00001, 0.001, 0.0001	Trainscg, Trainscg, Trainscg	00:03:27, 00:04:04, 00:07:10

The results of the first cross-validation are shown in Table 4, indicating out our best training for each integrator. The training data is shown in Table 3.

Table 4. Results of the first cross-validation with three different integrators

TRAINING	IDENTIFICATIÓN GATING NETWORK %	IDENTIFICATIÓN MEASURES AND SUGENO INTEGRAL %	IDENTIFICATIÓN FUZZY LOGIC SYSTEM %
1	95	95	93.3
2	92.5	94	89
3	95	93	90.8
4	92.5	89	87.5
5	88	87.5	91.6
6	94	85	89
7	90	85	84
8	95	92.5	83
9	92.5	88	86.6
10	88	89	86.6

Continuing with our results we now show the second cross-validation in table 5, underlining our best training for each integrator. The training data is shown in Table 3.

Table 5. Results of the second cross-validation with three different integrators

TRAINING	IDENTIFICATIÓN GATING NETWORK %	IDENTIFICATIÓN MEASURES AND SUGENO INTEGRAL %	IDENTIFICATIÓN FUZZY LOGIC SYSTEM %
1	88	86	81
2	87.5	88	84
3	90	90	83
4	89	88	88
5	92	91	83
6	92	90	82.5
7	91	92	84
8	92	89	89
9	88	90	88
10	92	91	89

And finally our results for the third cross-validation are shown in Table 6.

Table 6. Results of the third cross-validation with three different integrators

TRAINING	IDENTIFICATIÓN GATING NETWORK %	IDENTIFICATIÓN MEASURES AND SUGENO INTEGRAL %	IDENTIFICATIÓN FUZZY LOGIC SYSTEM %
1	94	91	81
2	94	93	85.8
3	94	93	86.6
4	92.5	91	87.5
5	91	90	85.8
6	96	94	93.3
7	91	87.5	87.5
8	96	92.5	93.3
9	89	87.5	89
10	94	86	87.5

8 Fuzzy Logic System

The fuzzy system consists of three input modules and one output, each module has three Gaussian functions, with a range of 0 to 1 for both entries of the output. See Figure 8.

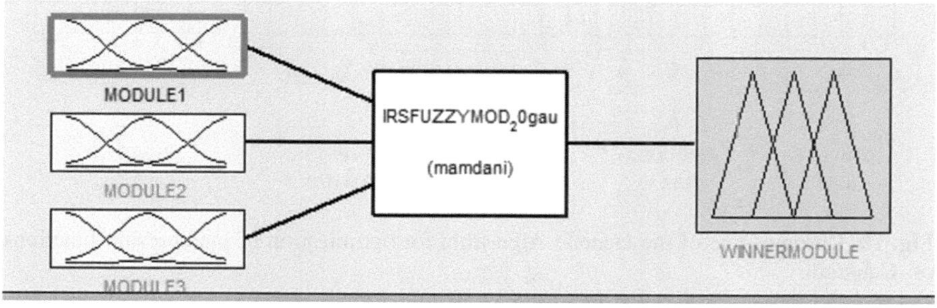

Fig. 8. Fuzzy Logic Design

Each input variable represents 3 modules, which are module1, module2, module3 and output variable, and each module consists of three membership functions activatiónmodule1baja, activationmodule2media, activatiónmodule3alta. See Figure 9.

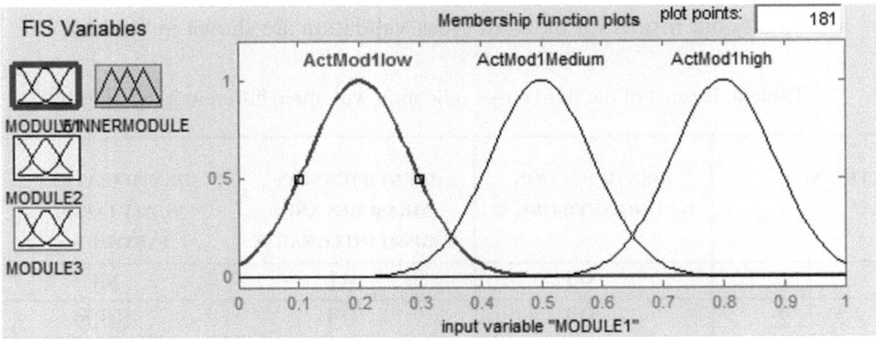

Fig. 9. Membership Functions.

9 Genetic Algorithm

Our genetic algorithm consists of 51 gens, 24 gens for the input and output membership functions parameters (Gaussians), these gens are moved between 0 and 1 and are real numbers. The remaining 27 gens are to reduce the rules of the fuzzy system, these are binary type. See Figure 10.

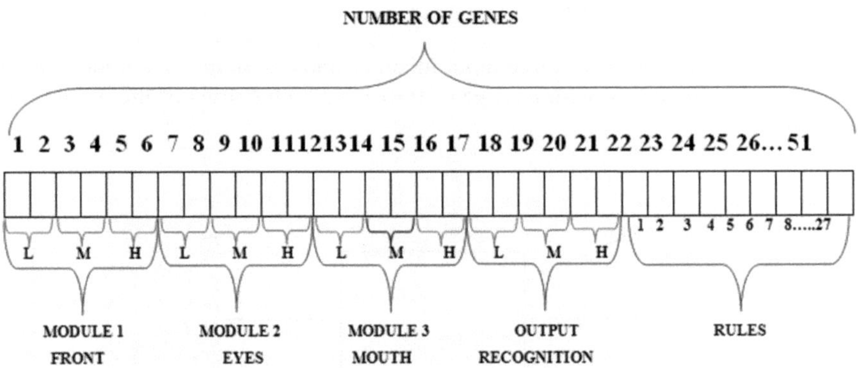

Fig. 10. Chromosome of the Genetic Algorithm for optimization of membership functions of Gaussian.

9.1 Objective Function

Our objective function is defined to minimize the error of recognition and the rules, where the most important objective is to minimize the recognition error and the second level of importance objective is to reduce the rules. This can be expressed as follows:

$$\text{Objective function} = \text{error recognition} + (\text{number rules}/27) \quad (2)$$

9.2 Results

We show the information of the 10 evolutions we performed for the best network with the ORL Database and the best network of the three cross-validations. See table 7.

Table 7. Content of evolutions of the Genetic Algorithm

Evolution	Crossing	Mutation	Generation	Individuals
1	0.7	0.06	100	100
2	0.6	0.05	60	70
3	0.8	0.05	60	70
4	0.5	0.07	60	70
5	0.9	0.04	60	70
6	0.4	0.08	60	70
7	0.6	0.04	60	70
8	0.6	0.55	60	70
9	0.6	0.45	60	70
10	0.6	0.40	60	70

There were 10 evolutions for the best network of the original Database, identifying the best development for that network. See Table 8.

Table 8. Results of the 10 evolutions in the genetic algorithm for the best network in the ORL Database.

Evolution	Number of Rules	Genetic Algorithm Error	Time	Genetic Algorithm D.B % Identification
1	22	15.8148	00:23:43	89
2	21	14.7778	00:22:18	88
3	23	11.8519	00:24:18	91
4	21	14.7778	00:18:49	88
5	23	13.8519	00:19:20	89
6	21	17.7778	00:18:19	86
7	24	21.8889	00:18:26	82.5
8	22	11.8148	00:18:33	91
9	23	12.8519	00:19:09	90
10	24	15.8889	00:18:03	87.5

Continuing with our results now we show the First cross-validation shown in Table 9, underlining our best evolution for that network. The evolutions information is shown in Table 7.

Table 9. Results of the 10 evolutions in the genetic algorithm for the best network in the first cross-validation.

Evolution	Number of Rules	Genetic Algorithm Error	Time	Genetic Algorithm D.B % Identification
1	20	13.7407	00:53:14	92.5
2	24	3.8889	00:22:25	**97.5**
3	27	14	00:27:49	89
4	25	14.8148	00:27:04	89
5	23	9.8148	00:23:29	92.5
6	27	15	00:25:59	87.5
7	26	15.8889	00:27.28	87.5
8	23	10.8519	00:24:14	92
9	25	8.8889	00:26:10	93
10	22	9.8148	00:23:02	92.5

Now we show in table 10 the results for the Second cross-validation, underlining our best evolution for that network. The evolution information is shown in Table 7.

Table 10. Results of the 10 evolutions in the genetic algorithm for the best network in the second cross-validation

Evolution	Number of Rules	Genetic Algorithm Error	Time	Genetic Algorithm D.B % Identification
1	22	10.8148	00:26:50	92.5
2	21	10.7778	00:28:10	92.5
3	22	10.8148	00:25:36	92.5
4	21	10.7778	00:26:56	92
5	22	10.8148	00:28:01	92
6	20	14.7407	00:27:46	88
7	27	10	00:27:57	92
8	23	10.8519	00:25:39	92
9	22	10.8148	00:24:20	92
10	22	10.8148	00:32:20	92

And finally our results for the third cross-validation are shown in Table 11. The evolution information is shown in Table 7.

Table 11. Results of the 10 evolutions in the genetic algorithm fcr the best network in the third cross-validation

Evolution	Number of Rules	Genetic Algorithm Error	Time	Genetic Algorithm D.B % Identification
1	20	9.7407	00:22:45	92.5
2	23	6.8519	00:22:25	95
3	20	21.7407	00:27.22	94
4	21	7.7778	00:25:01	94
5	22	7.8148	00:31:57	94
6	24	4.8889	00:23:22	97
7	21	6.7778	00:24:57	95
8	22	5.8148	00:26:39	96
9	27	7	00:26:40	95
10	22	5.8148	00:22:56	96

Having tested the best network for each database, we observe that good results are achieved testing a single network for the 4 databases, the information of the evolutions are in Table 7. Underlining our best evolution. See Table 12.

Table 12. Results for the ORL database, the 10 changes in the genetic algorithm with a single network for the four Databases.

Evolution	Number of Rules	Genetic Algorithm Error	Time	Genetic Algorithm D.B % Identification
1	25	33.8667	00:22:38	84
2	26	16	00:22:45	87.5
3	25	21.8889	00:30:44	82.5
4	26	21.8889	00:23:57	82.5
5	26	15	00:24:21	88
6	27	16	00:27:30	87.5
7	26	15	00:27:30	88
8	26	16	00:24:26	87.5
9	26	15.8889	00:23:20	87.5
10	26	11.7778	00:24:37	91

Continuing with our results now we show the First cross-validation shown in Table 13, underlining our best evolution. The evolutions information is shown in Table 7.

Table 13. Results of the First cross-validation, the 10 evolutions in the genetic algorithm with a single network for the four Database.

Evolution	Number of Rules	Genetic Algorithm Error	Time	Genetic Algorithm D.B % Identification
1	25	32.0000	00:19:15	82
2	26	20.7778	00:18:49	83
3	25	10.7778	00:18:04	92
4	26	11	00:18:40	92
5	26	13.7778	00:18:38	89
6	27	14	00:18:15	88
7	26	8.7778	00:18.46	93
8	26	9.8519	00:18:56	92.5
9	26	14.8148	00:18:31	89
10	26	9.8519	00:18:45	91

We show the results of the Second cross-validation in Table 14, underlining our best evolution. The evolution information is shown in Table 7.

Table 14. Results of the Second cross-validation, the 10 evolutions in the genetic algorithm with a single network for the four Database.

Evolution	Number of Rules	Genetic Algorithm Error	Time	Genetic Algorithm D.B % Identification
1	25	0.9259	00:00:20	100
2	26	0.9630	00:00:17	100
3	25	0.9259	00:00:24	100
4	26	0.9630	00:00:20	100
5	26	0.9630	00:00:20	100
6	27	0.9630	00:00:17	100
7	26	0.9630	00:00:19	100
8	26	0.9630	00:00:17	100
9	26	0.9630	00:00:18	100
10	26	0.9630	00:00:17	100

And finally our results for the third cross-validation are shown in Table 15. The evolution information is shown in Table 7.Underlining our best evolution.

Analyzing the previous results, we observe that the results of the second cross-validation are excellent and we obtained a 100% of identification, now we tested the best fuzzy integrator we obtained with the genetic algorithm of ORL Database. The results are as follows, see Table 16.

Table 15. Results of the Third cross-validation, the 10 evolutions in the genetic algorithm with a single network for the four Database.

Evolution	Number of Rules	Genetic Algorithm Error	Time	Genetic Algorithm D.B % Identification
1	25	25.7778	00:23:08	92
2	26	31.3778	00:25:15	86
3	25	29.0667	00:25:07	89
4	26	32.3556	00:30:03	87
5	26	31.4667	00:22:03	87
6	27	32.3556	00:23:01	87
7	26	29.8667	00:00:19	89
8	26	28.4444	00:28:21	92
9	26	32.3556	00:32:36	87
10	26	29.3333	00:23:46	92

Table 16. Results of the four databases with the same Fuzzy Integration Method.

Databases	Identificatión Results
ORL Database	90.8% 109/120
First cross-validation	100% 120/120
Second cross-validation	100% 120/120
Third cross-validation	100% 120/120
AVERAGE	**97.7 %**

Below is the best Fuzzy System we got from Database ORL, showing the Gaussian membership functions for the 3 input to the output. See Figures 11, 12, 13, 14 and 15.

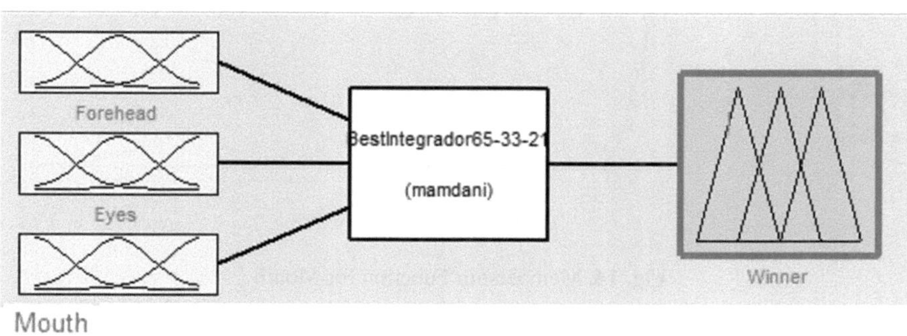

Fig. 11. Fuzzy System

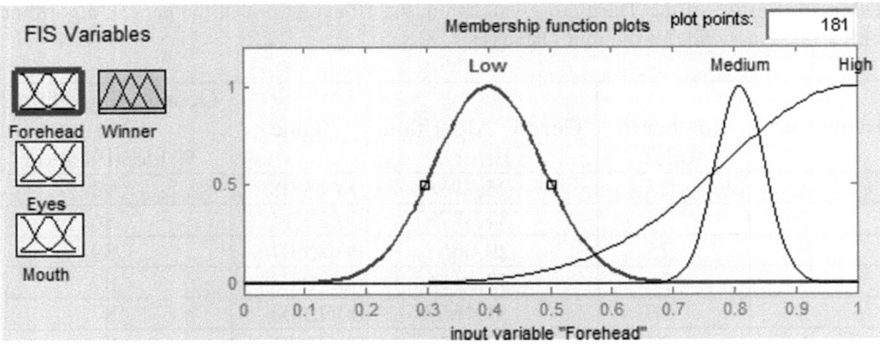

Fig. 12. Membership Function for forehead

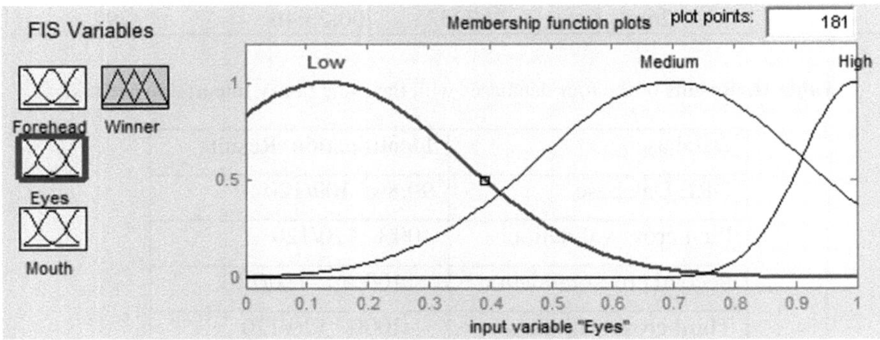

Fig. 13. Membership Function for Eyes

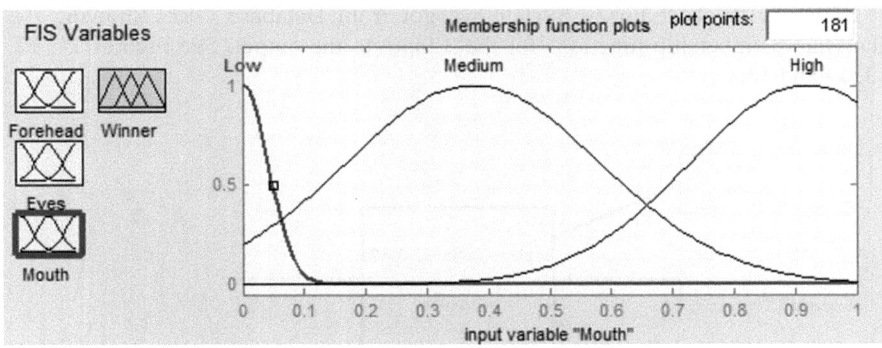

Fig. 14. Membership Function for Mouth

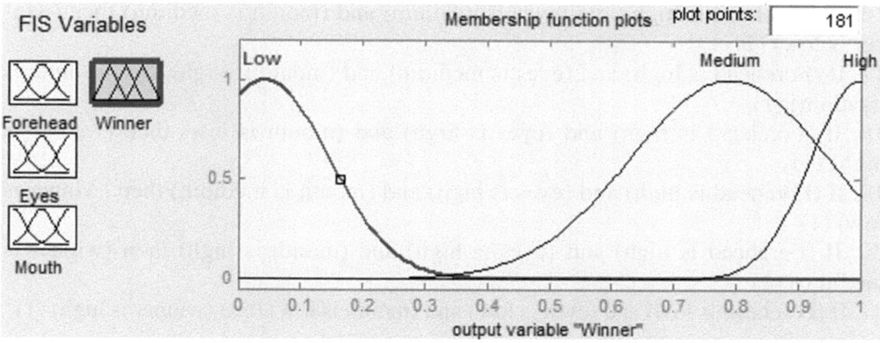

Fig. 15. Output Membership Function

And the rules that gave us the fuzzy system were as follows:

1. If (Forehead is low) and (eyes is low) and (mouth is medium) then (winner is low) (1)
2. If (Forehead is low) and (eyes is low) and (mouth is high) then (winner is medium) (1)
3. If (Forehead is low) and (eyes is medium) and (mouth is low) then (winner is high) (1)
4. If (Forehead is low) and (eyes is medium) and (mouth is medium) then (winner is low) (1)
5. If (Forehead is low) and (eyes is medium) and (mouth is high) then (winner is medium) (1)
6. If (Forehead is low) and (eyes is high) and (mouth is low) then (winner is high) (1)
7. If (Forehead is low) and (eyes is high) and (mouth is medium) then (winner is low) (1)
8. If (Forehead is low) and (eyes is high) and (mouth is high) then (winner is medium) (1)
9. If (Forehead is medium) and (eyes is low) and (mouth is low) then (winner is high) (1)
10. If (Forehead is medium) and (eyes is low) and (mouth is medium) then (winner is low) (1)
11. If (Forehead is medium) and (eyes is low) and (mouth is high) then (winner is medium) (1)
12. If (Forehead is medium) and (eyes is medium) and (mouth is low) then (winner is high) (1)
13. If (Forehead is medium) and (eyes is medium) and (mouth is medium) then (winner is low) (1)
14. If (Forehead is high) and (eyes is low) and (mouth is medium) then (winner is low) (1)
15. If (Forehead is high) and (eyes is low) and (mouth is high) then (winner is medium) (1)

16. If (Forehead is high) and (eyes is medium) and (mouth is medium) then (winner is low) (1)
17. If (Forehead is high) and (eyes is medium) and (mouth is high) then (winner is medium) (1)
18. If (Forehead is high) and (eyes is high) and (mouth is low) then (winner is high) (1)
19. If (Forehead is high) and (eyes is high) and (mouth is medium) then (winner is low) (1)
20. If (Forehead is high) and (eyes is high) and (mouth is high) then (winner is medium) (1)
21. If (Forehead is low) and (eyes is low) and (mouth is low) then (winner is high) (1)

And finally we made and test of a genetic algorithm to optimize triangular membership functions, the chromosome to achieve this optimization is with 54 gens, the first 27 gens are for the input and output parameters of the membership functions, 27 genes because we have three inputs each with three triangular membership and these have three parameters (a, b, c), these genes are moved between 0 and 1 and are real numbers. And the remaining 27 genes are to reduce the rules of the fuzzy system, these are binary gens. See Figure 16.

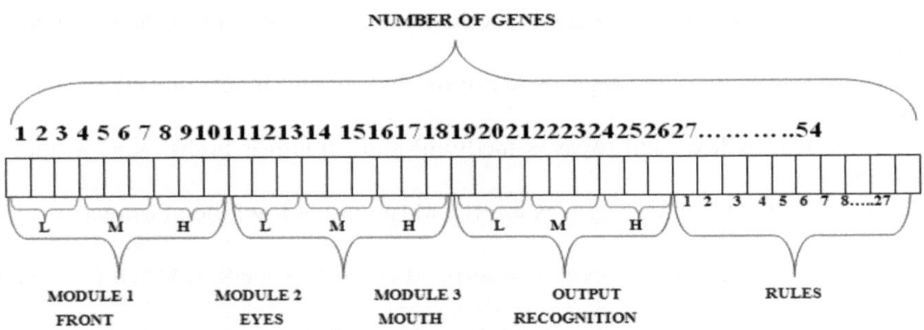

Fig. 16. Chromosome of the Genetic Algorithm for optimization of Triangular membership functions

The objective function is to minimize the error of recognition and the rules, where the most important objective is to minimize the recognition error and the second level of importance objective is to reduce the rules. but in this case uses weights for different objectives and can be expressed as follows:

$$\text{Objective function} = (\text{error recognition} * 0.80) + (\text{number rules} * 0.20) \qquad (2)$$

We tested the best fuzzy integrator obtained with the genetic algorithm for the ORL Database with triangular membership functions, and we also tested with the 3 cross validations. The results are as follows. See Table_17.

Table 17. Results of the four databases with the same Fuzzy.

Databases	Identificatión Results
ORL Database	96.67% 116/120
First cross-validation	100% 120/120
Second cross-validation	100% 120/120
Third cross-validation	99.15% 119/120
AVERAGE	**98.95 %**

Below is the best Fuzzy System we got from Database ORL, showing the Triangular membership functions for the 3 input to the output. See Figures 17, 18, 19, 20 and 21.

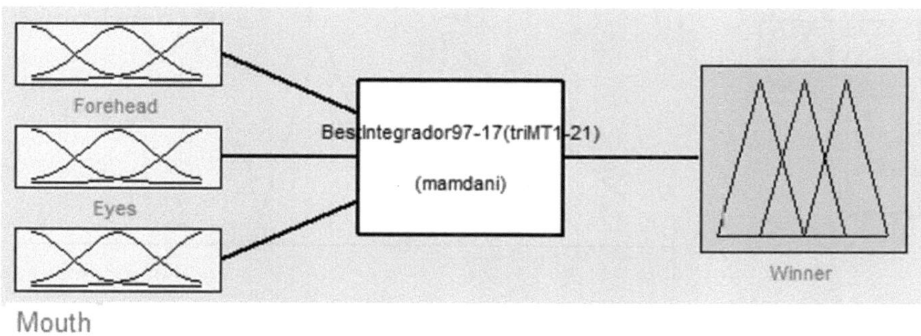

Fig. 17. Fuzzy System

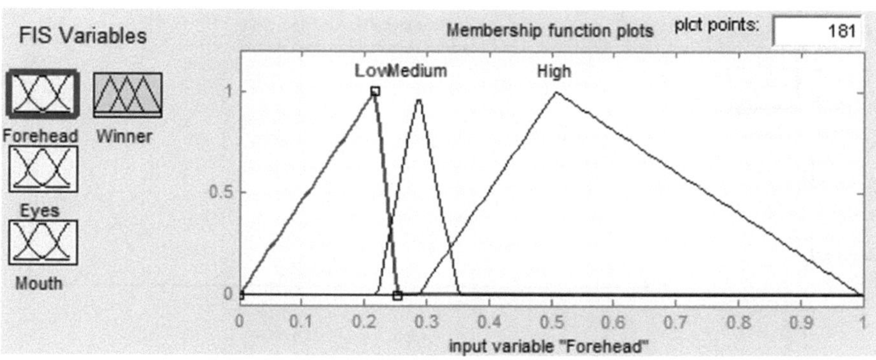

Fig. 18. Membership Function for forehead

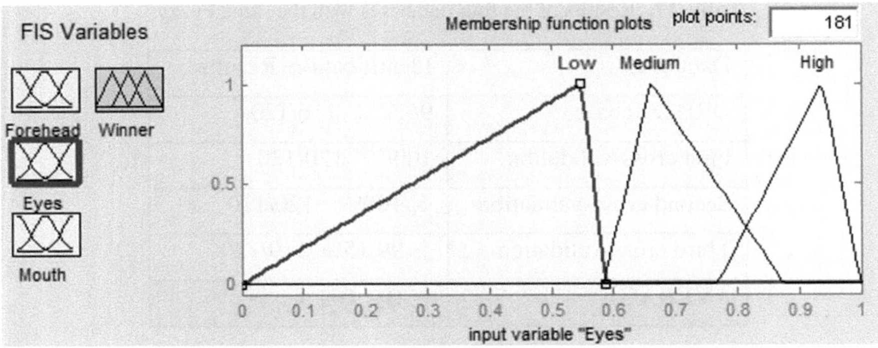

Fig. 19. Membership Function for Eyes

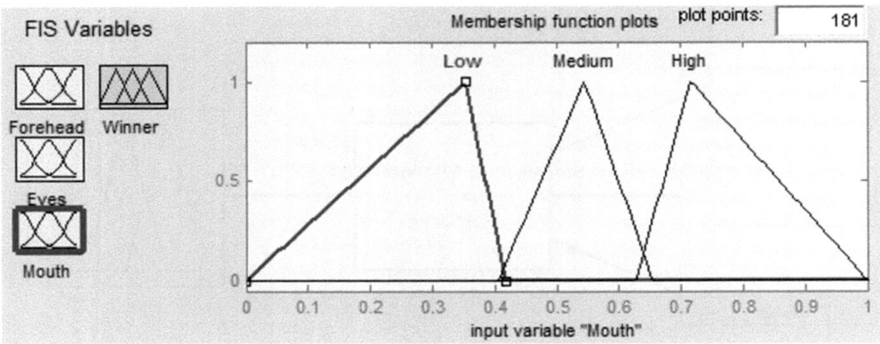

Fig. 20. Membership Function for Mouth

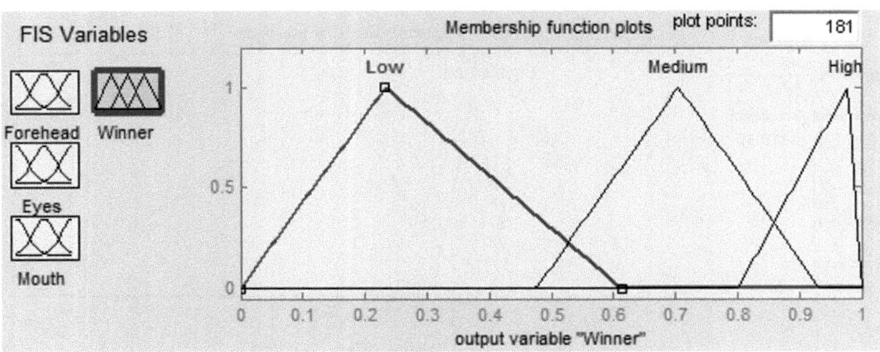

Fig. 21. Output Membership Function

And the rules that gave us the fuzzy system were as follows:

1. If (Forehead is low) and (eyes is low) and (mouth is low) then (winner is high) (1)
2. If (Forehead is low) and (eyes is low) and (mouth is medium) then (winner is medium) (1)
3. If (Forehead is low) and (eyes is low) and (mouth is high) then (winner is medium) (1)
4. If (Forehead is low) and (eyes is medium) and (mouth is low) then (winner is medium) (1)
5. If (Forehead is low) and (eyes is medium) and (mouth is medium) then (winner is high) (1)
6. If (Forehead is low) and (eyes is medium) and (mouth is high) then (winner is medium) (1)
7. If (Forehead is low) and (eyes is high) and (mouth is low) then (winner is high) (1)
8. If (Forehead is low) and (eyes is high) and (mouth is medium) then (winner is high) (1)
9. If (Forehead is low) and (eyes is high) and (mouth is high) then (winner is low) (1)
10. If (Forehead is medium) and (eyes is low) and (mouth is low) then (winner is medium) (1)
11. If (Forehead is medium) and (eyes is low) and (mouth is medium) then (winner is high) (1)
12. If (Forehead is medium) and (eyes is medium) and (mouth is low) then (winner is high) (1)
13. If (Forehead is medium) and (eyes is medium) and (mouth is medium) then (winner is medium) (1)
14. If (Forehead is medium) and (eyes is high) and (mouth is low) then (winner is low) (1)
15. If (Forehead is medium) and (eyes is high) and (mouth is medium) then (winner is high) (1)
16. If (Forehead is medium) and (eyes is high) and (mouth is high) then (winner is high) (1)
17. If (Forehead is high) and (eyes is low) and (mouth is low) then (winner is low) (1)
18. If (Forehead is high) and (eyes is low) and (mouth is medium) then (winner is medium) (1)
19. If (Forehead is high) and (eyes is low) and (mouth is high) then (winner is high) (1)
20. If (Forehead is high) and (eyes is medium) and (mouth is low) then (winner is low) (1)
21. If (Forehead is high) and (eyes is high) and (mouth is medium) then (winner is low) (1)

And finally a graphic showing the percentages obtained from the ORL Database and the three cross-validations, we got from integrators that uses non-optimized and optimized. See Fig. 22.

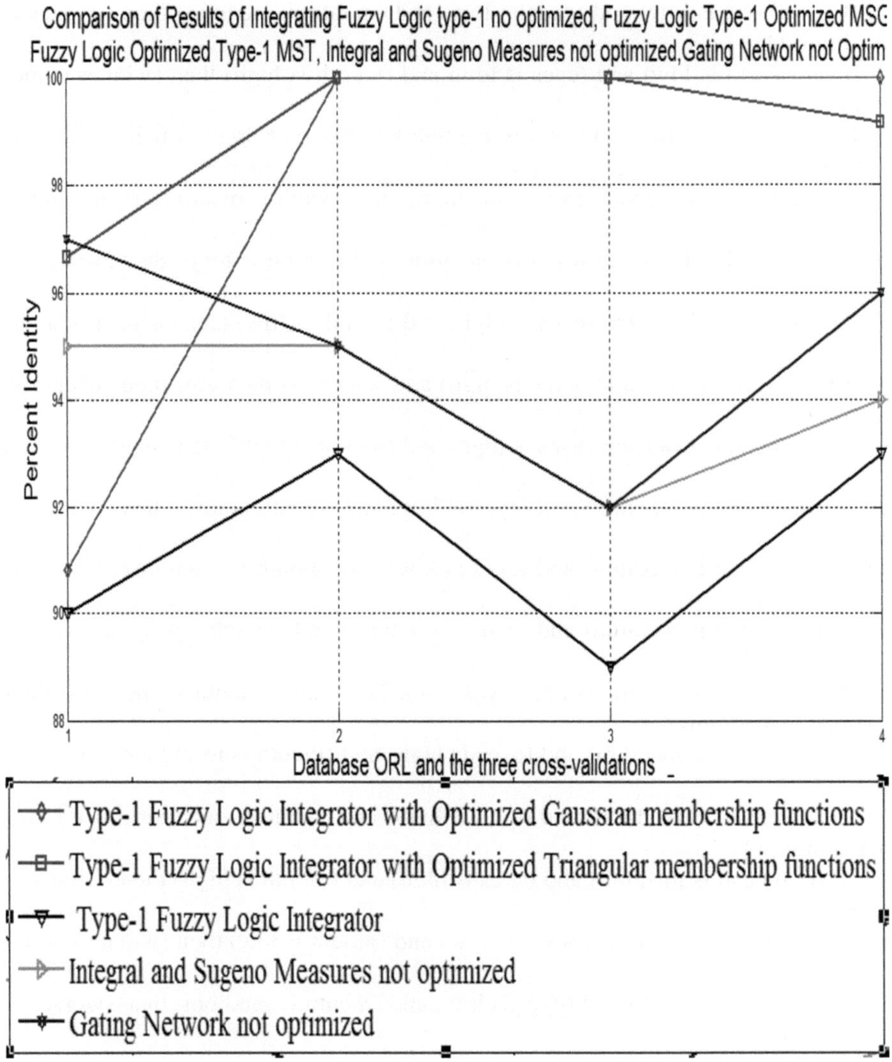

Fig. 22. Results of different response integrators

10 Conclusions

As a conclusion we can say that working with face as biometric measure is more reliable than other forms of authentication available.

Through the results shown in this work, we can see the results of response integrations: Gating Network with identification rate 97%, Sugeno integrals measures with identification rate 95%, type-1 fuzzy logic not optimized with identification rate 90.8%, type-1 fuzzy logic using genetic algorithms to optimize Gaussian membership function and fuzzy rules with identification rate 97.15 and type-1 fuzzy logic using genetic algorithms to optimized Triangular membership function and fuzzy rules with identification rate 98.95 (the best result of all integration methods).

We can also say that making a combination of two or more methods of artificial intelligence we can achieve a significant improvement in pattern identification, in this paper we used a hybrid system (neuro-fuzzy-genetic).

11 Future Work

Identification could be improved with pre-processing. Parallel processing could also be implemented to reduce response times. Tipe-2 fuzzy logic could be used to test if it helps improve identification.

Acknowledgment

We would like to express our gratitude to CONACYT, and Tijuana Institute of Technology for the facilities and resources granted for the development of this research.

References

[1] Hilera, J.R., Martínez, V.J.: Artificial Neural Networks (2000), `http://www.monografias.com/trabajos12/redneuro/redneuro.shtml`

[2] Skapura, D.M.: Building Neural Networks (1996), `http://www.une.edu.ve/electronica/neurona.htm`

[3] Adam, F.: Biologically Inspired Modular Neural Networks. Blacksburg, Virginia (2000)

[4] Holland, J.: Genetic algorithms. Scientific American, Julio de, 66–72 (1992)

[5] Serrano, E.P.: Introduction to Wavelet Transform and Its Applications to Signal Processing of Acoustic Emission. School of Science and Technology. National Unity of General San Martín

[6] Zadeh, L.A.: Outline of a new approach to the analysis of complex systems and decision processes. IEEE Transactions on Systems, Man, and Cybernetics 31(6), 891–901 (2001)

[7] Mamdani, E.H., Assilian, S.: An experiment in linguistic synthesis with a fuzzy logic controller. International Journal of Man-Machine Studies 7(1), 1–13 (1975)

[8] Mamdani, E.H.: Advances in the linguistic synthesis of fuzzy controllers. International Journal of Man-Machine Studies 8, 669–678 (1976)

[9] Poli, R., Langdon, W.B., McPhee, N.F.: A Field Guide to Genetic Programming, freely available via Lulu.com, Global Optimization Algorithms - Theory and Application (2008), ISBN 978-1-4092-0073-4
[10] Mitchell, M.: An Introduction to Genetic Algorithms. MIT Press, Cambridge (1996)
[11] Copyright© AT&T Laboratories, Cambridge (2002), http://www.cl.cam.ac.uk/research/dtg/attarchive/facedatabase.html
[12] Castillo, O., Melín, P.: Hybrid intelligent systems for time series prediction using neural networks, fuzzy logic, and fractal theory. IEEE Transactions on Neural Networks 13(6), 1395–1408 (2002)
[13] Faros, A.: Biologically Inspired Modular Neural Networks. Blacksburg, Virginia (May 2000); Kennedy, J., Eberhart, R.: Particle swarm optimization. In: Proceedings of IEEE International Conference on Neural Networks, Piscataway, NJ, pp. 1942–1948 (1995)
[14] Melín, P., Gonzalez, F., Martinez, G.: Pattern Recognition Using Modular Neural Networks and Genetic Algorithms. In: Proc. IC-AI 2004, Las Vegas, Nevada, USA, pp. 77–83 (2004)
[15] Nuñez, R.: ITT, Parallel Modular Neural Networks for Recognition of Persons (February 2008)
[16] Sugeno, M.: Theory of Fuzzy integrals and its application, Tokyo Institute of Technology (1974)
[17] Salinas, R.: Department of Electrical Engineering, University of Santiago de Chile, Neural Network Architecture in Parametric Face Recognition University Diego Portales, http://cabierta.uchile.cl/revista/17/articulos/pdf/paper4.pdf (August 2008)

A Modular Neural Network with Fuzzy Response Integration for Person Identification Using Biometric Measures

Magdalena Serrano and Patricia Melin

Tijuana Institute of Technology, Tijuana México
epmelin@hafsamx.org

Abstract. This paper describes an intelligent system for person identification with biometric measures such as signature, fingerprint and face. We describe the neural network architectures used to achieve person identification based on the biometrics measures. Simulation results show that the proposed method provides good recognition. Fuzzy integration of the three modules is tested on a single computer and also in a distributed environment.

1 Introduction

At the moment, systems based on biometric recognition have gained importance in applications that require the identification of users or restricted access. Compared with conventional methods based on using keys, we have the advantage that the biometric features may not be provided, copied or stolen.

These kinds of systems are usually easy to maintain. A biometric system is essentially a pattern recognition system that operates in the following manner: capturing a biometric measure, a set of features are extracted and compared with another group the features.

Biometric identification techniques are very diverse, since any element of a person is potentially usable as a biometric measure. Even with the diversity of existing techniques, to develop a biometric identification system, it is an entirely separate scheme from the technique used. The most used biometric measures are: fingerprint, iris, voice, signature, face, ear, hand geometry, vein structure, retina, etc.

The need for a way to identify the human being in a unique way, has led researchers to implement a wide range of methods.

Until today biometric methods have been implemented using different devices for the creation of patterns and generate the code that identifies the individual biometric measures. This is why, in this work we consider one of the most used biometric measures throughout history, which is the fingerprint, and other ones such as the face and signature.

O. Castillo et al. (Eds.): Evolutionary Design of Intelligent Systems, SCI 257, pp. 159–183.
springerlink.com

2 Neural Networks

A neural network is a model to perform a computational simulation of parts of the human brain using replication behavior in small-scale patterns that it performs for producing results from the perceived events.

An artificial neural network (ANN), often just called a "neural network" (NN), is a mathematical or computational model based on biological neural networks. It consists of an interconnected group of artificial neurons and processes information using a connectionist approach to computation. In most cases an ANN is an adaptive system that changes its structure based on external or internal information that flows through the network during the learning [2].

1. Biological neural networks are made up of real biological neurons that are connected or functionally related in the peripheral nervous system or the central nervous system. In the field of neuroscience, they are often identified as groups of neurons that perform a specific physiological function in laboratory analysis.
2. Artificial neural networks are made up of interconnecting artificial neurons (programming constructs that mimic the properties of biological neurons). Artificial neural networks may either be used to gain an understanding of biological neural networks, or for solving artificial intelligence problems without necessarily creating a model of a real biological system. The real, biological nervous system is highly complex and includes some features that may seem superfluous based on an understanding of artificial networks.

2.1 Structure of an Artificial Neural System

The artificial neural system is composed by several components that are necessary to structure the system. In figure 1 we show these components: the neuron, layers, and networks.

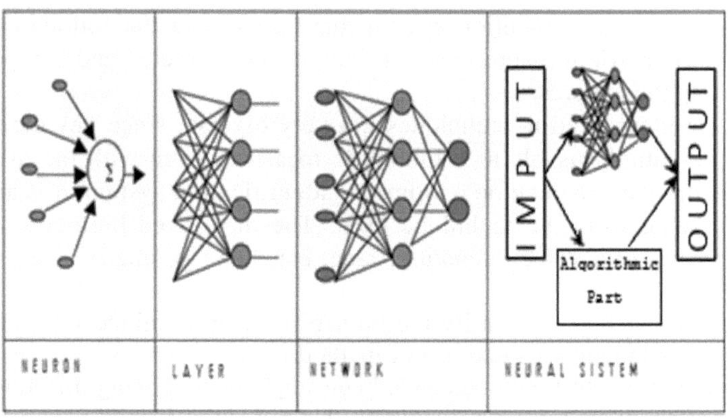

Fig. 1. Hierarchical structure of a system based on artificial neural networks

2.2 Modular Neural Networks

A modular neural network is a neural network characterized by a series of independent neural networks moderated by an intermediary. Each independent neural network serves as a module and operates on separate inputs to accomplish some subtask of the task the network is intended to perform.

The intermediary takes the outputs of each module and processes them to produce the output of the network as a whole. The intermediary only accepts the modules' outputs—it does not respond to, nor otherwise signal, the modules. Also the modules do not interact with each other [4].

The advantage is that if the model supports naturally a breakdown into more simple functions, the application of a modular network translates into faster learning. Each module can be built differently, in a way that meets the requirements of each subtask. A modular neural network can be represented by the scheme shown in figure 2.

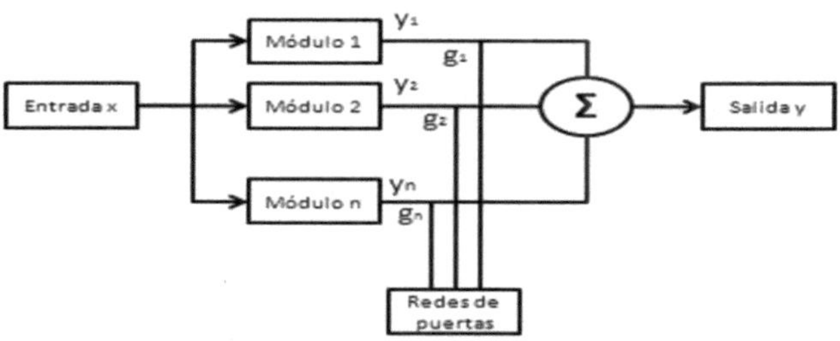

Fig. 2. A modular neural networks architecture.

3 Characteristics of a Biometric Measure

A biometric measure is a feature that can be used to make an identification. Whatever the measure, it must meet the following requirements:

1. Universality: means that anyone should have that characteristic.
2. Uniqueness: the existence of two people with identical characteristics has a very small probability.
3. Permanent Characteristics: the characteristic does not change over time
4. Quantification: the characteristic can be measured in a quantitative form.

3.1 Architecture of a Biometric System for Personal Identification

A biometric system has three basic components: The first is responsible for the acquisition of any analog or digital biometric feature of a person, such as the acquisition of a fingerprint image using a scanner. The second handles the compression, processing, storage and comparison of data acquired with the stored data. The third component provides an interface to applications on the same or another system. In figure 3 we show the phases of a biometrical identification system.

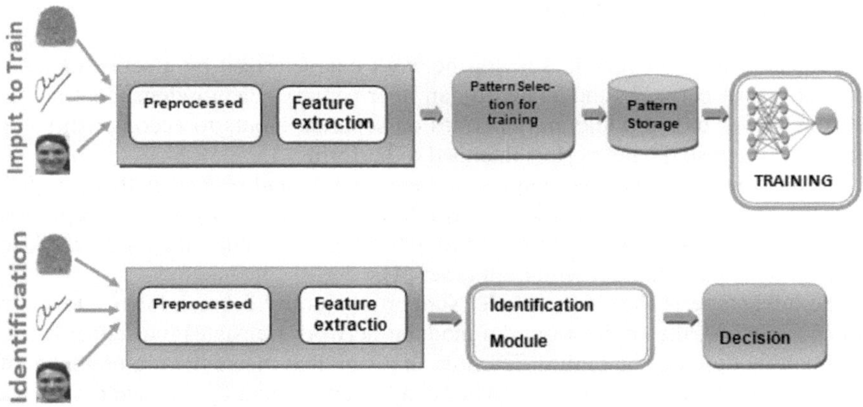

Fig. 3. Phases of a biometrical identification system.

4 Problem Statement and Proposed Method

The problem is to develop a hybrid system for person identification using parallel processing implementation determined by the biometric measures. We must develop a module for each of the biometric measures in the neural network architecture to implement:

- Fingerprint.
- Signature.
- Face.

Starting from the 3 above-mentioned modules it is important to make the unification of the 3 biometric measures to make the system obtain a higher percentage of identification, which aims at making the integration of the 3 systems into one, as shown in Figure 4.

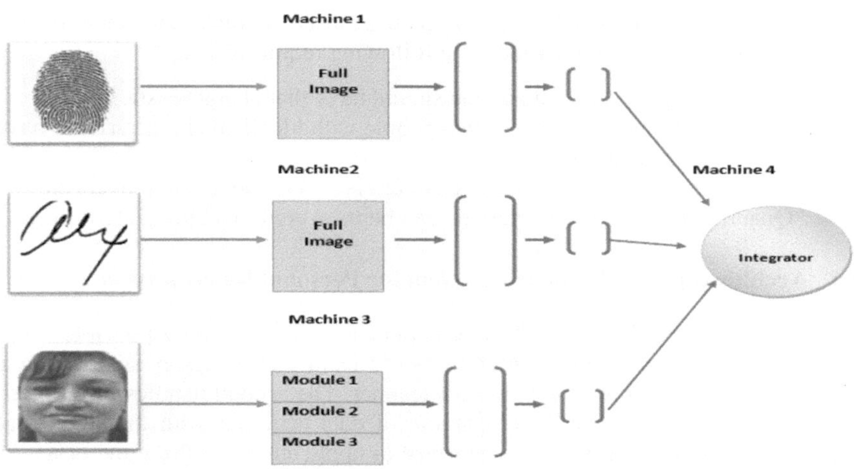

Fig. 4. Proposed scheme for the development of the system.

The implementation was done in 4 core computers, i.e, with 4 computers with 4 processors to activate 16 processors in parallel.

4.1 Biometric Databases

The databases were obtained from students and professors of the Master in science in computer science from Tijuana Institute the Technology.

The database consists of 10 Samples of signature, fingerprint and face, taken from 30 people, giving a total of 300 samples. We used for training the first 7 images and leaving the last 3 for identification. In total we trained the network with 210 images and 90 are left for the identification for face, fingerprint and signature.

The databases are shown in figures 6, 9, 10 and 11.

4.1.1 Signature Database

For the normalization of this database the images were cut to limit the signature, this produced a resizing of the images. We also applied Wavelets and the Sobel operator to preprocess the image, as shown in Figure 5.

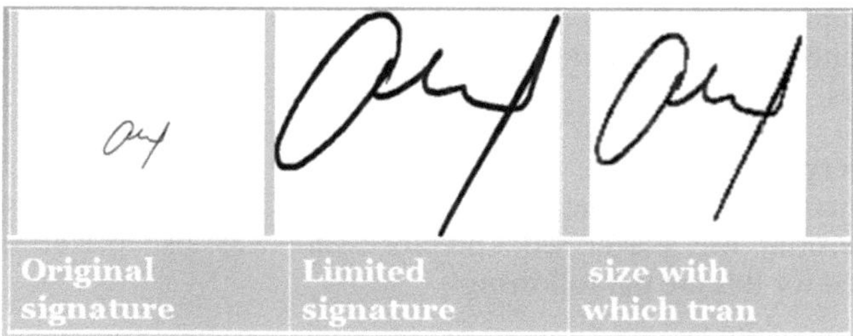

Fig. 5. Preprocessed database

Fig. 6. Database with examples of signatures.

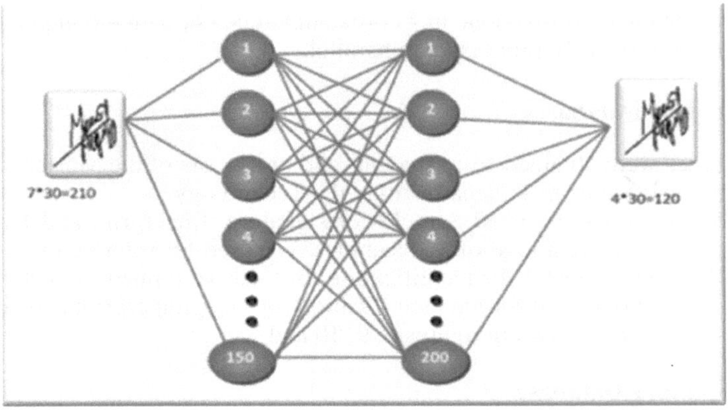

Fig. 7. Architecture of the net for signature.

After the preprocessing phase, the database can be visualized in Figure 6 with several examples of signatures .

The architecture for training the net is shown in Figure 7, were we have 2 hidden layers.

4.1.2 Fingerprint Database

For the normalization of the database, a cut of the standard image of the limitation of the fingerprint was performed, preprocessing was also performed, and the methods used for this are the following:

- Gradient magnitude
- Sobel edge masks
- Skeleton.
- Wavelets.

In figure 8 we show an example of fingerprint preprocessing. We show the gradient magnitude and skeleton methods applied to the fingerprint.

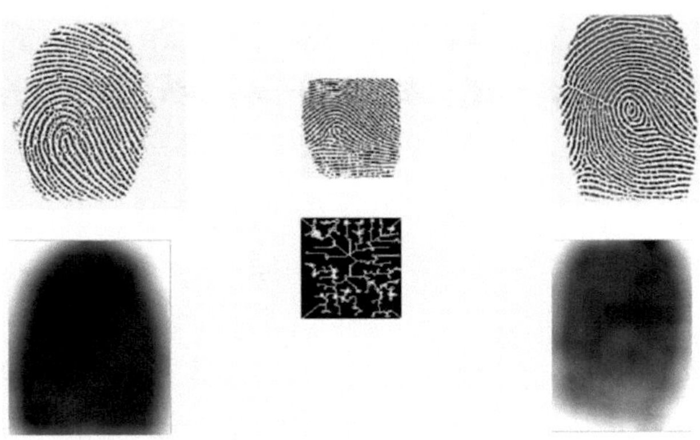

Fig. 8. Preprocessing of the fingerprint.

After preprocessing the database of images, they go in to the network with a dimension of [75 * 50] as shown in Figure 9.

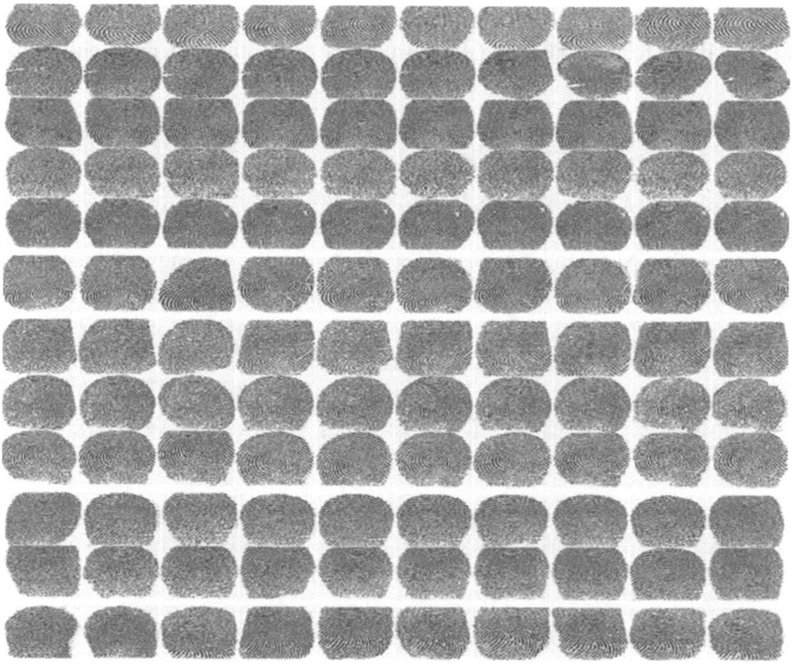

Fig. 9. Database of fingerprints

The architecture for training the net is shown in figure 10 and it has 2 hidden layers.

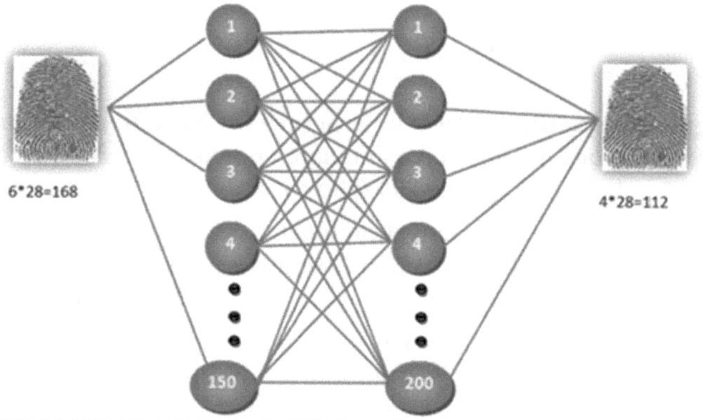

Fig. 10. Architecture of the neural network for fingerprint recognition.

4.1.3 Face Database

The face data base consist of a mixture of the ORL database and data acquired with students and professors of the institution, and has a different treatment for this reason, we first converted the images to gray tones and made a resizing to [92*112] in BMP format. The face database of the institution is shown en figure 11.

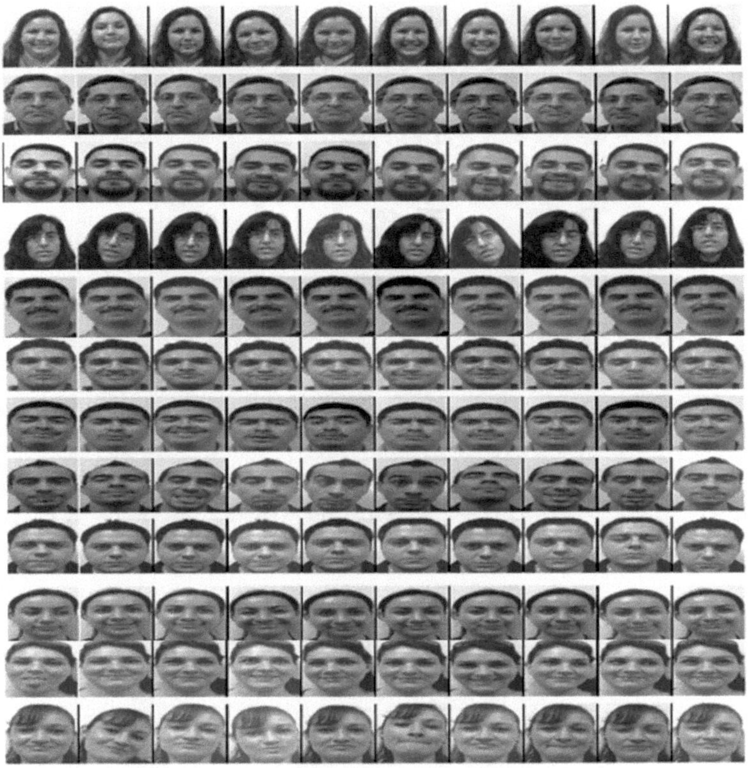

Fig. 11. Database of faces.

4.1.4 Design of the Neural Network Structure for Face Recognition

The number of neurons is allocated using an empirical expression created by Renato Salinas[8] and can be explained as follows:

- Input neurons: 2* (k+m), the activation function is a Tangent sigmoid.
- Hidden neurons: (k+m), the activation function is a Tangent sigmoid.
- Output neurons: the activation function is a Tangent logarithm sigmoid.

Where k is the number of individuals to train and m is the number of samples to train for each individual. In this case, these were 40 individuals for the ORL database , and 30 for the database of the institute and 7 samples therefore, k = 40, m = 7 and k = 30, m = 7.

Taking into account the previous data the architecture to train the modules is shown in Figure 12.

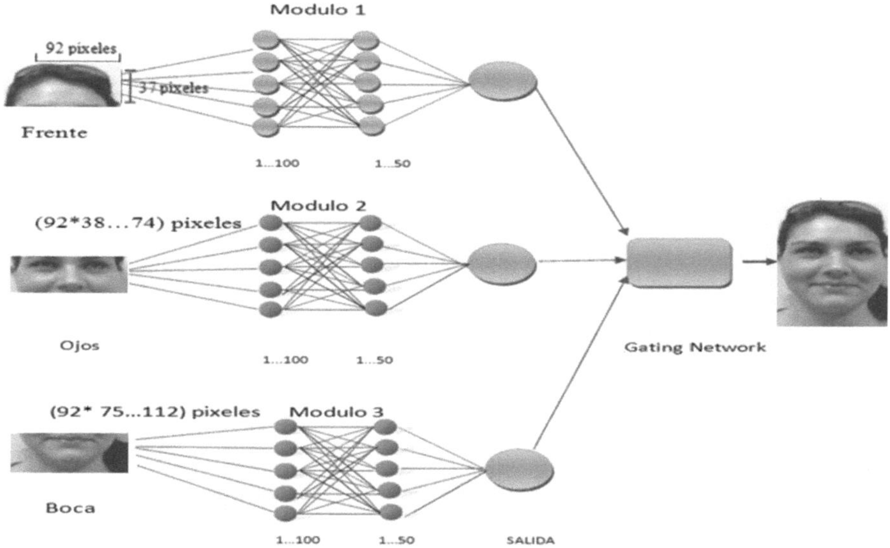

Fig. 12. Architecture of the neural network for face recognition.

5 Fuzzy Integration of Face, Signature and Fingerprint

We used a fuzzy logic module for the integration of the neural networks outpus. This method provides better performance in the subjective assignation of imputs from each of the individual networks;we show a model where one can see the integration of neural networks with fuzzy logic, so that the outputs of the neural network are processed by a fuzzy inference [11] mechanism, which can be seen illustrated in figure 13.

This phase consisted of a fuzzy system that can integrate the results of three different biometric measures (face, signature and fingerprint) and using this response as a result.

This system is designed with three input variables because each input belongs to biometric measure,this will increase if both the input variables, and the granularity of the number of rules would increase in a quantitative way.

Now we describe the design of the fuzzy integrator, since the granularity of the input variables, which are given as low, medium and high and the result is given by module1, module2 and module3 the we have the architecture shown in Figure 14. This type of fuzzy inference is of Mamdani form, which uses both the inputs

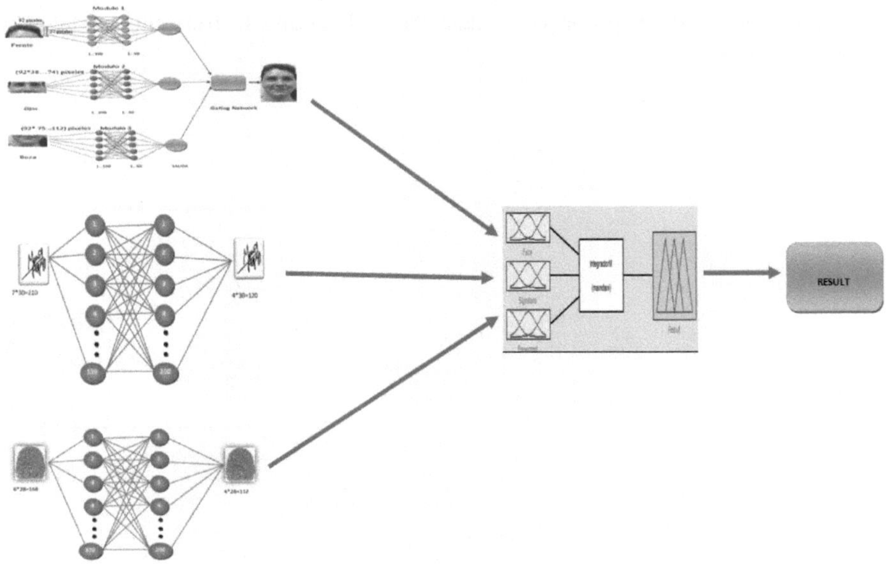

Fig. 13. Architecture of integration using fuzzy logic for integration of three biometrics measures.

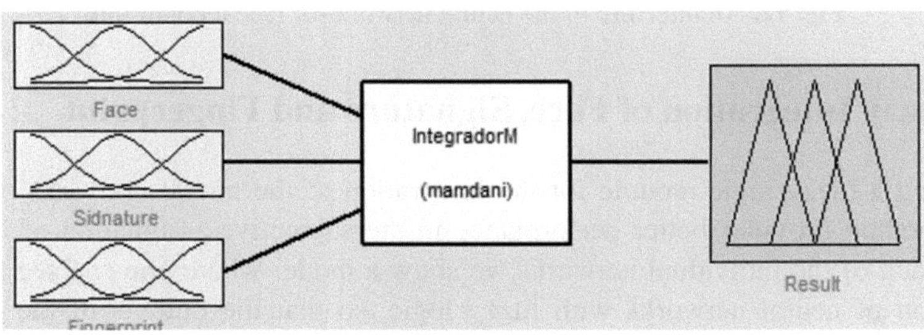

Fig. 14. Architecture of fuzzy Inference system.

as outputs of linguistic form. The gaussian membership functionare defined by the following equation.

$$\text{Gaussian}\ (x;\ c,\ \sigma) = e^{-\frac{1}{2}\left(\frac{x-c}{\sigma}\right)^2} \tag{1}$$

For the granularity of the variables we used the "medium"," low" and "high" membership functions, where these three linguistic values, distributed between 0 and 1.

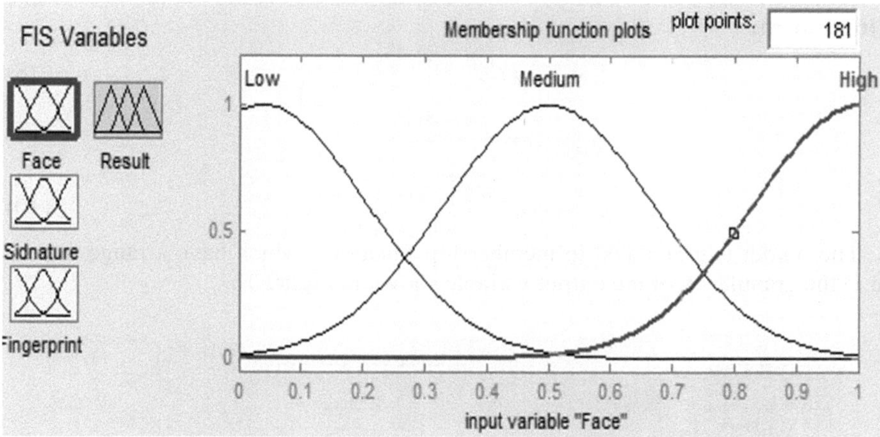

Fig. 15. granulation of the variable membership Face.

As shown in figure15, the face input variable has 3 membership functions of Gaussian type, which have the following parameters:

Low: parameter that ranges from [0,0.5]
Media: parameter that ranges from [0,1]
High: parameter ranges from [0.5, 1]

The Gaussian membership function for each of the linguistic values is defined as follows:

$$\mu_{Low}=e^{\left(\frac{X-0.02704}{0.27}\right)} \tag{2}$$

$$\mu_{Medium}=e^{\left(\frac{X-0.050}{0.1699}\right)} \tag{3}$$

$$\mu_{High}=e^{\left(\frac{X-1}{0.1699}\right)} \tag{4}$$

The membership functions for the variables of the signature and fingerprint have the same granularity as the one given to the face Gaussians and is defined in the same way as shown in the following equations:

Signature:

$$\mu_{Low}=e^{\left(\frac{X-0.02704}{0.27}\right)} \tag{5}$$

$$\mu_{Medium}=e^{\left(\frac{X-0.050}{0.1699}\right)} \tag{6}$$

$$\mu_{High}=e^{\left(\frac{X-1}{0.1699}\right)} \tag{7}$$

Fingerprint:

$$\mu_{Low}=e^{\left(\frac{x-0.01704}{0.47}\right)} \quad (8)$$

$$\mu_{Medium}=e^{\left(\frac{x-0.050}{0.1699}\right)} \quad (9)$$

$$\mu_{High}=e^{\left(\frac{x-1}{0.1699}\right)} \quad (10)$$

The output is granulated in membership functions, which have a range from 0 to 1, the granularity of the output variable shown in Figure 16.

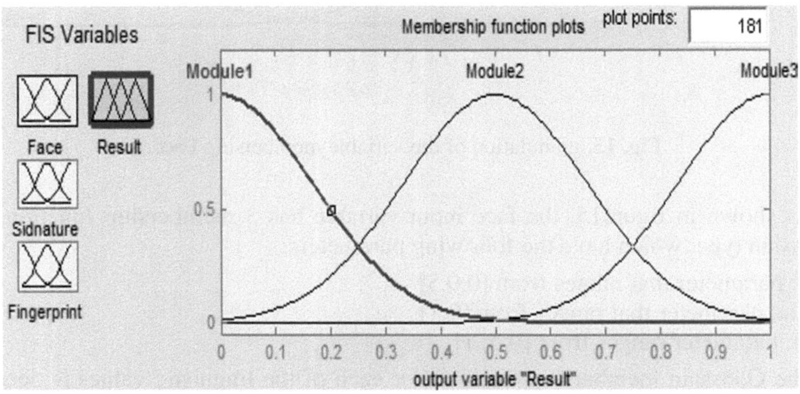

Fig. 16. Granulation of the output in the variable Result.

For the Gaussian outputs it is important to mention the ranges which are given as follows:

Module1: [0, 0.5]
Module2: [0, 1]
Module3: [0.5,1]

If we replace both the standard deviation and the mean in the equation of the Gaussian then we have:

$$\mu_{Module1}=e^{\left(\frac{x-6.939e-018}{0.1699}\right)} \quad (11)$$

$$\mu_{Module2}=e^{\left(\frac{x-0.050}{0.1699}\right)} \quad (12)$$

$$\mu_{Module3}=e^{\left(\frac{x-1}{0.1699}\right)} \quad (13)$$

Granulation of the output Variable is given as Modulo1, Modulo2, and Modulo3, which belong to the biometric face, fingerprint and signature respectively. This is important because when choosing a result response it is necessary to check which of the three modules is the one that has generated the correct answer. For this reason it is necessary that the rules cover all possible combinations for the solution of the problem.

Now we consider the optimization of the fuzzy inference system, the optimization can be in the type of system which can be Mamdani or Sugeno, and the membership function, that can be Gaussian, trapezoidal, triangles, etc.. In our case for the optimization, we use the 27 possible rules and leave the system with linguistic variables.

The rules were made taking into account all possible cases that may be present as shown in Table1.

Table 1. Rules for the fuzzy inference system.

Rules	Face	Signature	Fingerprint	Result
1	Low	Low	Low	Module1
2	Low	Low	Medium	Module2
3	Low	Low	High	Module3
4	Low	Medium	Low	Module2
5	Low	Medium	Medium	Module2
6	Low	Medium	High	Module3
7	Low	High	Low	Module2
8	Low	High	Medium	Module2
9	Low	High	High	Module2
10	Medium	Low	Low	Module1
11	Medium	Low	Medium	Module1
12	Medium	Low	High	Module3
13	Medium	Medium	Low	Module1
14	Medium	Medium	Medium	Module1
15	Medium	Medium	High	Module3
16	Medium	High	Low	Module2
17	Medium	High	Medium	Module2
18	Medium	High	High	Module2
19	High	Low	Low	Module1
20	High	Low	Medium	Module1
21	High	Low	High	Module1
22	High	Medium	Low	Module1
23	High	Medium	Medium	Module1
24	High	Medium	High	Module1
25	High	High	Low	Module1
26	High	High	Medium	Module1
27	High	High	High	Module1

Based on the fuzzy rules of table 1 we can simulate the performance of the fuzzy system, which is illustrated in figure 17.

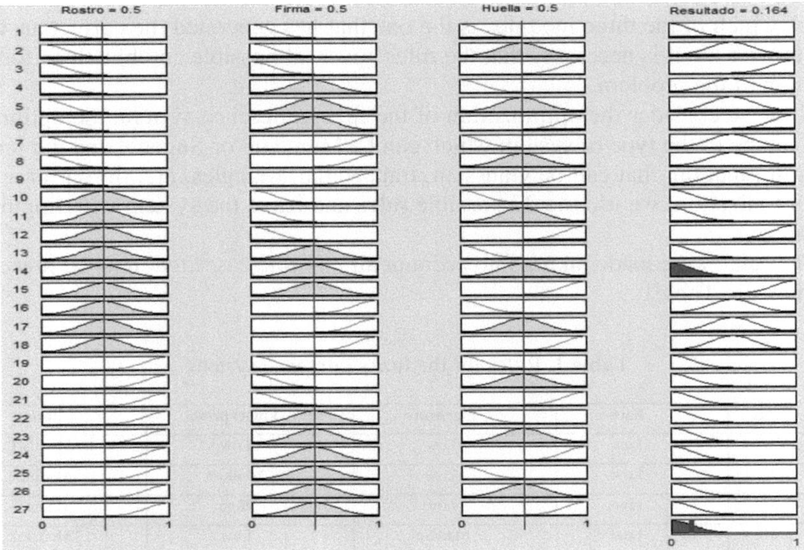

Fig. 17. Rule viewer

The general behavior of the fuzzy can be appreciated with non_linear surface generated by the fuzzy module. This is illustrated in three dimensions, with two biometric measures in each case.

Other fuzzy systems were made with triangular and trapezoidal membership functions, with a structure similar to the above mentioned inference system, which

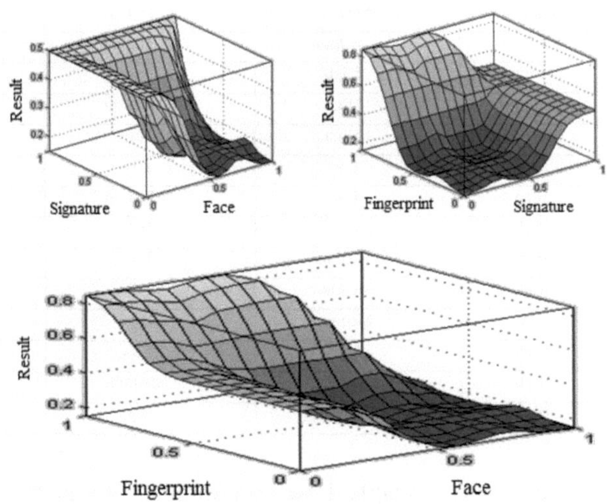

Fig. 18. Surface visor

is only referred to its realization, for the granulation with triangular membership functions we how use the following equation:

$$Triangle\,(x_i, a, b, c, d) max\left(\min\left(\frac{x-a}{b-a}, \frac{c-x}{c-b}\right), 0\right) \tag{14}$$

In implementing the fuzzy inference system with a Mamdani-type model with trapezoidal membership function, we used the formula of trapezoids which establishes the granularity of the values of variables.

$$\text{Trapezoid}\,(x_i, a, b, c, d) max\left(\min\left(\frac{x-a}{b-a}, 1, \frac{c-x}{c-b}\right), 0\right) \tag{15}$$

The results of these fuzzy inference systems were similar to the case of Gaussians and we have decided to use that experience and those who have achieved better results in identification of persons

6 Parallel Processing

Distributed or parallel computing speeds up the MNN execution of a program through its division into fragments that can be run simultaneously, each on one processor. Thus a program running on "n" processors might execute n times faster than using a single processor.

We chose to assign a fixed IP to each computer, so, to control the order of the machines in the network according to how they are located, and in this way the only work that the switch has will be to transfer the information between the processors, so that the computers are configured with the following assignments:

192.168.1.	Module1
192.168.1.	Module2
192.168.1.	Module3
192.168.1.	Module4

Figure 19 shows how the groups were placed on their shelves containers. If more computers are added in the cluster it is only necessary to assign the following IP address.

6.1 Distributed Computing Toolbox and MATLAB Distributed Computing Engine (MDCE)

The mdce service ensures that all other processes are running and that it is possible to communicate with them. Once the mdce service is running, you can use the nodestatus command to obtain information about the mdce service and all the processes it maintains.

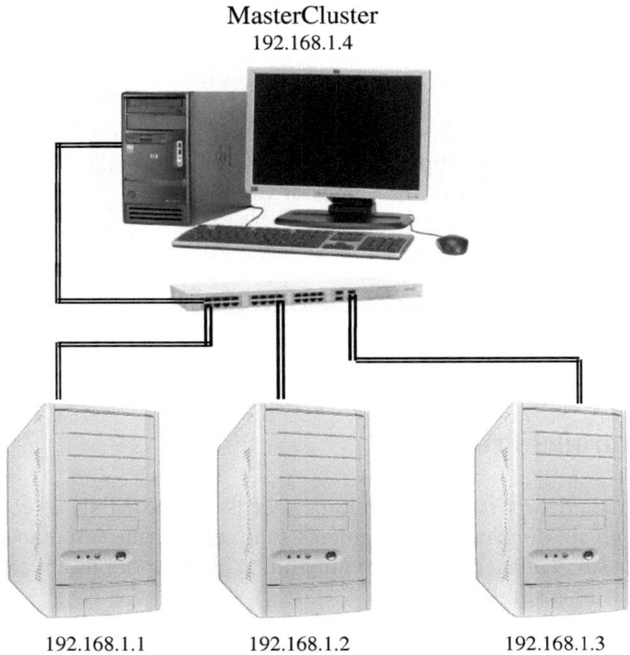

Fig. 19. Order of the machines

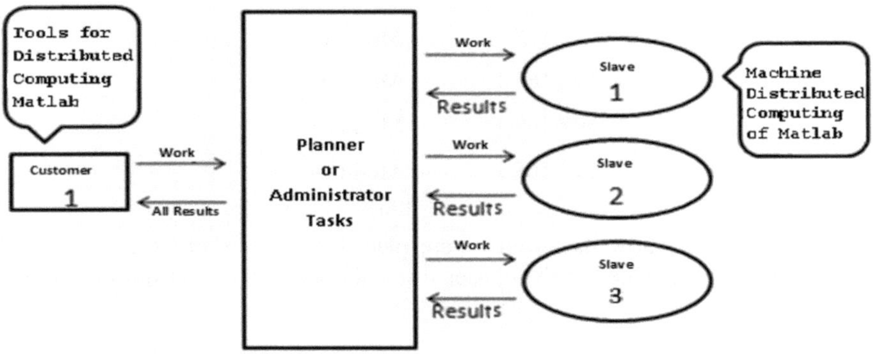

Fig. 20. Overview of Matlab as a computing distributed system

The distributed computing toolbox and MDCE allow us to coordinate and execute operations simultaneously on a cluster of computers to speed up execution of work (jobs) in MATLAB.

Figure 20 shows the interaction between machine, which uses the distributed computing toolbox to define jobs and tasks, and the MDCE. The planner can be

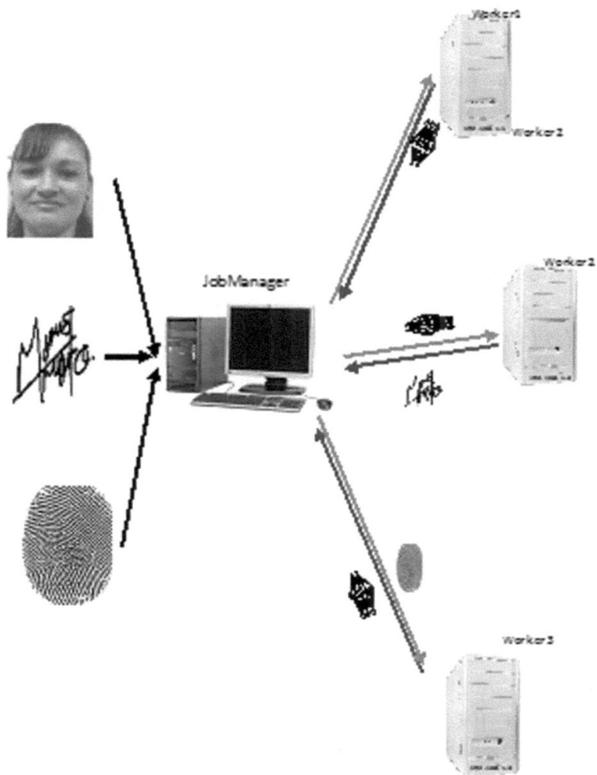

Fig. 21. Structure of the distributed algorithm in the Cluster

the manager of The Math Works jobs, included as part of MATLAB Distributed Computing Engine, a planner or a third party.

Management meetings conducted by the MDCE are shown in Figure 20. Processing requests of different clients are handled separately, a request at a time, for a single instant of time is carried out the task by the client 1,then 2 and so on for the n clients that can connect to the planner of tasks assigned to each period of time for processing.

6.2 Configuration Process

On each cluster node open a screen command line (MS-DOS) and type the following:

Cd Program files\ MATLAB\R2008a\Toolboox\distcomp\bin

1. Stop the execution of earlier versions of MDCE
   mdce stop

2. Install the new version of MATLAB and MDCE, after you continue with the next steps.

mdce install

3. MDCE starts the service on all cluster nodes

mdce start

4. Start the JobManager typing the following command.

startjobmanager -name <MyJobManager>
-remotehost <job manager hostname> -v

Verify that the Manager is running on the host specified.

5. Start the slaves (workers)

startworker -jobmanagerhost <job manager hostname> -jobmanager <MyJobManager> -remotehost <worker hostname> -v

6. Verify that the slaves are running on each computer:

nodestatus -remotehost <worker hostname>

It should be mentioned that the program receives the JobManager, which is the computer that controls all processes and this is where all tasks are carried out for pre-processing for biometric measures. As shown in Figure 21. This is how to distribute the work in the processor to do the biometric processing, after that the results are based on the fuzzy integrator and the Gating Network.

7 Simulation Results

In this section, we only show the results obtained for each of the biometric measures taking the tests with 5 methods of training and choosing the one with the better results. In table 1 we show the different methods of training used with the data bases and the method for integration, which was the gating network.

We perform several trainings in order to arrive to these results, although these are on average the best results that we have obtained for each of the biometric measures.

The results are given in three importants stages, the first is where the results of each of the biometric measures are presented separately, the second is where the three biometric measures are integrated through a fuzzy inference and tested on a single computer The third stage is the integration of the three biometric measures using response integration on distributed computing in parallel. We used different methods of training to check which one was the most appropriate biometric measurement. In Table 2 we shown the training methods used.

Table 2. Training methods for the neural networks.

Abbreviation	Training method	Integrator
TRAINGDX	Gradient descendent with momentum and adaptive learning rate backpropagation	Gating Network
TRAINGDA	Gradient descendent with adaptive learning rate bagpropagation	Gating Network
TRAINSCG	Scaled conjugate gradient backpropagation	Gating Network
TRAINRP	Resilient backpropagation	Gating Network
TRAINCGB	Conjugate gradient backpropagation with Powell-Beales restarts	Gating Network

7.1 Results of the Three Biometric Measures Considered Separately

In this section we show separate results for the biometric measures.

7.1.1 Signature Results

The results for the measurement of biometric signatures were very good, we achieved an average recognition of 99%, which is considerably good for the development of the system, and these results are shown below in Tables 3 and 4.

Table 3. Best simulation results without learning rate.

Training	Error Goal	Training Methods	Identification	Error	% Identification
1	0.0000001	Traingdx	87	3	87/90(97.50%)
2	0.0000001	Traingda	87	3	87/90(97.50%)
3	0.0000001	Trainscg	86	4	86/90(96.66%)
4	0.0000001	Trainrp	50	40	50/90(66.66%)
5	0.0000001	Traincgb	68	22	68/90(81.66%)

Table 3 shows the best training without the use of learning rate, as can be seen, we have a percentage of recognition of 97.50 with the training methods "traingdx" and "traingda".

Table 4. Best simulation results with learning rate of 0.001.

Training	Error Goal	Training Methods	Identification	Error	% Identification
1	0.00000001	Traingdx	89	1	89/90(99.16%)
2	0.00000001	Traingda	87	3	87/90(97.50%)
3	0.00000001	Trainscg	86	4	86/90(96.66%)
4	0.00000001	Trainrp	50	40	50/90(66.66%)
5	0.00000001	Traincgb	68	22	68/90(81.66%)

Table 4 shows the best results that were obtained for signature with a learning rate of 0.001, increasing our percentage of identification here to a 99.16% and the best Method of training was "traingdx".

7.1.2 Fingerprint Results

Several tests were made to reach the results shown in tables 5 and 6. In Table 5 we show the results of tests made without learning rate and a target error of 0.00000001 of 86.60%.The best results were obtained with the traingdx and traingda methods of training.

Table 5. Best results without learning rate

Training	Error Goal	Training Methods	Identification	Error	% Identification
1	0.00000001	Traingdx	75	15	75/90(86.60%)
2	0.00000001	Traingda	97	15	75/90(86.60%)
3	0.00000001	Trainscg	95	17	73/90(84.82%)
4	0.00000001	Trainrp	34	56	56/90(50%)
5	0.00000001	Traincgb	61	29	83/90(74.10%)

Better results were obtained by adding a learning rate of 0.001 with the same goal error is achieved with this increase above the percentage of identification in a 89.28% as shown in table 6, the best training method was trainscg.

Table 6. Best results with learning rate of 0.001

Training	Error Goal	Training Methods	Identification	Error	% Identification
1	0.0000001	Traingdx	99	13	92/112(88.39%)
2	0.0000001	Traingda	97	15	97/112(86.60%)
3	0.0000001	Trainscg	100	12	100/112(89.28%)
4	0.0000001	Trainrp	50	62	50/112(44.64%)
5	0.0000001	Traincgb	86	26	86/112(76.78%)

7.1.3 Face Results

For the face identification we have results for the data base of the institution and for the data base of the ORL, in both situations we performed different tests. In table 6 we show results that include a case of 100% identification for the data base from the institution by means of the training method traingdx.

Table 7. Results of original database take in the Tijuana Institute of Technology

Training	Error Goal	Training Methods	Epoch	Time	% Identification
1	0.0000001	Traingdx	500,500,500	2:11	90/90(100%)
2	0.0000001	Traingda	500,500,500	2:03	89/90(98.33%)
3	0.0000001	Trainscg	500,500,500	8:21	88/90(96.6%)
4	0.0000001	Trainrp	500,125,240	2:15	82/90(86.66%)
5	0.0000001	Traincgb	209,134,126	3:9	86/90(93.33%)

In order to verify the effectiveness of the result of the 100% of identification, tests of cross validation were realized where the positions of the images of the data base were alternated to verify that the percentage of identification persists, verifying if there is a result that fails as it is possible to be appreciated in table 7, another difference that can be appreciated with the cross validation is that the training method varies since with the normal data base we have the method of training traingdx and for these tests with which better results were achieved with trainscg, with a 95% of identification.

Table 8. Simulation results of cross validation

Train	Error Goal	Training Methods	Epoch	Time	% Identification
1	0.0000001	Traingdx	500,500,500	2:11	85/90(91.6%)
2	0.0000001	Traingda	500,500,500	2:03	84/90(90%)
3	0.0000001	Trainscg	500,500,500	8:21	87/90(95%)
4	0.0000001	Trainrp	500,125,240	2:15	82/90(86.66%)
5	0.0000001	Traincgb	209,134,126	3:9	86/90(93.33%)

Table 9. Best results of Three Biometrics measures with better training methods.

Biometric Measure	Training Method	Error Goal	% Identification
Face	Traingdx	0.0000001	90/90=100%
	TRainscg		88/90=97%
	Trraingda		89/90=99%
Face with Cross-Validation	Traingdx	0.0000001	85/90=94%
	TRainscg		87/90=96 %
	Trraingda		84/90=93%
Signature	Traingdx	0.00000001	87/90=96%
	TRainscg		87/90=96%
	Trraingda		86/90=96.66%
Signature with Cross-Validation	Traingdx	0.00000001	89/90=99%
	TRainscg		86/90=95%
	Trraingda		87/90=96%
Fingerprint	Traingdx	0.00000001	77/90=88.39%
	TRainscg		73/90=84.82%
	Trraingda		70/90=86.60%
Fingerprint with Cross-Validation	Traingdx	0.00000001	82/90=92.85%
	TRainscg		70/90=86.60%
	Trraingda		78/90=89.28%

For each of the biometric measures we have had very good results in an independent manner. It is important to consider the fact that in the future, when making the integration of the 3 biometric measures the percentage of identification will increase, because of the modularity and thus having a greater system reliability and performance.

Table 10. Results of Fuzzy integrator for the biometrics in set.

Type of Fuzzy Integrator	% of Identification	Identification Error	Biometric Measure	Range of Measure	Activations in each module
Mandami, Gausianas	100%	0	Face	0.3333	16
			Signature	0.6666	12
			Fingerprint	1	2

Table 11. Results of Integration in Distribute Compute with Fuzzy System, fuzzy integrals and measures Sugeno and gating network.

Image	Gating Network		Fuzzy Integral and Sugeno Measures		Fuzzy Logic Type- 1		Workers	Training Time
	% Rec	% Ident	% Rec	% Ident	% Rec	% Ident		
60	60/60 100%	60/60 100%	60/60 100%	60/60 100%	60/60 100%	60/60 100%	3 Rem	1 min 37 seg
80	80/80 100%	78/80 97%	80/80 100%	72/80 9%	80/80 100%	79/80 98.7%	3 Rem	2 min 27seg
90	90/90 100%	90/90 100%	90/90 100%	89/90 96%	90/90 100%	90/90 100%	3 Rem	3 min 58seg
120	120/120 100%	118/120 98%	120/120 100%	110/120 91%	120/10 100%	120/120 100%	3 Rem	5 min 49seg

The best results of the three biometric measures with the best training methods of are show in the table 8, where you can view face and face with cross-validation, signature and signature with cross validation as that fingerprint.

7.1.4 Best Results

In table 9 we show the best results of the three biometric measures with the three best training methods for each biometric measure.

7.2 Results of Fuzzy Response Integrator

For the three biometric measures a Mamdani type fuzzy integrator was used,which was developed using Gaussian type membership functions.
Table 10 shows that fuzzy integrator for the combination of the measures in a single value has achieved a 100% identification.

With 0 errors and is reached to verify that the three modules are activated to obtain this result, so the efficiency of a module compensates for other deficiencies. For the fuzzy inference system takes values between 0 and 1, as follows: from 0.3333 to the module face 0.6666 for Release 1 of the signature and fingerprint.

7.3 Results Using a Cluster

A consequence of the positive results in the previous stage it was decided to launch processes to remote nodes that are in the cluster of computers, just to verify that the results persist even when the database grows, the components can be seen in table 11, we use 7 training images per person and leaving 3 is to identify, 90 pictures to 30 persons and 60 to 20 persons, although the first tests were conducted for a number of 20 persons, a same operation to train with 6 images and allow for the identification is 4 a total of 120 for 30 persons and 80 to 20 persons.

In the table we show both the percentage of appreciation that is what is done with the images that were training as the percentage of identification that is what interests us is that the testing is done with the images that were not previously trained, It can be seen that the results are good.

8 Conclusions

We can conclude that working with biometric measures is much more reliable than with other forms of authentication and the fact that combining several biometric measures helps to have a greater percentage of reliability because measures cover the deficiencies of the others within a system.

At this moment, we have good results with the independent biometric measures, so it is possible to say that good results may be obtained after having of the integration of 3 measures. We will try this in the near future.

It is important to note that the goal of using parallel computing is to ensure that if the databases increase in thousands, the execution time for the identification of a person and the percentage of identification are reliable. Therefore we note that our research work with regard to biometric measures was much more effective with the "trainscg" training method and the best integrator were thegatting network and type-1 fuzzy systems.

Acknowledgment. We would like to express our gratitude to CONACYT, and Tijuana Institute of Technology for the facilities and resources granted for the development of this research.

References

[1] Castillo, O., Melin, P.: Hybrid intelligent systems for time series prediction using neural networks, fuzzy logic, and fractal theory. IEEE Transactions on Neural Networks 13(6), 1395–1408 (2002)

[2] Roger Jang, J.-S., Sun, C.-T., Mizutani, E.: Neuro-fuzzy and Soft Computing a Computational Approach to Leatning and Machine Inteligence

[3] Faros, A.: Biologically Inspired Modular Neural Networks. Blacksburg, Virginia (May 2000); Kennedy, J., Eberhart, R.: Particle swarm optimization. In: Proceedings of IEEE International Conference on Neural Networks, Piscataway, NJ, pp. 1942–1948 (1995)

[4] Melin, P., Gonzalez, F., Martinez, G.: Pattern Recognition Using Modular Neural Networks and Genetic Algorithms. In: Proc. IC-AI 2004, Las Vegas, Nevada, USA, pp. 77–83 (2004)
[5] Romero, M.: Sistemas Modulares, Mezcla de expertos y sistemas Hibridos (March 2004)
[6] Nuñez, R.: ITT, Redes Neuronales Modulares en Paralelo para el Reconocimiento de Personas (February 2008)
[7] Pérez, A.: Reconocimiento y verificación de firmas manuscritas off-line basado en el seguimiento de sus trazos componentes, eslova.com (2002/2003)
[8] Fu, H.-C., Lee, Y.-P., Chiang, C.-C., Hsiao-Tien: Divide and conquer learning and modular perceptron networks in IEEE transaction on neural networks. In: Goldberg, D. (ed.) Genetic Algorithms, vol. 12. Addison Wesley, Reading (1988)
[9] Salinas, R.: Departamento de Ingeniería Eléctrica, Universidad de Santiago de Chile, Red Neuronal de Arquitectura Paramétrica en Reconocimiento de Rostros, Universidad Diego Portales (August 2008), http://cabierta.uchile.cl/revista/17/articulos/pdf/paper4.pdf
[10] Romero, L.: Dpto. de Informática y Automática. Universidad de Salamanca- España. Calonge Cano Teodoro, Dpto de Informática. Universidad de Valladolid. España, Redes Neuronales y Reconocimiento de Patrones, http://lisisu02.fis.usal.es/airene/capit1.pdf
[11] Sugeno, M.: Theory of Fuzzy integrals and its application. Tokyo Institute of Technology (1974)
[12] The matworks Inc., MATLAB Distributed Computing Engine Systems Administrator's Guide (2007)

Signature Recognition with a Hybrid Approach Combining Modular Neural Networks and Fuzzy Logic for Response Integration

Mónica Beltrán, Patricia Melin, and Leonardo Trujillo

Tijuana Institute of Technology, Tijuana, México
epmelin@hafsamx.org

Abstract. This chapter describes a modular neural network (MNN) with fuzzy integration for the problem of signature recognition. Currently, biometric identification has gained a great deal of research interest within the pattern recognition community. For instance, many attempts have been made in order to automate the process of identifying a person's handwritten signature; however this problem has proven to be a very difficult task. In this work, we propose a MNN that has three separate modules, each using different image features as input, these are: edges, wavelet coefficients, and the Hough transform matrix. Then, the outputs from each of these modules are combined using a Sugeno fuzzy integral and a fuzzy inference system. The experimental results obtained using a database of 30 individual's shows that the modular architecture can achieve a very high 99.33% recognition accuracy with a test set of 150 images. Therefore, we conclude that the proposed architecture provides a suitable platform to build a signature recognition system. Furthermore we consider the verification of signatures as false acceptance, false rejection and error recognition of the MNN.

1 Introduction

Recently, there has been an increased interest in developing biometric recognition systems for security and identity verification purposes [2]. Such systems usually are intended to recognize different types of human traits, which include a person's face, their voice, fingerprints, and specific handwriting traits [2].

Particularly, the handwritten signature that each person posses is widely used for personal identification and has a rich social tradition. In fact, currently it is almost always necessary in all types of transactions that involve legal or financial documents.

However, it is not a trivial task for a computational system to automatically recognize a person's signature for the following reasons. First, there can be a great deal of variability when a person signs a document. This can be caused by different factors, such as a person's mood, free time to write the signature, and the level of concentration during the actual act of signing a document. Second, because signatures can be so diverse it is not evident which type of features should be used in order to describe and effectively differentiate among them. For instance, some

O. Castillo et al. (Eds.): Evolutionary Design of Intelligent Systems, SCI 257, pp. 185–201.
springerlink.com

signatures are mostly written using straight line segments, and still others have a much smoother form with curved and circular lines. Finally, many signatures share common traits that make them appear quite similar depending on the types of features that are analyzed.

In this work, we present a handwritten signature recognition system using Modular Neural Networks (MNNs) with the Sugeno fuzzy integral. We have chosen a MNN because they have proven to be a powerful, robust, and flexible tool, useful in many pattern recognition problems [13, 14]. In fact, we only extract simple and easily computed image features during our preprocessing stage, these features are: image edges, wavelet transform coefficients, and the Hough transform matrix. The MNN we propose uses these features to perform a very accurate discrimination of the input data used in our experimental tests. Therefore, we have confirmed that a MNN system can solve a difficult biometric recognition problem using a simple set of image features.

2 Problem Statement and Outline of Our Proposal

The problem we address in this chapter is concerned with the automatic recognition of a person's signature that is captured on a Tablet PC. We suppose that we have a set of N different people, and each has a unique personal signature. The system is trained using several samples from each person, and during testing it must determine the correct label for a previously unknown sample.

The system we are proposing consists on a MNN with three separate modules. Each module is given as input the features extracted with different feature extraction methods: edge detection, wavelet transform, and Hough transform. The responses from each of the modules are combined using a Sugeno fuzzy integral, which determines the person to whom the input signature corresponds. A general schematic of this architecture is shown in Figure 1, where all of the modules and stages are clearly shown.

In the following section we present a brief review of some of the main concepts needed to understand our work.

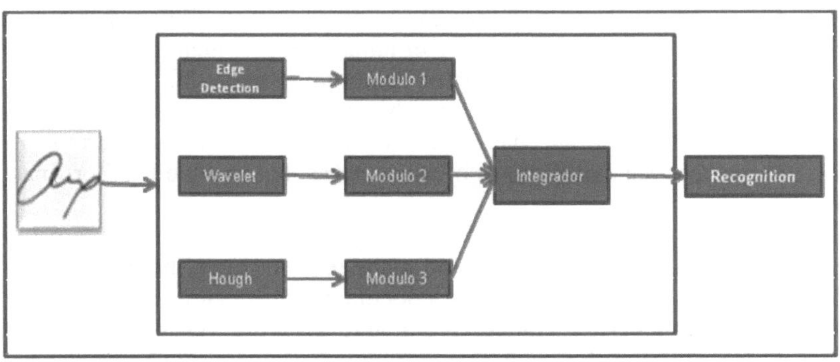

Fig. 1. General architecture of the proposed Modular Neural Network for signature recognition.

3 Background Theory

In this section we provide a general review of artificial neural networks and modular architectures, we discuss how the output from the modular system can be integrated using Sugeno fuzzy integrals, and we describe the feature extraction methods that provide the input for each of the modules in our MNN.

3.1 Modular Neural Networks

Artificial Neural Networks (ANNs) are information processing systems that employ a conceptual model that is based on the basic functional properties of biological neural networks. In the past twenty or thirty years, ANN research has grow very rapidly, in the development of new theories of how these systems work, in the design of more complex and intricate models, and in their application to a diverse set problem domains. Regarding the latter, application domains for ANN include pattern recognition, data mining, time series prediction, robot control, and in the development of hybrid methods with fuzzy logic and genetic algorithms, to mention but a few examples [7,13, 14].

In canonical implementations, most systems employ a monolithic network in order to solve the given task. However, when a system needs to process large amounts of data or when the problem is highly complex, then it is not trivial, and sometimes unfeasible, to establish a good architecture and topology for a single network that can solve the problem. For instance, in such problems a researcher might attempt to use a very large and complex ANN. Nevertheless, large networks are often difficult to train, and for this reason they rarely achieve the desired performance [13].

In order to overcome some of the aforementioned shortcomings of monolithic ANNs, many researchers have proposed modular approaches [11]. MNNs are based on the general principle of divide-and-conquer, where one attempts to divide a large problem into smaller sub-problems that are easier to solve independently. Then, these partial solutions are combined in order to obtain the complete solution for the original problem.

MNNs employ a parallel combination of several ANNs, and normally contain two main components: (1) local experts; and (2) an integrating unit. The basic architecture is shown in Figure 1 [16].

Each module consists of a single ANN, and each is considered to be an expert in a specific task. After the input is given to each module it is necessary to combine all of the outputs in some way, this task is carried out by a special module called an *integrator*. The simplest form of integration is given by a gating network, that basically switches between the outputs of the different modules based on simple criteria, such as the maximum level of activation. However, a better combination of the responses from each module can be obtained using more elaborate methods of integration, such as the Sugeno fuzzy integral [16].

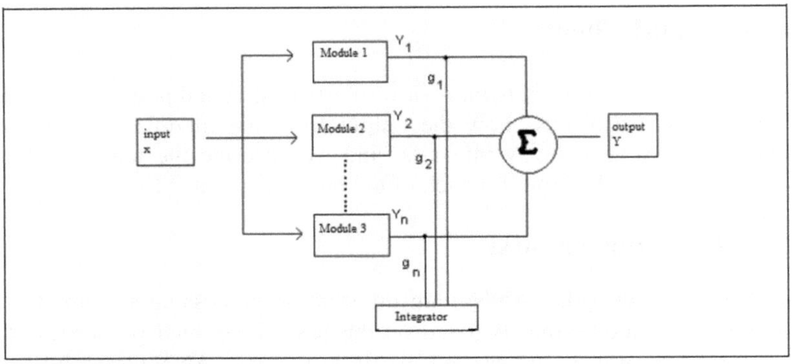

Fig. 2. Architecture of a Modular Network.

3.2 Sugeno Fuzzy Integral

The Sugeno fuzzy integral is a nonlinear aggregation operator that can combine different sources of information [4, 8, 9]. The intuitive idea behind this operator is based on how humans integrate information during a decision making process. In such scenarios it is necessary to evaluate different attributes, and to assign priorities based on partially subjective criteria. In order to replicate this process on an automatic system, a good model can be obtained by using a fuzzy representation [5, 8, 12]. Finally, several works have shown that the use of a Sugeno fuzzy integral as a MNN integrator can produce a very high level of performance [6, 8, 11], and for these reasons we have chosen it for the system we describe here.

3.3 Fuzzy Systems

Fuzzy theory was initiated by Lotfi A. Zadeh in 1965 with his seminal paper "Fuzzy sets". Before working on fuzzy theory, Zadeh was a well-respected scholar in control theory.

A big event in the 70's was the birth of fuzzy controllers for real systems. In 1975, Mamdani and Assilian established the basic framework of fuzzy controller and applied the fuzzy controller to control a steam engine. Their results were published in another seminal paper in fuzzy theory "An experiment in linguistic synthesis with a fuzzy logic controller". They found that the fuzzy controller was very easy to construct and worked remarkably well [3, 15].

The fuzzy inference system is a popular computing framework based on the concepts of fuzzy set theory, fuzzy if- then rules, and fuzzy reasoning. It has found successful applications in a wide variety of field, such as automatic control, data classification, decision analysis, experts systems, times series prediction, robotics, and patter recognition [3, 15].

The basic structure of a fuzzy inference system consists of three conceptual components: a rule base, which contains a selection of fuzzy rules; a database, which defines the membership functions used in the fuzzy rules; and a reasoning mechanism, which performs the inference procedure upon the rules and given facts to derive a reasonable output or conclusion [3].

3.4 Feature Extraction

In this work, we employ three individual modules, and each receives different image features extracted from the original image of a person's signature. Each of these feature extraction methods are briefly described next.

3.4.1 Edge Detection

For images of handwritten signatures, edges can capture much of the overall structure present within, because people normally write using a single color on a white background. Hence, we have chosen to apply the Canny edge detector to each image that generates a binary image of edge pixels, see Fig 3.

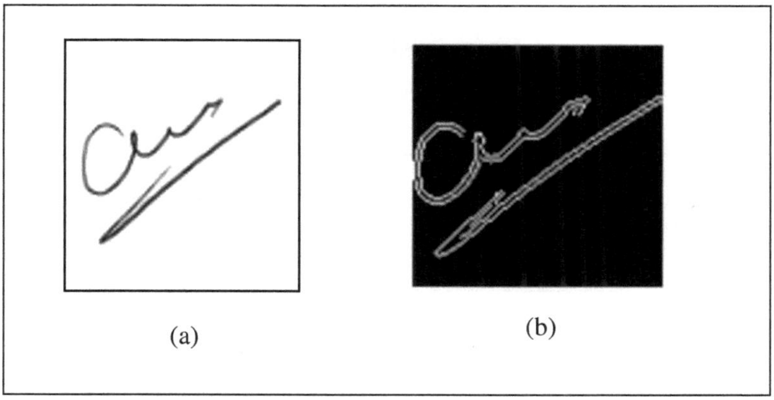

Fig. 3. (a) Original image of a signature. (b) Image edges.

3.4.2 Wavelet Transform

The wavelet transform decomposes a signal using a family of orthogonal functions, it accounts for both the frequency and the spatial location at each point. The most common application is the Discrete Wavelet Transform (DWT) using a Haar wavelet [17]. The DWT produces a matrix of wavelet coefficients that allows us to compress, and if needed reconstruct, the original image. In Figure 4 we can observe the two compression levels used in our work.

3.4.3 Hough Transform

In the third and final module we employ the Hough transform matrix as our image features [1]. The Hough transform can extract line segments from the image. In Figure 5 we show a sample image of a signature and its corresponding Hough transform matrix. Finally, in order to reduce the size of the matrix, and the size of the corresponding ANN, we compress the information of the Hough matrix by 25%.

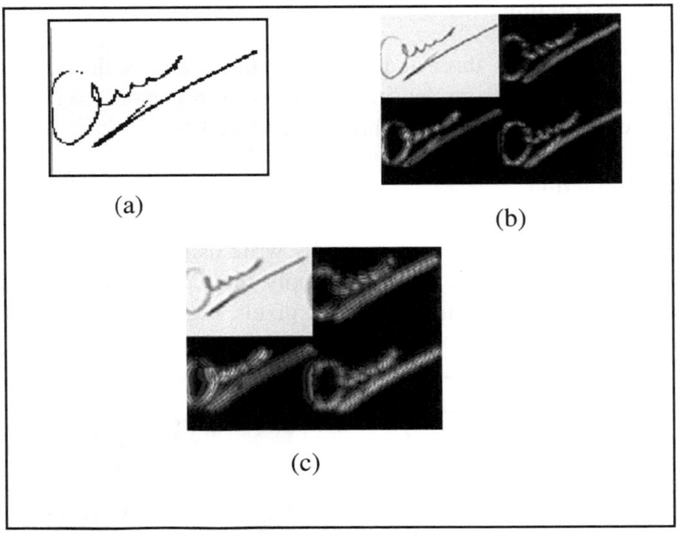

Fig. 4. (a) Original image of a signature. (b) First level of decomposition. (c) Second level of decomposition.

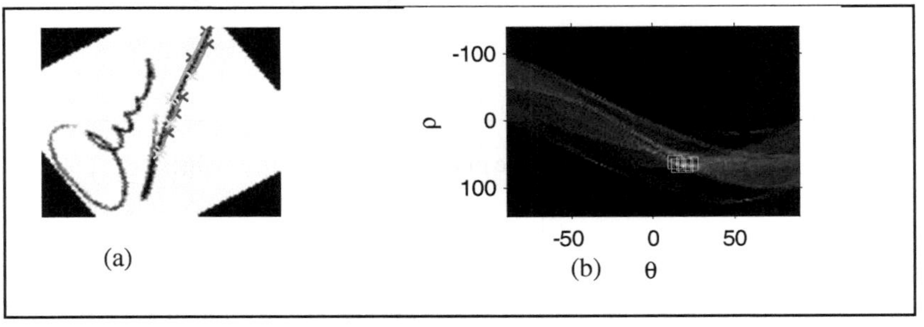

Fig. 5. (a) Sample of a signature image with some of the lines found by the Hough transform (b) The Hough transform matrix.

3.4.4 Verification of Signatures

Currently, security practice always involves PIN number, password, and access card. However, these signs are not very reliable, since it can be forgotten or lost [2].

Automatic signature verification is one of the most practical ways to verify human´s identify. Signature verification can be used in many applications such as security, access control, or financial and contractual matters.

The process of signature verification often consists of a learning stage and a testing stage, as shown in figure 6. In the learning stage, the verification system uses the feature extracted from one or several training samples to build a reference signature database. In the testing stage the user inputs the signature into input device. Then the system uses this information to extract the reference in the database, and compares the features extracted from the input signature with the reference. Finally the verification process out whether the test signature is genuine or not.

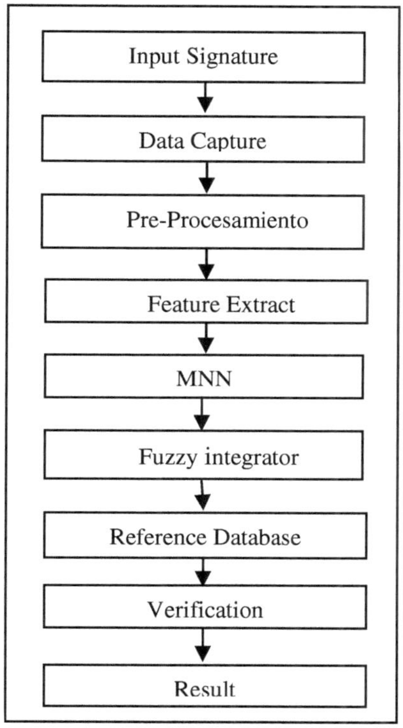

Fig. 6. Signature verification process

In the research area of signature verification, a type I error rate and type II error rate are usually called false reject rate (FRR) and false acceptance rate (FAR) respectively. To minimize the type II errors, which represent the acceptance of the counterfeited signatures will normally increase the type I errors, which are the rejections of genuine signature. In most case, type II error rate is considered to be more important, but it is not a must. This will depend on the purpose, design, characteristics and application of the verification systems. If the system requests a high security, false accept rate should reduced to its lowest; if the security is not so strict, the system can be adjust to its lowest average false rate.[2]

The fuzzy system will answer the greater activation of the 3 modules signing, taking the form of higher activation winner; this means the 27 rules in the system are considered fuzzy.

Once the winner module did a signature verification process to know whether the signature that shows fuzzy integrator corresponds to the person.

For this we also conducted 30 trainings with 150 different samples of genuine signatures, to make activation and get an average of this activation. The average of the activations is used as to whether a signature is forged or genuine.

Typically when the signature is authentic we obtain a high activation and when the signature is false the activation is low, although not necessarily, as may happen if there is a high activation but the signature is false activation or a low but firm is true, so we take into account four different cases:

1. False acceptance (FRA).
2. False rejection (FRR).
3. Error.
4. Signature Authentic.

After taking as reference the average of activations, 85 samples were collected from forged signatures of 17 persons and 65 authentic samples of 13 persons, giving a total of 150 samples between false and authentic signatures of 30 persons. In total 210 images of signatures of each module for the training signatures are authentic.

Table 1 shows the case that can be given upon verification of signatures, taken as a basis the average activations.

Table 1. Signature verification procedure

Recognizes	Overcome threshold	Original signature	Result
Yes	Yes	No	False Acceptance
Yes	No	Yes	False Rejection
No	Yes	Yes	Error
No	No	Yes	Error
No	No	No	Correct
No	Yes	No	Correct

4 Experiments

In this section we present our database of signature images, describe our experimental set-up, and detail the experimental results we have obtained using monolithic and modular networks.

4.1 Image Database

For this work we build a database of images with the signatures of 30 different people, students and professors from the computer science department at the Tijuana Institute of Technology, BC, México. We collected 12 samples of the signature from each person; this gives a total of 360 images in total. Sample images from the database are shown in Figure 7.

Fig. 7. Images from our database of signatures. Each row shows different samples from the signature of the same person.

4.2 Experimental Setup

In this work, we are interested in verifying the performance of our proposed MNN for the problem of signature recognition. Therefore, in order to obtain comparative measures we divide our experiments into four separate tests.

1. First, we use each module as a monolithic ANN for signature recognition. Therefore, we obtain three sets of results, one for each module, where in each case a different feature extraction method is used.
2. Second, we train our MNN using all three modules concurrently and the Sugeno fuzzy integral as our integration method.
3. Third, we train our MNN using all three modules concurrently and the Fuzzy System as our integration method.
4. Fourth, Signature verification: false acceptance, false rejection.

In all tests 210 images were chosen randomly and used for training, and the remaining 150 were used as a testing set. Additionally, after some preliminary runs it was determined that the best performance was achieved when the ANNs were trained with the Scaled Conjugate Gradient (Trainscg) algorithm, with a goal error of 0.001. Moreover, all networks had the same basic ANN architecture, with two hidden layers. In what follows, we present a detailed account of each of these experimental tests.

4.2.1 Monolithic ANNs

The results for the first monolithic ANN are summarized in Table 2. The table shows a corresponding ID number for each training case, the total epochs required to achieve the goal error, the neurons in each hidden layer, and the total time required for training. Recognition performance is shown with the number of correct recognitions obtained with the 150 testing images, and the corresponding accuracy

Table 2. Performance for a monolithic ANN using edge features; bold indicates best performance.

No	Epochs	Neurons	Time	Correct	Accuracy (%)
01	80	100-100	00:01:11	123/150	82
02	59	100-100	00:01:18	120/150	80
03	**78**	**100-100**	**00:01:07**	**131/150**	**87**
04	90	100-100	00:01:26	117/150	78
05	80	100-100	00:01:08	119/150	79
06	78	100-100	00:01:34	123/150	82
07	53	100-100	00:00:46	123/150	82
08	79	80-90	00:01:08	123/150	82
09	55	80-90	00:00:56	128/150	85
10	58	80-90	00:00:58	122/150	81

score. In this case, the best performance was achieved in the third training run where the algorithm required 78 epochs, and the ANN correctly classified 131 of the testing images.

The second monolithic ANN uses the wavelet features as input, and the obtained results are summarized in Table 3. In this case the best performance was obtained in the third training run, with a total of 5 epochs, and 144 correctly classified images. It is obvious that wavelet features provide a very good discriminative description of the signature images we are testing.

Table 3. Performance for a monolithic ANN using wavelet features.

Train	Epochs	Neurons	Time	Correct	Accuracy (%)
01	12	100-100	00:00:18	135/150	90
02	30	100-100	00:00:25	138/150	92
03	**05**	**100-100**	**00:00:08**	**144/150**	**96**
04	09	100-100	00:00:11	140/150	93
05	06	80-90	00:00:08	142/150	95
06	05	80-90	00:00:05	140/150	93
07	10	80-90	00:00:14	141/150	94
08	07	80-90	00:00:09	138/150	92
09	10	80-90	00:00:15	140/150	93
10	05	80-90	00:00:06	137/150	91

Finally, the third monolithic ANN uses the Hough transform matrix, and the corresponding results are shown in Table 4. The best performance is achieved in the fourth training run, with a total of 6 epochs and 141 correctly classified images.

Table 4. Performance for a monolithic ANN using the Hough transform.

Train	Epochs	Neurons	Time	Correct	Accuracy (%)
01	63	100-100	00:00:19	135/150	90
02	65	100-100	00:00:51	140/150	93
03	68	100-100	00:00:19	141/150	94
04	**06**	**80-90**	**00:00:08**	**141/150**	**94**
05	04	80-90	00:00:05	140/150	93
06	45	80-90	00:00:11	138/150	92
07	08	80-90	00:00:09	138/150	92
08	05	80-90	00:00:08	137/150	91
09	05	80-90	00:00:06	137/150	91
10	33	50-50	00:00:20	138/150	92

It is important to note that in all three cases, the monolithic methods did achieve good results. The best performance was obtained using wavelet features, and the Hough transform matrix also produced very similar results. On the other hand, the simple edge features produced a less accurate recognition than the other two methods.

4.2.2 Modular Neural Network with Sugeno Fuzzy Integral

The final experimental results correspond to the complete MNN described in Figure 1, and Table 5 summarizes the results of ten independent training runs. For the modular architecture, performance was consistently very high across all runs, and the best recognition accuracy of 98% was achieved in half of the runs. In fact, even the worst performance of 95% is better or equal than all but one of the monolithic ANNs (see Table 3).

4.2.3 Modular Neural Network with a Fuzzy System

We use a fuzzy systems integrator for the three modules of the network. The fuzzy systems are of Mamdani type, contain three inputs (module 1, module 2, module 3) output (winner module), and 27 rules. Several tests were performed with the

Table 5. Results for the Modular Neural Network with fuzzy Sugeno Integral

Trian	Epochs	Time	Correct	Accuracy (%)
01	**55**	**00:00:49**	**147/150**	**98**
02	150	00:01:34	146/150	97
03	180	00:01:53	144/150	96
04	300	00:02:20	147/150	98
05	150	00:01:30	147/150	98
06	155	00:01:45	146/150	97
07	320	00:02:49	147/150	98
08	310	00:02:38	148/150	98
09	285	00:01:58	145/150	96
10	03	00:00:02	143/150	95

fuzzy systems that have the same input, and output rules, but with different functions of membership: Triangular, trapezoidal and Gaussian.

In figures 8, 9, 10 we show the fuzzy systems with trapezoidal Membership functions, Triangular and Gaussian.

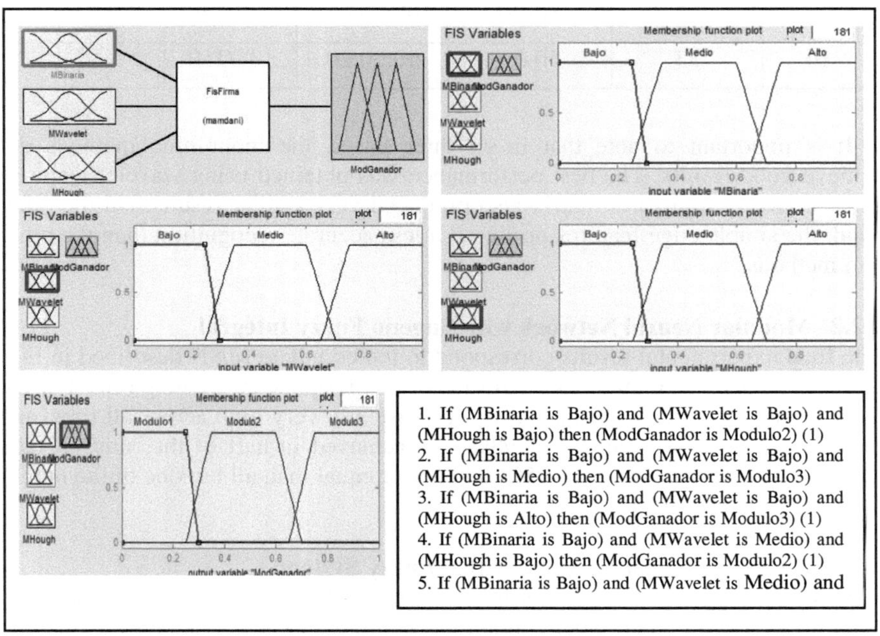

Fig. 8. Representation of fuzzy systems with trapezoidal membership functions.

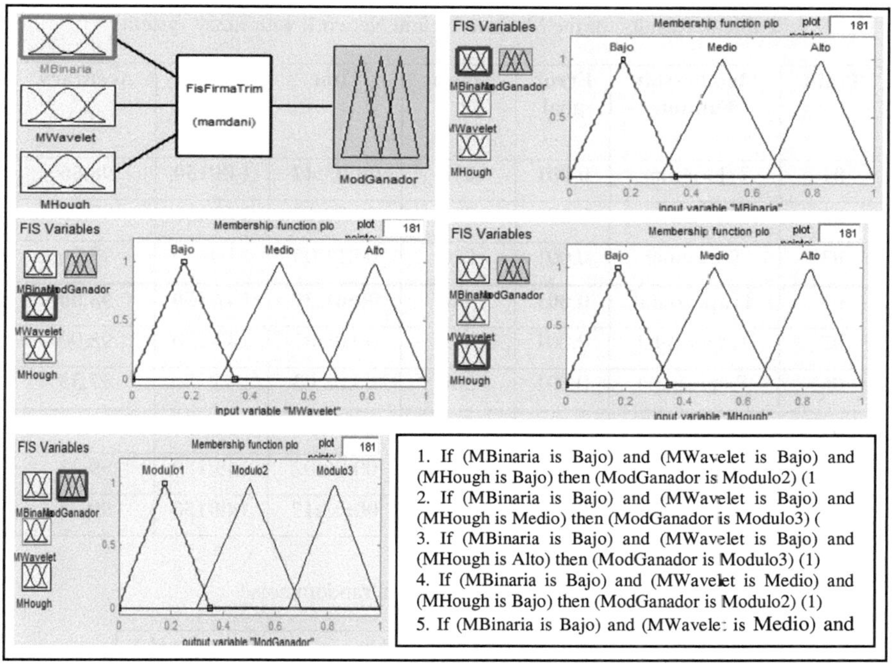

Fig. 9. Representation of fuzzy systems with Triangular membership functions.

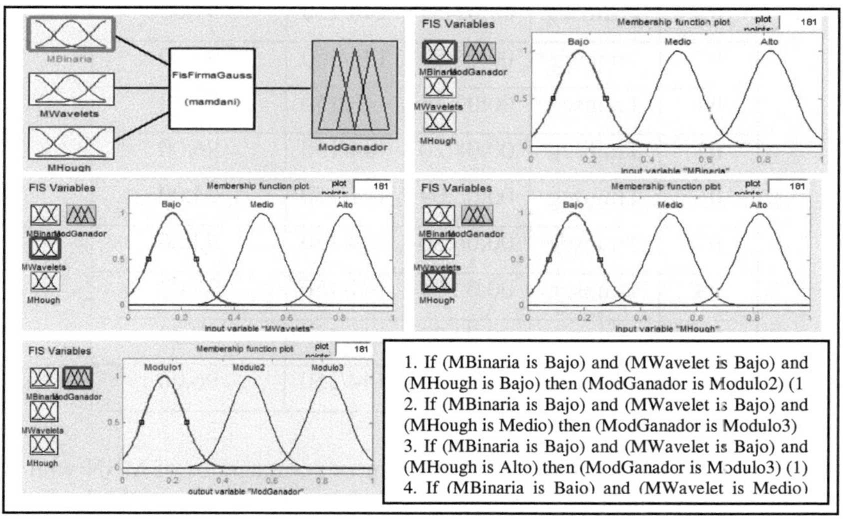

Fig. 10. Representation of fuzzy systems with Gaussian membership functions.

Table 6. Results for the Modular Neural Network with fuzzy system

Train	Membership Funtion	Error goal	Epochs	Time	Correct	Accuracy (%)
01	**Triangular**	**0.001**	**232**	**00:03:42**	**148/150**	**98.66**
02	Triangular	0.001	560	00:09:15	146/150	97.33
03	Triangular	0.001	710	00:11:05	148/150	98.66
04	**Trapezoidal**	**0.001**	**96**	**00:01:37**	**147/150**	**98.00**
05	Trapezoidal	0.001	302	00:08:01	147/150	98.00
06	Trapezoidal	0.001	304	00:08:03	146/150	97.33
07	Gaussian	0.001	150	00:02:50	146/150	97.33
08	Gaussian	0.001	257	00:06:02	149/150	99.33
09	**Gaussian**	**0.001**	**223**	**00:03:17**	**149/150**	**99.33**

Table 7. Result with uniform random noise

Train	Method	Time	Co-rrect	Accuracy (%)
01	Trainscg	00:02:56	141/150	92.66
02	**Trainscg**	**00:02:57**	**146/150**	**97.33**
03	Trainscg	00:04:50	144/150	96.00
04	Trainscg	00:03:25	143/150	95.33
05	Trainscg	00:04:20	144/150	96.00
06	Trainscg	00:03:09	141/150	94.00
07	Trainscg	00:06:50	139/150	92.66
08	Trainscg	00:02:54	141/150	94.00
09	Trainscg	00:03:33	146/150	97.33
10	Trainscg	00:05:56	144/150	96.00

The results obtained with the fuzzy system as an integrator of MNN with different membership functions (see table 6), were good, in this case the best result was obtained in the training with 9 Gaussian membership function, with a total of 223 epochs, and 149 images are classified correctly. The method of training is scaled conjugate gradient (Transcg). Overcoming the best result with fuzzy Sugeno integral (see table 5).

4.2.4 Modular Neural Network with a Fuzzy System Adding Uniform Random Noise

After multiple tests done with the fuzzy system, and taking into account that the best result was obtained with Gaussian membership functions, we applied noise to the images of signatures, using "uniform random noise". The noise level of 0.5 was applied. Table 7 shows the top 10 results. The best training is the one in the second row, with a total of 146 correctly classified images.

4.2.5 Results of Verification of Signatures

Table 8 shows the results as a percentage for each case: false acceptance, false rejection, error recognition and the percentage of correct signatures.

Table 8. Results from the verification of signatures

Train	Time	False Acceptance (%)	False Rejection (%)	Error Recognition (%)	Correct Sgnatures (%)
01	00:06:32	9.33	8.00	1.33	81.33
02	00:06:03	7.33	18.00	0.66	74.00
03	00:07:32	10.00	5.33	2.00	82.66
04	00:08:03	14.66	2.00	1.33	82.00
05	00:08:02	7.33	18.00	0.66	74.00
06	00:09:00	16.00	4.00	1.33	78.66
07	00:06:06	6.66	14.00	1.33	78.00
08	00:06:42	11.33	10.66	1.33	76.66
09	00:06:52	15.33	4.66	2.00	78.00
10	00:06:13	12.00	6.00	2.66	79.33

5 Summary, Conclusions and Future Work

In this chapter we have addressed the problem of signature recognition, a common behavioral biometric measure. We proposed a modular system using ANNs and three types of image features: edges, wavelet coefficients, and the Hough transform matrix. In our system, the responses from each module were combined using a Sugeno fuzzy integral and a fuzzy inference system. In order to test our system, we built a database of image signatures from 30 different individuals. In our experiments, the proposed architecture achieves a very high recognition rate, results that confirm the usefulness of the proposal.

In our tests, we have confirmed that the modular approach always outperforms, with varying degrees, the monolithic ANNs tested here. However, in some cases the difference in performance was not very high, only 3 or 2 percent. Nevertheless, we believe that if the recognition problem is made more difficult then the modular approach will more clearly show a better overall performance.

Furthermore, our results also show that even with the simple image features used in this work, each of the ANN modules is indeed capable of learning very good discriminating functions that can correctly differentiate between our set of image signatures.

Moreover, the fuzzy system as a unit exceeds the percentage achieved with recognition of Sugeno fuzzy integral. For this reason the last experiments were conducted with the fuzzy system integrator.

Finally, the results we have obtained suggest several possible extensions for our work, which include the following:

1. Test the system with a more challenging image database, using more signatures and a smaller set of training samples, in order to verify the robustness of our approach.
2. Optimizing the MNN architecture with a genetic algorithm.

References

[1] Ballard, D.: Generalizing the Hough transform to detect arbitrary shapes. Pattern Recognition 13(2), 111–122 (1981)

[2] Zhang, D.: Automated Biometrics Technologies and Systems, ch. 10, Hong Kong Polytechnic University, pp. 203–206. Kluwer Academic Publishers, Dordrecht (2000)

[3] Jang, J., Sun, C., Mizutani, E.: Neuro-Fuzzy and Soft Computing, pp. 1–70. Prentice-Hall, Upper Sanddle River (1997)

[4] Keller, J., Gader, P., Hocaoglu, A.: Integrals in image processing and recognition. In: Grabisch, M., et al. (eds.) Fuzzy Measures and Integrals: Theory and Applications, pp. 435–466. Physica-Verlag, NY (2000)

[5] Grabisch, M.: A new algorithm for identifying fuzzy measures and its application to pattern recognition. In: Proc. of 4th IEEE Int. Conf. on Fuzzy Systems, Yokohama, Japan, pp. 145–150 (1995)

[6] Grabisch, M., Murofushi, T., Sugeno, M.: Fuzzy Measures and Integrals: Theory and Applications, pp. 348–373. Physica-Verlag, NY (2000)

[7] Castillo, O., Melin, P., Kacprzyk, J., Pedrycz, W.: Hybrid Intelligent Systems Analysis and Design. Springer, Heidelberg (2007)

[8] Mendoza, O., Melin, P.: The Fuzzy Sugeno Integral as a Decision Operator in the Recognition of Images with Modular Neural Networks, Tijuana Institute of Technology. Springer, Heidelberg (2007)

[9] Melin, P., Mancilla, A., González, C., Bravo, D.: Modular Neuronal Networks with Fuzzy Sugeno Integral Response Integration for Face and Fingerprint Recognition. In: The International Multi Conference in Computer Science and Computer Enginnering, Las Vegas, USA, vol. 1, pp. 91–97 (2004)

[10] Melin, P., Mancilla, A., Lopez, M., Solano, D., Soto, M., Castillo, O.: Pattern Recognition for Industrial Security using the Fuzzy Sugeno Integral and Modular Neural Network, Soft Computing in Industrial Applications, vol. 39, pp. 4–9. Springer, Heidelberg (2007)

[11] Melin, P., Felix, C., Castillo, O.: Face Recognition using Modular Neural Networks and the Fuzzy Sugeno Integral for Response Integration. Journal of Intelligent Systems 20(2), 29–275 (2005)

[12] Melin, P., Gonzalez, C., Bravo, D., Gonzalez, F., Martinez, G.: Modular Neural Networks and Fuzzy Sugeno Integral for Pattern Recognition: Then Case of Human Face and Fingerprint, Tijuana Institute of Technology. Springer, Heidelberg (2007)

[13] Melin, P., Castillo, O.: Hybrid Intelligent Systems for Pattern Recognition Using Soft Computing: An Evolutionary Approach for Neural Networks and Fuzzy Systems, 1st edn. Studies in Fuzziness and Soft Computing. Springer, Heidelberg (2005)
[14] Melin, P., Castillo, O.: Hybrid Intelligent Systems for Pattern Recognition. Springer, Heidelberg (2005)
[15] Sepulveda, R., Montiel, O., Castillo, O., Melin, P.: Fundamentals of Fuzzy logic, pp. 115–124. ILCSA, ijuana B.C (2002)
[16] Kung, S., Mak, M., Lin, S.: Biometric Authentication A Machine Learning Approach. Prentice Hall Information and System Sciences Series, Kailath, T. (series ed.), pp. 27–49 (2005)
[17] Santoso, S., Powers, E., Grady, E.: Power quality disturbance data compression using wavelet transform methods. IEEE Trans. on Power Delivery 12(3), 1250–1257 (1997)

Part III
Learning and Social Simulation

A Hybrid Recommender System Architecture for Learning Objects

Mario García-Valdez and Brunett Parra

Tijuana Institute of Technology, Tijuana México
rparra_galaviz@hotmail.com, mariog@tectijuana.mx

Abstract. In this paper we present the architecture of a hybrid recommender system to support an adaptive hypermedia educational (AHE) system. Currently the instructor (using fuzzy rules) specifies the sequence in which learning objects are presented to students. The instructor can also give students a chance to choose from a pool of objects and helps them make their selection by assigning to each object a recommendation rating based on the student's profile. We propose a hybrid recommender system that uses collaborative filtering techniques together with fuzzy inference systems to provide recommendations, considering the instructor's experience as well as the ratings given by similar students.

1 Introduction

The goal of Adaptive Hypermedia systems (AH) is to enhance the functionality of hypermedia, tailoring to each user's needs the navigation and presentation of resources [7]. In a previous work an Adaptive Hypermedia Educational (AHE) system based on learning objects was proposed by the authors [1]. Learning objects in this context are reusable web based resources (i.e. a web page, a video or images) that support a certain learning activity, these resources are authored as components, so they can be combined with others. In a course each unit of instruction can be supported by many learning objects, and instructors normally define the sequence in which these learning objects are going to be presented to students. In this current implementation of the system, instructors can personalize to each student, the sequencing and selection of learning objects using a rule-based sequencing model based on the Simple Sequencing specification [8]. Instructors can specify rules that give permission to students so they can choose which objects they want to visit. In this paper we present a preliminary design of a hybrid recommender system to help students make their selections. The proposed recommender systems consider a personalized rating given to the learning object by the instructor and also the ratings given by other students. The aim of this paper is to present an overview of the main components of the recommendation process and the algorithms used. In section 1 a brief overview of the field of recommender systems is presented; a more in-depth survey can be found in [2]. An overview of

O. Castillo et al. (Eds.): Evolutionary Design of Intelligent Systems, SCI 257, pp. 205–211.
springerlink.com

the proposed recommender system is given in section 2, and in section 3 the details of the recommender algorithm are presented. Finally some conclusions are presented in section 4.

2 Recommender Systems

Recommender systems (RS) help users deal with information overload by providing personalized recommendations of content and services [2]. These systems are used by commercial websites and e-marketing tools to increase sales, by presenting to users those products that they more probably want to buy. The majority of RS use a *collaborative filtering* approach, which is the method of making automatic predictions (filtering) about the interests of a user by collecting information on the tastes of many users (collaborative) and also preferences he liked in the past. For example, a collaborative filtering system of musical taste, can make predictions about which music a user would want, given a partial list of preferences of other users and his own. Other RS use a *content-based* approach where items recommended to the user are similar to those the user liked in the past. Other RS use a hybrid approach. Formally the recommendation problem can be stated as follows [2]: Let C be the set of all users, and S the set of items that can be recommended, both sets can be very large. Let u be the utility function that measures the usefulness of item s to user u. i.e. u: $C \times S \rightarrow R$ where R is a totally ordered set. Then for each $c \in C$, we want to choose an item $s' \in S$ that maximizes the user's utility:

$$\forall c \in C, \quad s_c' = arg\,max_{s \in S}\, u(c, s)$$

Usually u is represented by a rating in a recommender system and is only defined for a subset of the $C \times S$ space, because not all users give a rating to all the items. A rating matrix of $C \times S$ has the ratings of items indicated by users, and to indicate that a user has not rated an item the symbol "$\emptyset$" can be used. The RS engine should be able to estimate these missing combinations and issue recommendations based in these predictions.

3 Recommender System Architecture

In this section the proposed recommender system (RS) is presented. This system is used in an Adaptive Hypermedia Educational system, in which students must complete certain learning activities previously specified by the instructor. These learning activities can have a recommended value based on the student's profile which has information about the student's performance and learning style. Instructors specify their recommendations using a Mamdani fuzzy inference system with rules which have a recommended fuzzy recommended value as their consequent:

```
IF Visual IS Strong AND Verbal IS Mild THEN
      Recommended IS Low
IF Visual IS Mild AND Verbal IS Strong THEN
      Recommended IS High
```

These rules are static, as the instructor has defined the membership functions i.e. `Strong`, `Medium` and `Mild` corresponding to `Visual` and `Verbal` linguistic variables. Learning activities are multimedia resources (i.e. video, text or audio) in this example a learning activity presented in text format has a higher recommended value for students with a strong verbal learning style. These heuristic based recommendations rely on the instructor's subjective appreciations about the preferences of students. One way to add an adaptive behavior to the system is by changing the parameters of membership functions in response to student's feedback. In this paper another approach is explored, adding another recommendation value to learning activities, this time the recommendation is given by a collaborative filtering algorithm. Now each learning activity has a recommended value that takes into account the instructor and also their peers' ratings. In this algorithm the same features as the heuristic recommendation are considered:

- **Learning style.** The learning style is previously assessed by a test, which gives students a grade from 0 to 20 in each of the learning styles (visual, verbal, aural, physical, logical, social, solitary).
- **Student Performance.** Learning activities record the performance of students, and the percentage of objectives reached by students.

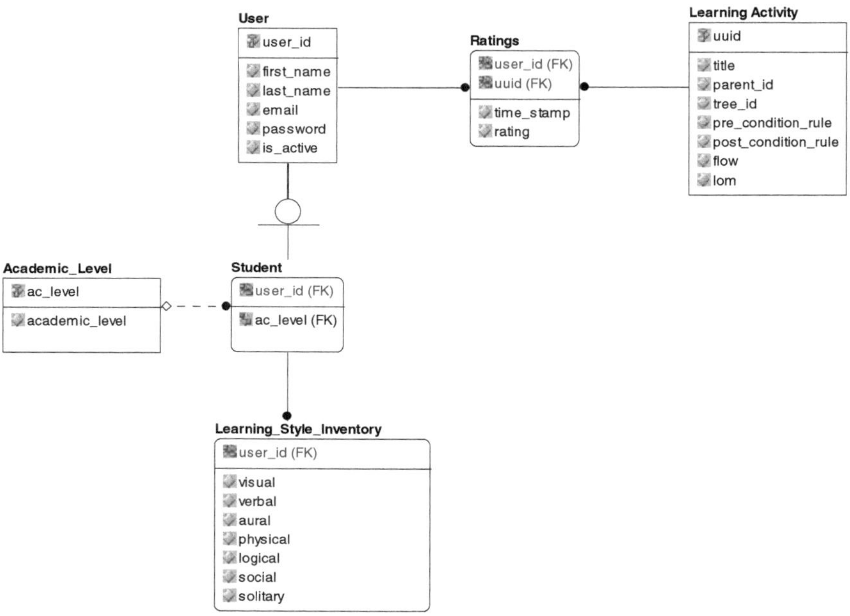

Fig. 1. Recommender System Data Model

Each student is given an opportunity to rate each learning activity they complete, giving it a rating from 1 to 5. These ratings are also considered by the collaborative filtering algorithms. The data model of the recommender system is illustrated in Figure 1.

Each student is represented by a vector of features in this case their learning style inventory; there is also a rating matrix with the ratings of all users. These two sets of data are used separately by two k-NN algorithms, one based on Euclidean distance for selecting students with a similar profile, and another based on Pearson Correlation to find students with similar preferences as there is a positive correlation between their ratings. The recommender algorithm, considers three special cases:

1. **A new student is added.** In order to make accurate recommendations collaborative filtering algorithms need to know the student's previous preferences, based on values of the rating matrix. When a new user is added to the system or he has not made certain number of ratings, there is not enough rating data to give an accurate recommendation. If a student is "new" in the system then the instructor's recommendation value and the ratings of similar students with respect to their learning styles is considered. The level of newness can be seen as a fuzzy variable based on the minimum number of ratings needed to make accurate predictions. This value can be estimated on-line.
2. **A new learning activity is added.** As in the previous case, there is also a problem when a new learning activity is added to the system, as again it does not have enough ratings. Each learning activity has standard metadata [3] indicating among other things: their intended audience, difficulty level, format, authors and version. This information can be used to make a content-based recommendation, when a new learning activity is added to the system, also the instructors recommendation is considered. As the learning activity is receiving ratings, the collaborative filtering recommendations become more accurate and gain more weight in the overall recommendation.
3. **Sparse rating matrixes.** This is a typical problem in other recommender systems and happens when there are a high number of users and items, and users only rate a small fraction of the items. In this case, students must rate learning activities as they finish them, but they only rate those learning activities in their path. This is illustrated in Figure 2, where only Student 1 and Student 2 rated the same learning activities, in this particular application the rating matrix not only has which students rated the same learning activities, but to some extent; also which students followed similar paths. As an option the proposed recommender system can also consider the performance evaluation of the student as an indirect rating value.

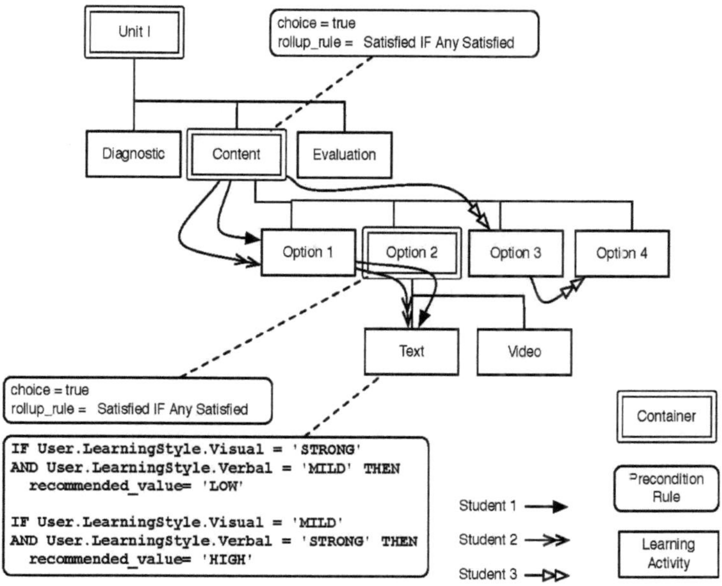

Fig. 2. Paths of different students

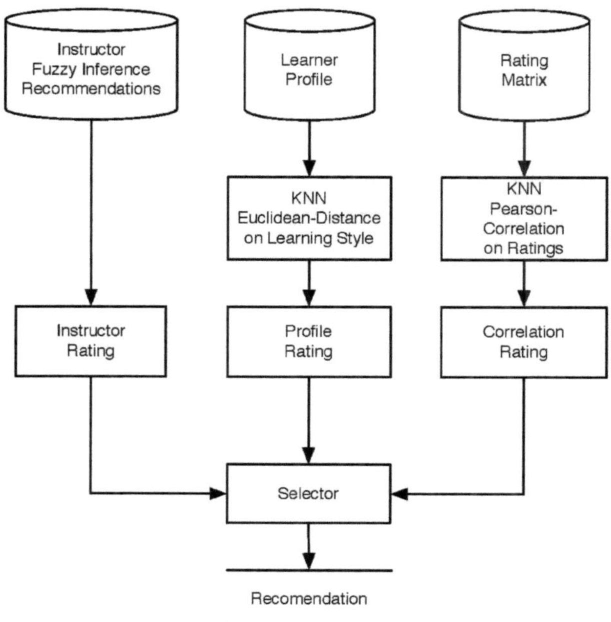

Fig. 3. Recommender System Process

An overview of the recommender process is illustrated in Figure 3, there are three different predictions of ratings: one that considers the instructor's rule-based system, other that considers the similarity of students regarding their learning styles and finally one that considers the correlation between student's ratings. These three predictions are then integrated by a fuzzy inference system which gives the final recommendation value.

4 Memory-Based RecommenderAlgorithm

Memory-Based algorithms have been proposed [4] for collaborative filtering. The objective of these algorithms is to predict the vote (rating) a particular active user is going to give to items, based on a sample or population of the voting of other users. Each of these votes are represented by v_{ij} which is the vote of user i on item j. This algorithm considers the mean vote of other users $\bar{v}_i$ and the active user $\bar{v}_a$ and assumes that the predicted vote $p_{a,j}$ of active user a to item j, is the weighted sum of the votes of other users:

$$p_{a,j} = \bar{v}_a + k \sum_{i=1}^{n} w(a,i)(v_{ij} - \bar{v}_i)$$

Where n is the number of users with non zero weights. Here the weights $w(a,i)$ correspond to the similarity between the active user a and user i (i.e. Pearson's correlation or Euclidean distance), k is a normalization factor such that the absolute value of the weights sum to unity. In this particular implementation the number of users considered are the K-Nearest Neighbors, and the weight function used are: Euclidean distance for the Profile Rating and Pearson's for the Correlation Rating. These ratings are then integrated by a Fuzzy Inference system that considers the number of items and ratings (given or received) to determine the newness of the user or learning activity, with this information a final recommendation value is finally assigned to the Learning Object.

5 Conclusions

In this paper we have presented the design of a hybrid recommender system, considering some of the problems associated with this type of systems, multiple recommendations are computed and then a selector module chooses the appropriate value for the intended item. Fuzzy inference is used for making heuristic recommendations and for the final selection. The implementation is still work in progress, and refinements can be made as a result of further experiments.

References

[1] Breese, J.S., Heckerman, D., Kadie, C.: Empirical analysis of predictive algorithms for collaborative filtering. In: Proceedings of the 14th Conference on Uncertainty in Artificial Intelligence, pp. 43–52 (1998)

[2] Sarwar, B., Karypis, G., Konstan, J.: Analysis of recommendation algorithms for e-commerce. In: Proceedings of ACM E-Commerce (2000)
[3] Memletics, Memletics® Learning Styles Inventory (2008), http://www.learning-styles-online.com/inventory/
[4] García Valdez, J.M.: Aprendizaje colaborativo basado en recursos adaptativos (2008)
[5] Adomavicius, G., Tuzhilin, A.: Towards the Next Generation of Recommender Systems: A Survey of the State-of-the-Art and Possible Extensions. IEEE Trans. Knowl. Data Eng. 17(6), 734–749 (2005)
[6] Delgado, J., Ishii, N.: Memory-Based Weighted-Majority Prediction for Recommender Systems. In: ACM SIGIR 1999 Workshop on Recommender Systems: Algorithms and Evaluation (1999)
[7] Brusilovsky, P.: Methods and techniques of adaptive hypermedia. User Modeling and User-Adapted Interaction 6(2-3), 88–129
[8] IMS Global Learning Consortium, I. IMS Simple Sequencing Information and Behavior Model (Version 1.0 Final Specification) (2003b)

TA-Fuzzy Semantic Networks for Interaction Representation in Social Simulation

Dora-Luz Flores, Antonio Rodríguez-Díaz, Juan R. Castro, and Carelia Gaxiola

Universidad Autónoma de Baja California, Tijuana, B.C. México
dflores@uabc.mx, ardiaz@uabc.mx,
jrcastror@uabc.mx, cgaxiola@uabc.mx

Abstract. The need to model interactions between people of different cultures, religions and ethnic groups is evident. In Social Simulation, the combination of Artificial Intelligence and Multi-agent Systems has proven to be a good tool for modeling social groups, however much remains to achieve a model which represents a society with differences between individuals. Our proposal is to combine fuzzy logic, semantic networks and transactional analysis for representation of social interactions, taking into account the perception and a psychosocial profile of each individual. This model will facilitate the implementation of socially intelligent agents.

1 Introduction

Simulations as a research tool have gained more attention by researchers as a possibility for study and understand phenomena. Several disciplines have adopted it as a regular tool with success to generate data close to the real phenomena. In [1] and [5] presented two different types of formalizations, one using UML another using Logic Agent-Based to present tools for Social Simulation.

Traditionally, social sciences use statistical methods to develop and the study of models that describe the social phenomena, but with an emergent systemic approach, the possibility of developing software is more and more appealing. The social researchers coming from different fields, such as, sociology, psychology, anthropology, among others, have developed qualitative tools to take them to the practice in social simulations, but this is not sufficient to develop simulations that are resembled the reality more accurately. In order to carry out these simulations it is precise to define the scope in which they will be developed, from the point of view of the multi-agents systems (MAS) is an ideal frame to carry out the simulation.

A MAS is a distributed system in which the nodes or elements are artificial intelligence systems, that are called agents, or, it is a distributed system where the conduct of these agents produces an intelligent result altogether. It is necessary to notice that

O. Castillo et al. (Eds.): Evolutionary Design of Intelligent Systems, SCI 257, pp. 213–225.
springerlink.com

the agents are not necessarily intelligent; two approaches exist to construct them; the classic one, where the agent is equipped with greater possible intelligence; and the constructivist, which persecutes the idea to offer intelligence to the assembly of all the agents. This type of approach is habitually called emergent behaviour.

An interaction assumes the presence of agents able to act and/or to communicate to achieve a goal by means of individual and/or cooperative work; in this sense, the necessity to take into account the intentions of the agents in the interactions is fundamental, this can be obtained considering the psychological part of the same agents by means of the transactional analysis theory (TA).

TA theory [2] has been used anywhere in the world as a successful tool to understand the bases of the behaviour and feelings of the people as well as for the conflict detection in the interactions. Dr Berne introduced it for the first time at the beginning of the 60's and it is considered a tool that can be used to explain the human interaction [8].

With the purpose of describing the social interactions, one sets out to construct a representation mechanism that will take into account the attitude, implicit social interrelations, and the communication, assigning roles based on cultural scripts. As well as the inclusion of the fuzzy logic to represent the knowledge base of the agents by a fuzzy inference system.

2 Objective of the Model

This innovative proposal for the representation of the model is in which the theory of TA is modified to define the interactions among agents within a MAS. In this sense, the role of each agent is determined according to a set of elements that are comprised by the TA.

The transactional analysis provides to the people a rational method to analyze and to understand the human behaviour. [3] propose in their model the structure of the personality in a systemic and consistent sense. The TA sees the "personality" as the result of the interactions of forces within an individual. TA not only tries to formalize the human interaction, in addition detect conflicts too. The transactions happen between the people when they send messages. A transaction consists of a stimulus and an answer. The transaction can be obvious and verbal. Some times are not verbal nor are so obvious, for example a smile or a pitching.

Structural analysis is within TA theory; it is based on the concept of the ego state (ES). An ES is defined as a pattern of feelings, thoughts, and experiences and they are related to a behaviour profile. This personality is conformed of three ego states: Parent (P), Adult (A), and Child (C). Each ES perceives the environment of totally different form. The Parent ego state, tries to use the learned patterns (imitation), these patterns are called "culture". The Adult ego state, changes the environments stimulate into information, this is processed and archived in the previous experience. Finally the Child ego state reacts of a steep way based in a pre-logical mechanism and distorted perceptions.

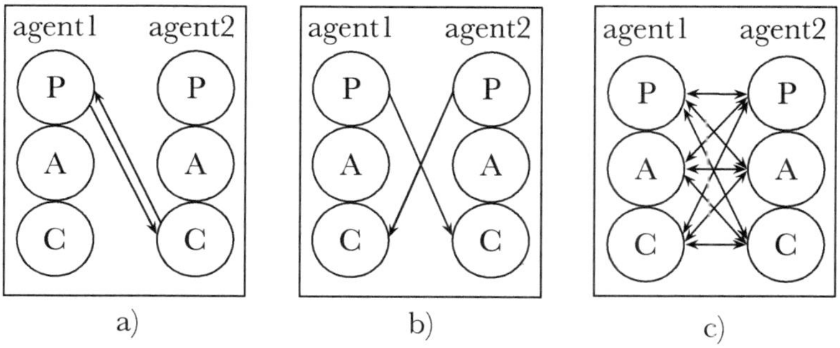

Fig. 1. Types of transactions. a) complementary, b) crossed, and c) ulterior.

In order than a transaction exists, a stimulate must be generated by a particular ES of one person and an answer that comes given from ES of another person. Stroke is named in the TA language when a person recognizes another one with a smile or a pitching, for example. Two or more strokes become a transaction. All the transactions can be classified in complementary, crossed, and ulteriors.

A complementary transaction is when the message initiates in a transmitters ES and is received by the appropriate receivers ES. In this case, cooperation and understanding of both parties is possible, that is to say, it exists a dialog. A crossed transaction happens when there is an unexpected answer to the stimulus, an unsuitable ES is activated and the lines of the transaction are crossed, when this happens appears a conflict and the communication is interrupted. The ulterior transactions are the most complex ones. These transactions involves more than one ES and the message is sent from more than one ES at the same time. These transactions can be seen in a graphic way in Fig. 1.

Another concept of the TA that taken into account is the analysis of the game. The games in which we participated, each person takes its own role. The Triangle of Drama of Karpman [12] defines three roles, persecutor, victim, and rescuer, which always are presents in all the games in which we participated. The drama generates the sensations that happen when people hide its intentions and secrets and later manipulates them to obtain personal advantage.

Once a role is defined, this one participates within a script. A script is a particular sequence of actions played by a role. The scripts have the particularitity that are repeated several times. The scripts have names and the way in which they will be developed can be calculated, then it is possible to think about them as small theater play where each participant repeats its lines over and over again.

In order to complete the proposal, it is taken into account the four positions from life developed by [6] that we shall call psychological posture. Ever since we are born we are taking some of these positions, when a child is defenseless and dependant takes the I'm not ok –You're ok ($-+$) position and it happens in the first year of life. The second position (I'm not ok –You're not ok, $--$) it is the one that is adopted when a child is in the second year of its life, where there is not more comfort and it receives scolds. When a child is mistreated in his childhood and survives it can assume the third called position I'm ok –You're not ok (+−) and always it is recriminating to the society assuring that it is always right and that everybody in the world is against its. Finally, the fourth position is I'm ok –You're ok (++), unlike the first three which they are in unconscious form, this position is not a feeling, is based in thoughts, faith and actions.

Including to this proposal the TA theory provides an approach to the reality because the agents must have major similarity to the human behaviour [4], but this is not enough since these concepts do not take into account the fuzzy concepts. Soft computing is an approach to computing which parallels the remarkable ability of the human mind to reason and learn in an environment of uncertainty and imprecision [14]; this is a characteristic of human being.

3 Fuzzy MAS Formal Description

In this section, the formal description of representation model of interactions among agents in a MAS is made. In order to describe a situation in adapted way, simple and easy to understand, representations can be used. A semantic network is a type of representation and consists of four parts, a lexical part, a structural part, one of procedure and a semantic part [11]. The lexical part, determines the symbols that are allowed in the vocabulary of the representation. The structural part describes the restrictions on the form in which the symbols can be become ordered. The operative part specifies the access procedures that allow to create descriptions, to modify them and to respond to questions. The semantic part establishes a form to associate the meaning with the descriptions.

Many schools of thought exist about the meaning of the semantics. But after all, the meaning always seems to have roots in the human perception and the intuition. A unknown object is identified with a idealized object if their characteristics are similar and not necessarily identical.

3.1 Formal Definition of Multi-agent System

A finite automaton is used to formally define the multi-agent system model. As it's defined in [10] a finite automaton is formed by several parts; this finite automaton has a set of states and rules and it goes from one state to another following the input symbol. It has an input alphabet that indicates the symbols that can store. It has an initial state and a set of acceptance states. The formal definition states that a finite automaton is a list of these five objects: set of states, input alphabet, rules for moving, start state, and accept states. In mathematical language a list of five elements is often called a 5-tuple. A finite automaton is defined as a tuple of five parts. Transition function is called to the rules for moving and it is denoted by δ. If a finite automaton has an arrow from the state x to the state y labeled with the input symbol 1 means that, if the automata is in the state x and reads a symbol 1 this automaton moves to the state y. This can be indicated of the form $\delta(x,1) = y$.

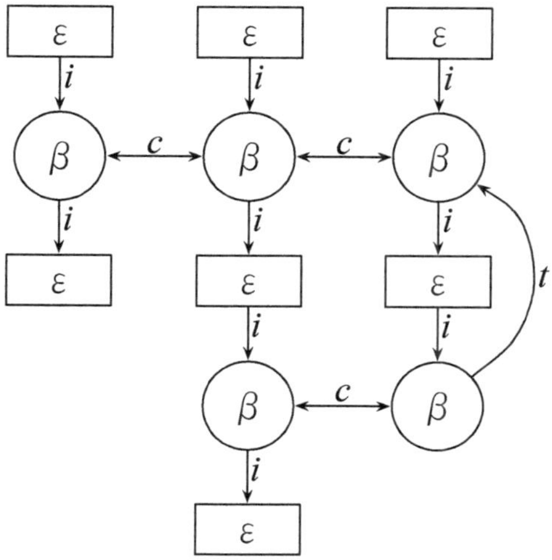

Fig. 2. Graphic representation of a MAS

Definition 1. A MAS is defined as a finite automaton with six elements

$$MAS = (\beta, \varepsilon, \alpha, \delta, q_0, \phi)$$

where:

1. β is a set of finite nodes that describes an agent, denoted by BP,
2. ε is a set of finite nodes that describes the events that happen in the MAS, denoted by E,
3. α is the finite set named alphabet, where $\alpha = \{i, t, c\}$ represents a link label between a node and another one. A labeled i link can initiate a type-BP node or type-E node; a labeled t link can terminate a type-BP node, finally a labeled c link corefers two type-BP nodes,
4. $\delta : \eta \times \alpha \rightarrow P(\eta)$ is the transition function, where $\eta = \beta \cup \varepsilon$ and it is defined by

$$\delta(\eta, \alpha) = \begin{cases} P(\beta) & \eta \in \beta \text{ and } \alpha \in \{c, t\} \\ P(\varepsilon) & \eta \in \beta \text{ and } \alpha = i \\ P(\eta) & \eta \in \varepsilon \text{ and } \alpha = i \end{cases},$$

5. $q_0 \in \varepsilon$ is the start state, and
6. $\phi \subseteq \beta$ is the set of accept states.

Fig. 2 shows a graphic representation of the semantic network of a MAS.

Definition 2. An agent is defineted as a 5-tuple

$$\beta = (\lambda, ESi, ESu, \pi i, \pi u)$$

where:

1. λ represents the perception of an event for an agent. It is defined in section 3.2 as a linguistic variable.
2. $ESi = \{P, A, C\}$ represents the ego state of an agent. Where **P** = Parent, **A** = Adult, and **C** = Child.
3. $ESu = \{P, A, C\}$ represents the perception of the ego state of another agent. Where **P** = Parent, **A** = Adult, and **C** = Child.
4. πi represents the psychological posture of an agent. It is defined in section 3.2 as a fuzzy inference system.
5. πu represents the perception of the psychological posture of the other agent. It is defined in section 3.2 as a fuzzy inference system.

Fig. 3 shows a graphic representation of an agent.

A link labels are *i*, an acronym for initiates, meaning that a type-BP node or a type-E node at the tail of an *i* link leads to the one at the head of the link; *t*, for terminates, meaning that a type-BP node at the tail turns off that happened in a previous time at the head; and *c* for corefers, meaning that two type-BP nodes are related in the same time of two or more agents. Links labeled with *c* have two heads to indicate that these nodes are related in both directions.

With two kind of nodes and three kind of links, $2 \times 3 \times 2$ combinations can be created of the type node-link-node; some combination from these 12 possibilities are shown in Fig. 4.

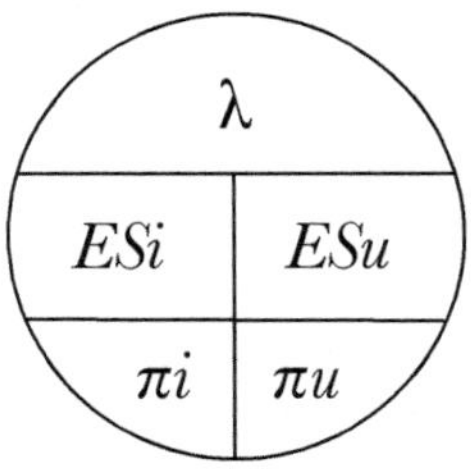

Fig. 3. Graphic representation of an agent.

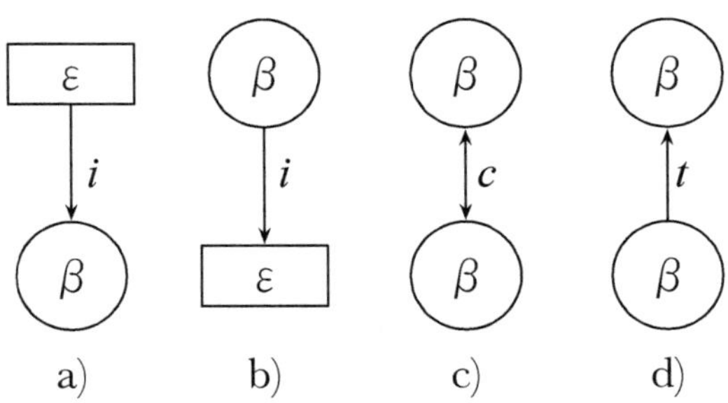

Fig. 4. Some cases of combinations in a semantic network. a) Enablement, b) Reaction, c) Perseverence, and d) Change of mind.

3.2 Definition of the Fuzzy Inference System

With the purpose to represent the perception of the event (λ) for an agent, a membership functions (MF) is defined. The MF maps each element of a collection of objects to a membership grade (or membership value) between 0 and 1 [7]. With this techniques of fuzzy logic [13] can be applied to turn the perceptions to linguistic values.

Let Λ = "event perception." The fuzzy sets are defined "not_important," "important," and "very_important" that are characterized by MFs $\mu_{not\_important}(\lambda)$, $\mu_{important}(\lambda)$, and $\mu_{very\_important}(\lambda)$ respectively. If "event perception" assumes the value of "not_important," then exists the expression "event perception is not important," and so forth for the others values, these linguistic values are displayed in Fig. 5, where the universe of discourse Λ is totally covered by the MFs and the transition from one MF to another is smooth and gradual.

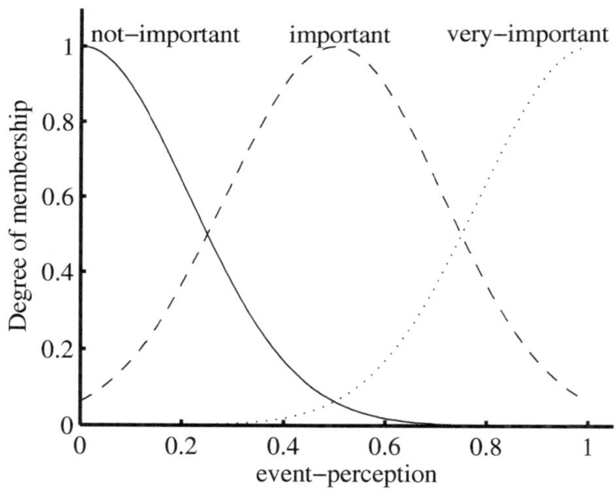

Fig. 5. Membership functions for event perception

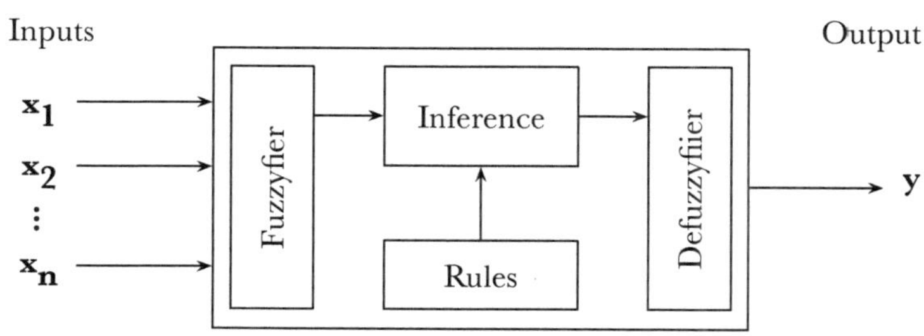

Fig. 6. Structure of fuzzy inference system

Table 1. Linguistic variables πi and πu and their linguistic values

Linguistic variables	linguistic values	
πi	NOK	OK
πu	NOK	OK

Table 2. Linguistic variables ρi and ρu and their linguistic values

Linguistic variables	linguistic values			
ρi	No-role	Persecutor	Rescuer	Victim
ρu	No-role	Persecutor	Rescuer	Victim

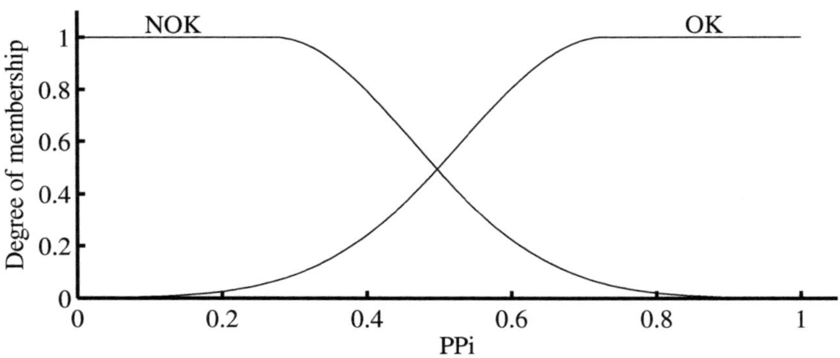

Fig. 7. Antecedent MF of Mamdani fuzzy model

In order to represent the roles that participate in a script uses a Mamdani fuzzy inference system (FIS). A FIS is based on fuzzy set theory, fuzzy if-then rules, and fuzzy reasoning [7]. Fig. 6 shown the blocks of the FIS. The rule base contains a selection of fuzzy rules; the database (or dictionary) defines the membership fuctions used in the fuzzy rules; and the reasoning mechanism, which performs the inference procedure upon the rules and given facts to derive a reasonable output or conclusion.

Table 1 shows the linguistic values for the input variables to the FIS, they are defined by πi (psychological posture of the agent) and πu (perception of psychological posture of another agent) as linguistic variables. Table 2 shows the linguistic values for the output variables, which are ρi (role of the agent) and ρu (perception of the role of another agent).

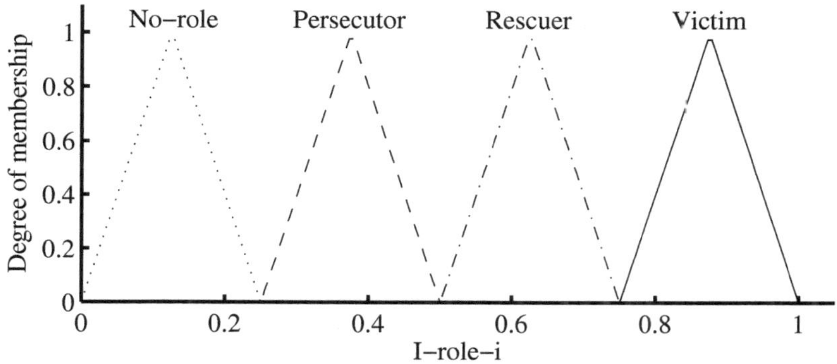

Fig. 8. Consequent MF of Mamdani fuzzy model

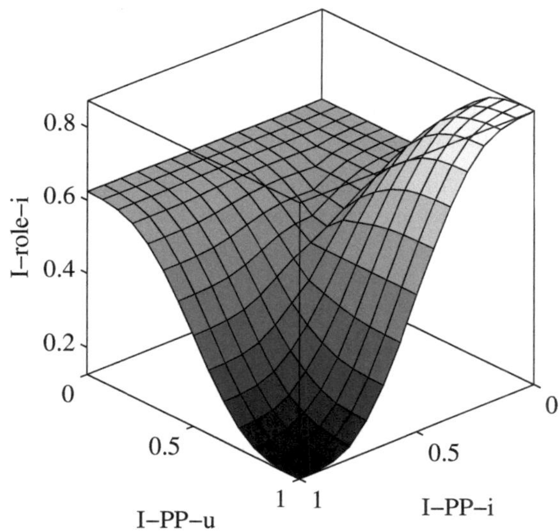

Fig. 9. Overall input-output surface

These variables and their linguistic values can be represented by membership functions as it is shown in the Fig. 7 and Fig. 8.

In Fig. 9 can be observed that while the psychological posture of each agents is more closed to linguistic value OK, they do not take any role. If an agent approaches linguistic value OK and the other agent approaches more linguistic value NOK, the role that takes the agent is Rescuer and the perception of role of the other agent is Victim. This is shows the overall input-output surface with the max-min composition and the defuzzyfication type centroid.

4 Case of Study

In order to use the model presented in the previous section, an example taken from [9] is used, that describes a psychological profile called "Don Juan" syndrome and presents a characterization of psychological script that a man who still is in this situation. The sequence of steps that follows in this psychological game is presented next:

1. Don Juan uses flattery and promises to present himself as the Rescuer of a woman who is prey to her own need to be free and to feel appreciated (the Victim).
2. The work of seduction continues until she capitulates (the response).
3. The moment "Don Juan" reneges on any further demands for emotional closeness, the woman remains bewildered and shocked.
4. As soon as she realizes how gullible she has been, she turns into a Persecutor seeking revenge, and Don Juan, in turn, becomes the Victim of female voracity–ready to start fresh anew as another woman's Rescuer (this move is the switch).
5. The game's playoff is for Don Juan to prove once again how voracious women are and to feel "all set and raring to go" in a new attempt to win over the ideal woman; the woman's payoff is to confirm man are untrustworthy.

In this game different types of messages can be identifed and a collection of them could be made, in the Table 3 some classifications of these messages with an example are shown.

Table 3. Classification of messages in a psychological game

Type of message	Example of message
Social	"Hello!"
Flattery	"You are beautiful"
Promise	"I'll do it next week!"
Reneges	"You ask too much"
Demand	"I want to see you right now!"
Revenge	"You're just like all the rest."

4.1 Scenario A. One Agent Don Juan

In this subsection we describes a scenario where there are 3 agents, one of them has the profile of "Don Juan," the two kind of agents using the proposed model are defined.

Definition 3. "Don Juan" is defineted as a 5-tuple $\beta_{agent1} = (\lambda, ESi, ESu, \pi i, \pi u)$ where:

1. $\lambda = very - important$,
2. $ESi = \{P\}$,
3. $ESu = \{C\}$,

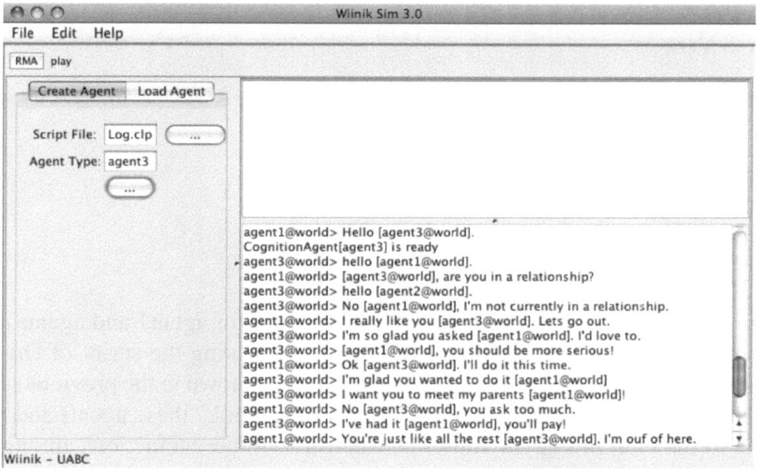

Fig. 10. Screenshot shows to the script of "Don Juan."

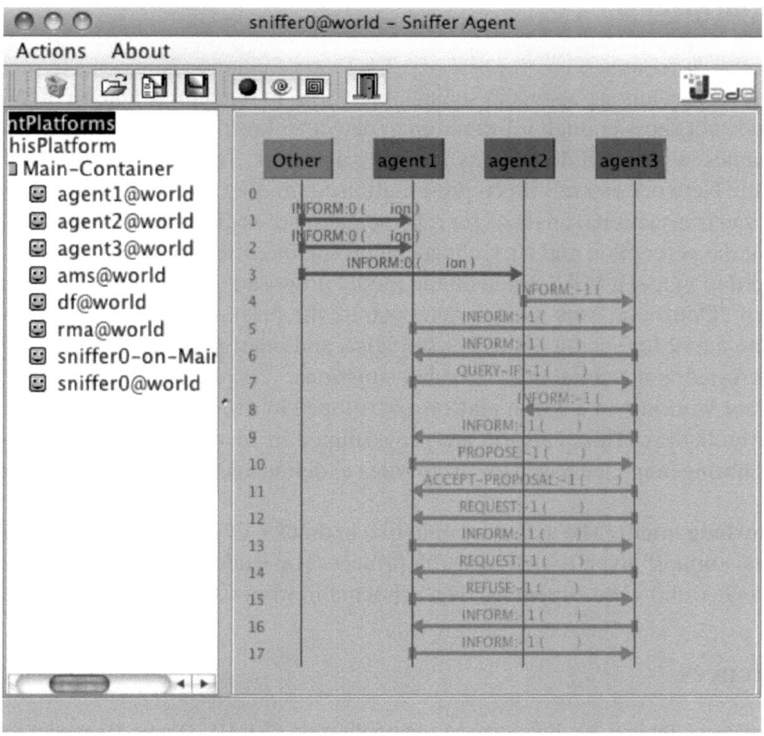

Fig. 11. Screenshot shows interactions among agents who exist in the world.

4. $\pi i = OK$,
5. $\pi u = NOK$.

Definition 4. agent2 is defineted as a 5-tuple $\beta_{agent2} = (\lambda, ESi, ESu, \pi i, \pi u)$ where:

1. $\lambda = very - important$,
2. $ESi = \{C\}$,
3. $ESu = \{P\}$,
4. $\pi i = NOK$,
5. $\pi u = OK$.

With this information establishing the role of Rescuer for agent1 and agents agent2 and agent3 are positioned in the role of Victim. Following the script of Don Juan must exist these two roles for agents to begin the steps shown in the previous section. In Fig. 10 shows a screenshoot of the way in which "hook" these agents and follow the dash to the end. In Fig. 11 shows a screenshot of the interactions among these agents, which can display the kind of message that is sent and received by the agents.

5 Conclusions

This research focuses on a proposal of a TA-Fuzzy Semantic Networks to represent the interactions among agents in social simulation. In the representation of Winston [11] does not show enough information to detect and resolve conflicts or to identify the manner in which a decision is taken by an agent. The proposal for TA-Fuzzy Semantic Network extends this representation combining fuzzy logic, semantic networks and transactional analysis for representation of social interactions, taking into account the perception and a psychosocial profile of each individual.

A tool to model a MAS based on the model proposed was presented. The case of study of "Don Juan" was used to demonstrate the proposed model. This approach could be a tool for sociologists, psychologists and other groups of researchers who are interested in the simulation of social situations.

We are working on a social platform developed by IMOX research group to perform simulations. This platform are programmed in Java language is also using a programming language based on fuzzy rules to define agents and their environment.

Acknowledgement. The authors would like to thank CONACYT and UABC for the financial support given to this research project. The students (Dora-Luz Flores and Carelia Gaxiola) were supported by a scholarship from CONACYT.

References

1. Bouabana-Tebibel, T., Belmesk, M.: Formalization of UML Object Dynamics and Behaviour. In: 2004 IEEE International Conference on System, Man and Cybernetics, pp. 4971–4976 (2004)
2. Berne, E.: Games people play: The psychology of human relationship. Penguin, London (1964)

3. Berne, E.: Principles of Group Treatment. Oxford University Press, New York (1964)
4. Cheng, Z., Capretz, M.A., Osano, M.: A Model for Negotiation among Agents based on the Transactional Analysis Theory. Autonomous Descentralized System, 427–433 (1995)
5. Frank Dignum, F., Edmonds, B., Sonenberg, L.: The Use of Logic in Agent-Based Social Simulation. Journal of Artificial Societies and Social Simulation 7(4) (2004), http://jasss.soc.surrey.ac.uk/7/4/8.html
6. Harris, T.: I'm OK - You're OK. Harper & Row, New York (1969)
7. Jang, J.-S.R., Sun, C.-T., Mizutani, E.: Neuro-Fuzzy And Soft Computing. In: A Computational Approach to Learning and Machine Intelligence, Prentice Hall, Englewood Cliffs (1997)
8. Kreyenberg, J.: Transactional analysis in organizations as a systemic constructivist approach. Transactional Analysis Journal 35(4), 300–310 (2005)
9. Novellino, M.: The Don Juan Syndrome: The Script of the Great Losing Lover. Transactional Analysis Journal 36(1), 33–43 (2006)
10. Sipser, M.: Introduction to the Theory of Computation. PWS Publishing Company, Boston (1997)
11. Winston, P.H.: Artificial Intelligence. Addison Wesley, Reading (1992)
12. Woollams, S., Brown, M.: TA: the total handbook of transactional analysis. Prentice Hall, London (1979)
13. Zadeh, L.A.: Fuzzy sets. Information and Control 8, 338–353 (1965)
14. Zadeh, L.A.: Fuzzy Logic, neural networks and soft computing. One-page course announcement of CS CS 294(4), the University of California at Berkeley (1992)

Fuzzy Personality Model Based on Transactional Analysis and VSM for Socially Intelligent Agents and Robots

Carelia Gaxiola[1], Antonio Rodríguez-Díaz[2], Susan Jones[3], Manuel Castañón-Puga[4], and Dora-Luz Flores[5]

[1] Universidad Autónoma Baja California, Tijuana, B.C. México
cgaxiola@uabc.mx
[2] Universidad Autónoma Baja California, Tijuana, B.C. México
ardiaz@uabc.mx
[3] University of Sunderland, UK
susan.jones@sunderland.uk
[4] Universidad Autónoma Baja California, Tijuana, B.C. México
puga@uabc.mx
[5] Universidad Autónoma Baja California, Tijuana, B.C. México
dflores@uabc.mx

Abstract. This paper presents a model to add personality features to robots, the Transactional Analysis (TA) is used to define a psychological profile and the Viable Systems Model (VSM) is used to help explain the decisions made by the robot and how this impacts on its viability.

1 Introduction

Representing personality in multi-agent system (MAS) has been a key issue in recent years; agent behaviour is expected to reproduce human behaviour and thus, social phenomena in a MAS must replicate social issues in real life. In the beginning of the 90's, agent based systems where considered a significant advancement in software development [14], and software evolution [12]. Today agents are of great interest in a variety of fields in Computer Science. Characteristics such as autonomy, collaboration, reasoning, adaptability, mobility and goal orientation are among the main features that make agent oriented systems a great tool for social simulation [17].

However, humans have more than just these characteristics, since they have a personality profile, which plays a central role as suggested in [8][1][2][10][15][6].

Unfortunately, one of the first problems encountered while trying to model personality is that the term itself refers to a concept, which can be interpreted in many ways, thus uncertainty can arise. Also, it can not be observed directly, but indirectly by actions performed by someone.

There are many issues that require interacting with the person in order to be able to describe them. Things like thought and feelings can be "talked about", but others

O. Castillo et al. (Eds.): Evolutionary Design of Intelligent Systems, SCI 257, pp. 227–241.
springerlink.com

are not at the "conscious level", so they need to be observed for a longer period of time. In other words, to study personality requires a long-term strategy.

Often, when we talk about someone, we talk about what makes this person different from others, or even what makes this person special. In [16] personality is defined as a style of behaviour.

In some Personality Theories, individual personality differences between persons are the main issue as stated in [5]. Nevertheless, personality theorists are also interested in what is common between people, i.e. their "internal structure", how a person can be "ensemble" and how a person "works". Most theorists try to explain personality and social dynamics in terms of a "soul", "consciousness", "super ego" or even "spirit", which are difficult to model and consequently difficult to program. So, if we are interested in programming a system in which some piece of software will represent a person with a personality, we must consider only those theories that use not so abstracts concepts.

One such model is Transactional Analysis (TA), which was created by Eric Berne [4] and it has been considered a very useful and practical tool by therapists all over the world [18]. Therefore, in this paper we focus in defining an internal structure of a robot based on TA that we will call Robot Personality (*RP*) which will be used to model personality.

Given that the trend in MAS research in general is to begin to model agents within a structural context, often using language and abstractions from the systems domain, one possible solution to this problem is to look into the systems domain itself for such 'well-defined' models and processes which could provide the framework required to examine macroscopic behaviour in MAS modeling and experimentation [9].

In [9] we propose the use of Stanford Beer's viable systems model (VSM) to complement modelling approaches. In this paper we present a combined TA & VSM model approach, where we use TA theory in order to define the psychological profile of the robot and the VSM is used to help explain the decisions made by the robot and how this impacts on its viability or its ability to stay alive.

This paper is organized as follows: next section presents previous work related to modelling personality, especially in multi-agent systems. In section 3 we present information about our proposed model. Section 4 presents the experiment done and finally, section 5 deals with our conclusions.

2 Similar Work

In [11] a computational model of personality is proposed, the purpose of the model is to implement non-intellectual functions of the human mind on computer systems. The personality model was formulated based on psychoanalysis.

In [7] the Transactional Analysis Theory is used in order to define the inner structure of an agent along with the negotiation process occurred among them until the cooperation is established.

In [8] a design of a rational agent based on decision theory is presented. Emotional states and personality are defined formally as a finite state automaton. Emotional states are considered as methods for decision making inside the agent.

Changes in external stimuli provoke changes of emotional state and, thus changes in agent's decision-making behaviour. Personality is defined as emotional states along with transition rules between states. Also, a probabilistic version is considered to model personality. An agent's personality can be predicted given an initial state and emotional inputs.

In [1][2] "affective agent"-user interface is considered. Personality is defined as a complex structure that distinguishes a person, nation or group. An emotion is defined as an affection that interrupts and re-directs attention (usually accompanied by a stimulus). In this model, the Five-Factor-Model (FFM) is used. The descriptive nature of FFM gives an explicit model of personality and makes it possible to concentrate in the use of the affective interface to express directly these characteristics. In this project, personality and emotions are used as filters that restrict decision-making process.

In [10] a basic structure for a dialog automata is described, which is used to model users. This structure tries to model psychological terms such as personality and emotions. This proposal is based on the analysis of finite states and offers a general perspective on which formal methods (algebraic in general) and results can be applied to a variety of problems.

In [15] a methodology is presented for the study of the mind as part of Artificial Intelligence. This paper presents an architecture for motivated agents; the architecture is composed by several modules that handle automatic processes in which reflexive administration of limited resources that include planning, decision-making, scheduling, etc. are involved, along with meta-administrative processes like internal perception and actions.

In [13] a synthetic character is built that generates emphatic behaviour based on cathexis flux.

In [6], an interaction model for decentralized autonomous systems based on Transactional Analysis is presented. Cooperation between agents is negotiated by stroke exchange. After collaboration is established, each autonomous system plays certain role trying to accomplish a specific goal. However, only parameter formalization and hints on how to use this structure are given, but no implementation is considered. On the next section, we consider this issue.

3 TA and VSM Personality Model

Definition 1. In our basic model a *robot's personality* is represented by a tuple

$$RP = < A, ES >$$

where:

1. A is the set of actions a robot can perform
2. $ES = \{P, A, C\}$. Ego States where P=Parent, A=Adult, C=Child

In the simplified model of TA used, each *ES* holds a series of rules that define how a *RP* should perceive the world under that particular *ES*. For each *ES*, rules are different inside *RP*, in the case of state *P*, rules are associated mainly with social behaviour and the way things should be done in each situation. The set of rules for state *P* can be considered the "cultural background" of the *RP*. For state *A*, rules are mainly about information gathering and problem solving mechanisms. Finally, for state *C*, rules are about postures as an individual, i.e. what makes this *RP* different from others. In this case we can consider things like favourite colour, games, etc. Pre-logical reasoning and spontaneous reaction are considered in this state (see Figure 1).

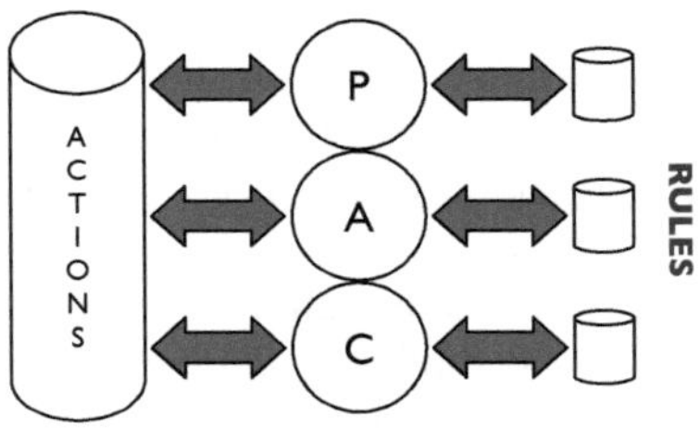

Fig. 1. Representing Rules-Actions and TA States

Another important concept is that of Personal Energy (*K*) which is equivalent (but not necessarily identical) to the cathexis concept in humans. Following Berne's model, *K* is divided in three parts: first, for each event that a *RP* perceives, the list of actions from *A* is scanned and assigned a weight for each state; this means that the same action will have three different weights, one for each state. This first value represents the Potential Energy (Kp) a particular event can give to an action for a *RP* in state X, i.e. how an event can trigger an action. As an example, consider action "To Play". For state *A*, Kp should be very low since playing is not an action a human would normally do when in this state. But for state *C*, Kp must be very high since this action is strongly related to this state. This assignment is independent for each state.

When considering which of the states will be the actual state that executes an action, we will consider the one that holds the maximum of Potential Energy, i.e. the highest Kp.

$$ES = max\{Kp(P), Kp(A), Kp(C)\} \tag{1}$$

Consider this: a *RP* perceives an event; this causes each of its *ES* to take the list *A* of actions and assign to each of them a weight. Then each *ES* sorts the list from higher to lower weight. After that, each *ES* takes the first action of its list and compares it with the action chosen by the other *ES*s. The one with highest Kp will be considered the actual *ES* (this means that the *RP* will be in that particular *ES*) and thus its action will be executed by *RP*. In Figure 2, action "To Work" of state *A* has the highest Kp and thus the *RP* will start working.

On the other hand, it must be taken into account that humans have a tendency to continue an activity, which we consider interesting or important, either because

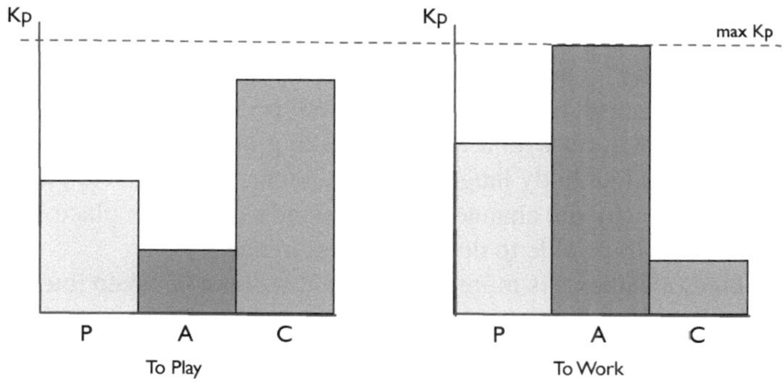

Fig. 2. Action "To Work" of state *A* has the highest Kp and thus the *RP* will start working

we feel obligated, it is convenient or we are just like it, and to discontinue activities that are no longer interesting or important. This kind of cathexis can be thought of as "the interest" a *RP* has and we will call it the Kinetic Energy Kc of the *RP* to executing an action. If we consider that, when a *RP* starts executing an action X, Kp(X) starts to be "consumed" (diminishing); but if there is an interest $Kc(X)$, this consumption will be compensated and thus, *RP* will execute X a longer period of time (see Figure 3a).

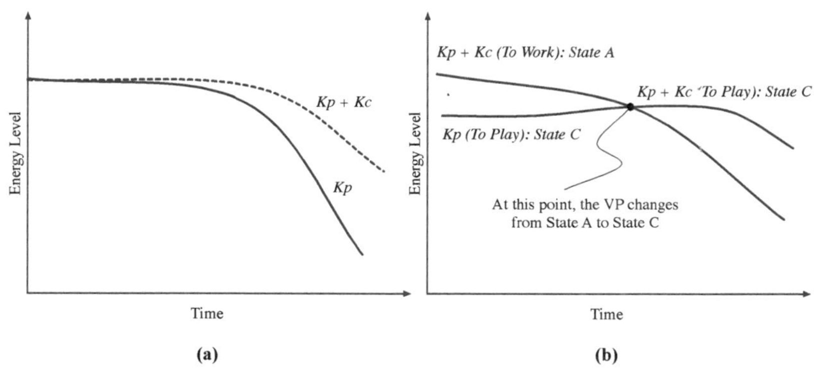

Fig. 3. Comparision between Potencial Energy (Kp) vs. Potencial Energy + Kinetic Energy ($Kp+Kc$) on action X

An interpretation of this can be as follows: each unit of time we must decide what to do, those actions that are most important will normally make us decide on them first, our interest on executing them will continue as long as we still consider them important. If for some reason the interest disappears, we will stop doing this action and turn to another. As can be seen in Figure 3b, as action "To Work" is executing, it is also being "consumed." When Kp("*ToPlay*") is higher than Kp("*ToWork*"), the *RP* will change from state *A* to state *C* and will start executing action "*ToPlay*".

Once a function to determine the "actual *ES*" is defined, it is possible to see the changes of *ES* of any *RP* and thus the possible action that will be executed, given a sense of "personality". On the other hand, the opposite can be considered, from events in the virtual world, to infer which *ES* a *RP* is in any moment. In fact, this is what a TA therapist will do with a client. By asking questions, hearing "the story" and some other signs like body language, expressions, attitudes, etc., he will try to infer when, how and why the changes in the patient's states take place. In this case, experience is crucial to be able to do a correct inference.

The third kind of cathexis is more complex and will not be taken into account for this paper.

3.1 Viable System Model

The Viable Systems Model looks at an organisation interacting with its environment. The organisation is viewed as two parts: the Operation which does all the basic work and the bits that provide a service to the Operation by ensuring the whole organisation works together in an integrated way. These bits are called the Metasystem.

Beer's first insight was to consider the human organism as three main interacting parts: the muscles & organs, the nervous systems, and the external environment.

These are generalised in the Viable Systems Model as follows:

- The Operation (O). The muscles and organs. The bits that do all the basic work. The primary activities. In RP the operation is *A*, and everything in the robot related to perform the actions in *A* correctly.
- The Metasystem (M). The brain and nervous systems.
- The Environment (E). All those parts of the outside world that are of direct relevance to the system in focus.

The following diagram illustrates the basic VSM.

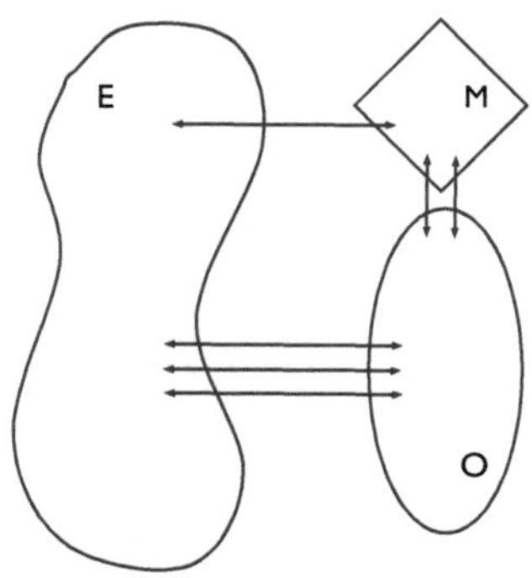

Fig. 4. Basic VSM

The arrows indicate the many and various ways that the three parts interact. The Operation will consist of a number of Operational units. The Operational units themselves must be viable, and thus can be looked at as smaller Viable Systems embedded in the larger system [3].

The main functions of the Metasystem are:

1. Look at the entire collection of Operational units and deal with ways of getting them to work together in mutually beneficial ways, and with the resolution of conflicts. This function is called "Internal Eye".
2. Look at the external environment, assess the threats and opportunities and make plans to ensure the organisation can adapt to a changing environment. This function is called "External Eye".
3. Establish the ground rules, which set the tone for the whole organisation. This is the Policy System. Is in the Policy System where the ES are located.

A need N_i takes the *RP* to the state S_i, i.e. the need of "food" takes the *RP* to the state "hungry". The Metasystem identify "needs" and following the rules imposed for the Policy System determine what action the Operation will perform. When the action is being executed another necessity can appear, this will take the robot to another state, i.e. in order to exit from "hungry" state the robot need food, the action related is "To Eat", but while the robot are eating the need to work appears, the robot will be then in the "Responsible" state.

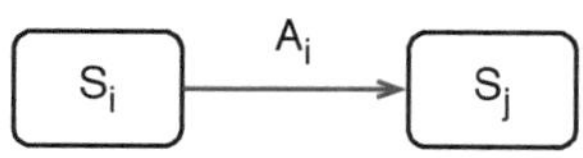

Fig. 5. Transition from state S_i to S_j due the execution of the action A_i

In the Figure 6 we can see an activities diagram of the action decision-making process in *RP*.

4 Case of Study

4.1 Experiment Protocol

The experiment begins with a new robot (*RP*) defined as "software being" that "lives" inside a virtual world. The robot needs to work to meet a specific goal. For work undertaken the robot earns the equivalent of monetary tokens, but the battery level decreases and also the performance level slowly depreciates. With the tokens earned, the robot can pay to recharge its batteries but must also pay for maintenance services to restore its performance levels to the state they were at the beginning of the simulation.

Besides parameters defined in section 3, we will use other parameters for these study cases:

$BatL(t) = \{1..100\}$: Battery Level. This parameter defines the amount of "physical energy" (different from cathexis) that a *RP* has a time t. This parameter will decrease by work and rest in order to simulate energy consuming, so $BatL(t) > BatL(t+1)$, if for some reason $BatL(t) = 0$, the *RP* will die, to avoid this the robot

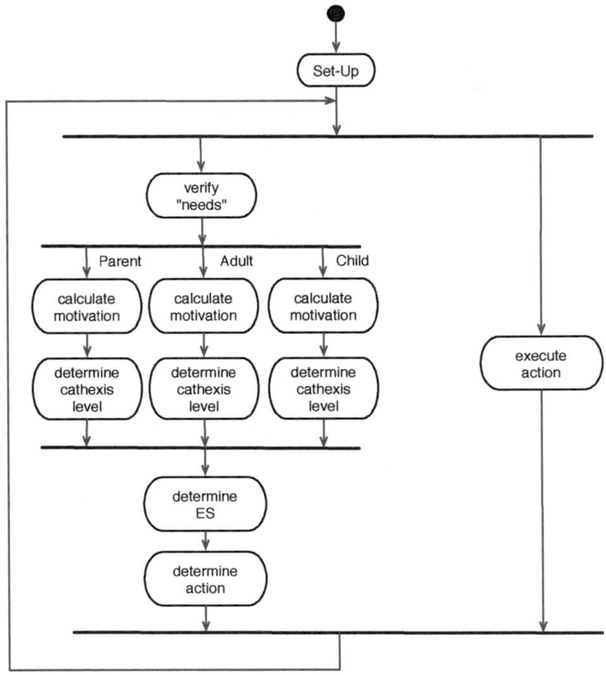

Fig. 6. Activities diagram

can pay to "eat", this means recharge its batteries. All *RP*s should estimate when they need to "eat" in order to keep alive.

$PerL(t) = 1..100$: Performance Level. This parameter defines the state of "Health Level" that a *RP* has a time t. This parameter will decrease by work and eat in order to simulate health depreciation, so $PerL(t) > PerL(t+1)$, as the performance level drops, the time taken to carry out the work to meet the goal takes longer, so the robot can pay to "rest", this means restore its performance level. The cost of the maintenance service is high and the time taken to undertake maintenance is lengthy, in comparison to charging the battery, which is relatively quick.

4.2 Goal of the Experiment

The goal of the experiment is to demonstrate that a robot with some kind of pathology, a lazy robot or a workaholic, will die earlier than a robot without such pathologies. The workaholic robot will not perform its own personal maintenance because of their intrinsic focus on their work. This will eventually lead to bad performance, which will ultimately affect the productivity of the robot because it is a workaholic who will spend more time working. Eventually the performance will be so low that the robot will not recharge its batteries and it will die.

4.3 Representation of Three Ego States

The program used to simulate the psychological states of the robot and to control the actions it takes based upon its decision-making, has been developed using LabView from National Instruments.

The psychological profile of each Ego State i.e. Parent, Adult and Child is defined by two physiological needs, namely the need to eat or acquire energy, the need to rest or undertake preventative maintenance and the social need to work or engage in meaningful activity.

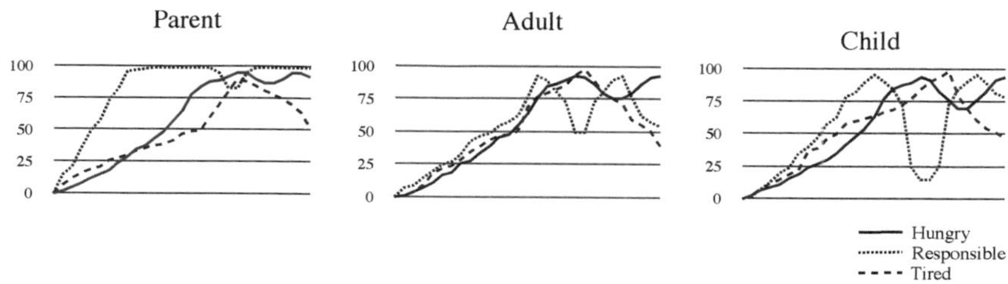

Fig. 7. Psychological profile of the robot

The user interface includes a set of controls to define the 'interest' that each ego state has for a particular need. The decision is taken in accordance to the "motivation" level at any particular moment in time. Each Ego State is plotted on a graph, which records the motivation level for undertaking an activity against time. At any one time, the level of each of the three states is presented to help the user to demonstrate the predominate state of the robot with respect to the three "needs" (Figure 8).

4.4 Control of Decision Making by the RP

The decision making process is controlled by the cathexis level, which are illustrated by three tanks (see Figure 8), each tank representing the cathexis level of each Ego State, that is the Parent, Adult and Child. The decision of which activity is to be performed is taken by the ego state with the highest level in the tank.

4.4.1 Fuzzy Inference System

Since the Kp is determinant in the decision making process of the *RP*, we define a FIS to calculate it, taking into account the personality profile (pp), the interest level in performing an action i and the need indicator that triggers action i, i.e. in the process of calculating the Kp for the ES Parent and the action "*ToEat*" the input is the pp, the interest level that the ES Parent has in 'eat' and the $BatL(t)$, then the output is the Kp for the action "*ToEat*".

Table 1 shows the linguistic values for the input variables of the FIS, they are $InterestLevel$ and $NeedLevel$, the personality profile (pp) is fixed for each simulation, describing the personality profile of a *RP*. Table 2 shows the linguistic variable Kp, the output of the FIS.

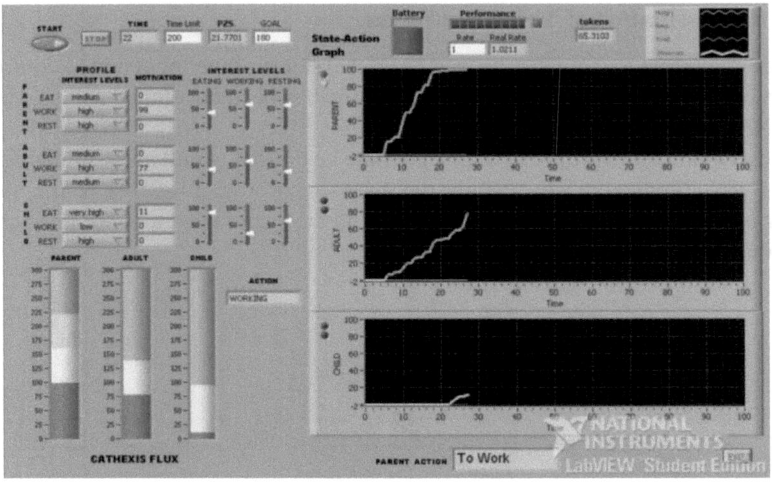

Fig. 8. Screenshot of the user interface

Table 1. Linguistic variables *InterestLevel* and *NeedLevel* and their Linguistic variables

Linguistic variables	Linguistic values				
InterestLevel	"veryLow"	"low"	"medium"	"high"	"veryHigh"
NeedLevel	"veryLow"	"low"	"medium"	"high"	"veryHigh"

Table 2. Linguistic variable Kp and their Linguistic variables

Linguistic variable	Linguistic values				
Kp	"veryLow"	"low"	"medium"	"high"	"veryHigh"

The membership functions of *InterestLevel*, *NeedLevel* and the output Kp are represented in Fig 9

For each action in each ES a FIS is defined. The rules for each one are the same. There are two cases when the *NeedLevel* determines the Kp without evaluating any other input, the first one is for the case when the *NeedLevel* becomes critical to the viability of the robot, e.g., the $BatL(t)$ is "veryLow", and the other one is for the case when the need becomes fulfilled, e.g. the $BatL(t)$ is "veryHigh".

If *NeedLevel* is "veryLow" then Kp is "veryHigh"
If *NeedLevel* is "veryHigh" then Kp is "veryLow"

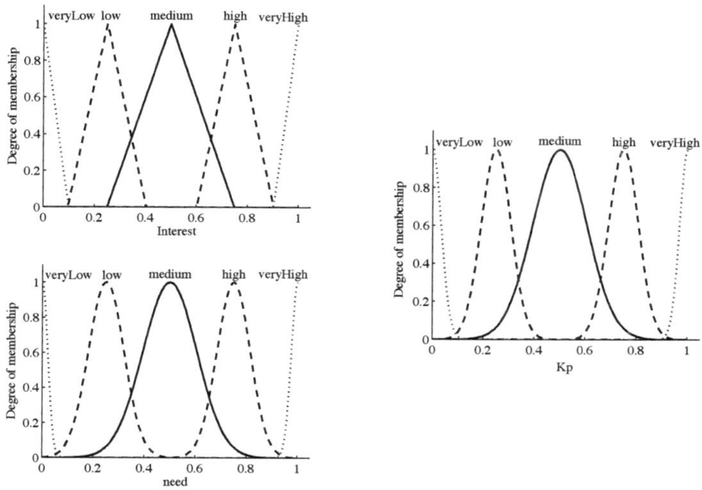

(a) (a)

Fig. 9. (a) antecedent MFs and (b) consecuent MF

An example of the rules that evaluate *InterestLevel* and *NeedLevel* are described below

If *NeedLevel* is "veryHigh" then *Kp* is "veryHigh"
If *NeedLevel* is "veryLow" then *Kp* is "veryLow"
If *InterestLevel* is "low" and *NeedLevel* is "medium" then *Kp* is "low"
If *InterestLevel* is "medium" and *NeedLevel* is "medium" then *Kp* is "medium"
If *InterestLevel* is "high" and *NeedLevel* is "medium" then *Kp* is "high"
If *InterestLevel* is "veryLow" and *NeedLevel* is "high" then *Kp* is "low"
If *InterestLevel* is "low" and *NeedLevel* is "high" then *Kp* is "medium"
If *InterestLevel* is "medium" and *NeedLevel* is "high" then *Kp* is "high"
If *InterestLevel* is "high" and *NeedLevel* is "high" then *Kp* is "high"

The response of the *RP* not only relies in the *InterestLevel* and *NeedLevel* but in the personality profile as well which is taken in to account in the *Kp* calculation. The Fig 10 shows the overall input-output surface.

4.5 Indicators of 'Need'

In addition to the controls to define the profile, the indicators to show the cathexis level and the graphs to show the motivation level, there are three indicators to show the battery level associated with the necessity to eat, the performance level associated with the necessity to rest and a counter to show the work undertaken by the robot. In addition, there is a goal indicator that is, how much work the robot needs to do for a given period of time, defined for the indicator labelled as Time Limit.

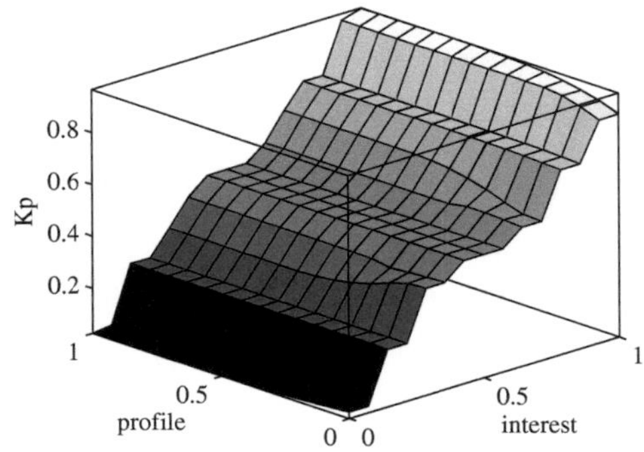

Fig. 10. Overall input-output surface

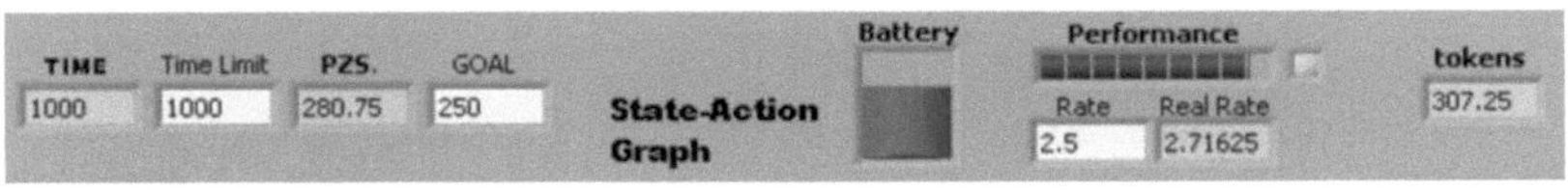

Fig. 11. Indicators of 'Need'

4.6 Robot Behaviour

In the Figure 12 we show a screenshot of the result of the simulation of a robot with a "high" concern in "work", and "medium" concern in "eat", as a result of this compulsion to "work" the robot die early due starvation.

We can identify this *RP* as a workaholic robot that keeps working without taking sufficient time out for maintenance, and as it keeps working the performance will gradually decrease, the time to achieve the goal will extend and the stress levels to

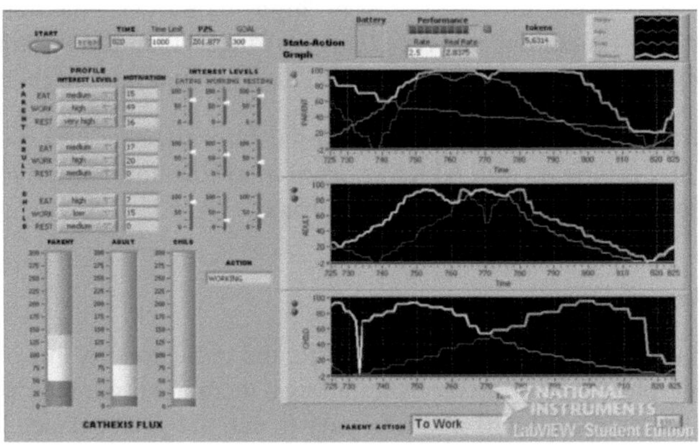

Fig. 12. Screenshot of the simulation of a "workaholic" robot

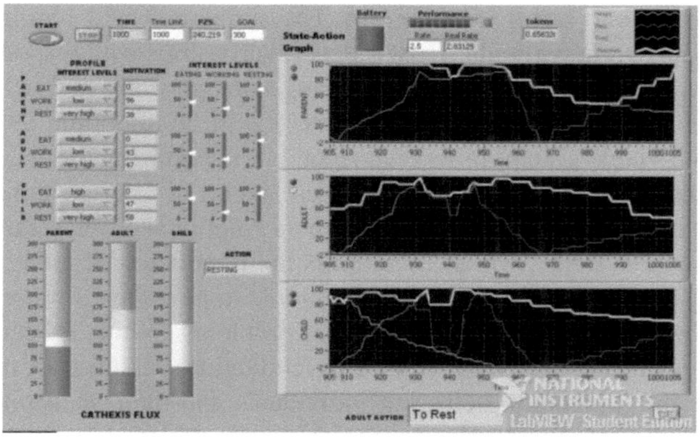

Fig. 13. Screenshot of the simulation of a "workaholic" robot

keep working will remain high. This behaviour will eventually cause the robot to collapse.

In Figure 13 we show a screenshot of the result of the simulation of a robot with a "very high" concern in "rest", and "low" concern in "work", as a result of this the robot only work when the necessity of "eat" or "rest" appear, in consequence the tokens earned at the end of the given period of time is extremely low. We can identify a robot with this *RP* as a lazy robot.

5 Conclusions and Future Work

With the working simulation the aim will be to explore how a psychological pathology as represented in the robot profile, will impact upon its decision making, its ability to achieve the goals set, its responses to critical viability criteria and ultimately its 'existence' as a working, self-sustaining robot and viable system.

In this experiment the psychological profile of the robot as defined by the three ego states described in TA is directly linked to its goal seeking behaviour as a viable system. Looking more broadly at the rationale behind this combination of a cognitive model with a systems model, the intention is to conduct experiments to help understand social phenomena.

Clearly the experiment attempts to simulate a cognate being which can achieve goals and maintain viability. However the model is relatively simple, the robot is only aware of its own needs and is functioning in a very simple environment that has limited perturbations and lacks the complexity of real social systems. The work will look to extend the simulation to gradually increase the complexity and in doing so, gradually improve on our understanding on how the decisions are taken in order to maintain a viable existence.

Conscious of the need to consider 'context' in MAS systems, future work must involve the interaction of robots in a constrained 'world', similar to a small family where individuals at different stages of maturity e.g. father and son whose manifestations of ego states are different. The simulation will require the robots to communicate and in particular must reflect relationships with responsibilities for others and how these interactions lead to changes in ego states and in the decisions that individual robots make relative to others. It will be important to consider the robot both as a viable system in its own right, but recognising the recursive nature of systems, the robot must also be considered as part of a larger family of robots with whom it has dependencies which ensure its emotional or physical well being.

In large social systems, which encompass the population of cities like Tijuana for example, social science researchers need tools to help them find or identify the mechanisms which drive and underpin social problems. While TA can help social researchers understand behaviour from a psychological perspective, the VSM and its more systemic paradigm, helps put individuals and their decision-making behaviour into a context with clearly recognisable characteristics that can be identified and analysed. This powerful combination of the cognitive with the systemic and structural should help represent social entities both at the level of the individual and at an organisational or social level - effectively a more natural simulation of a social system. Given the complexity of large urban populations like Tijuana we need an approach which can take an account of the context of an individual in a social system and at the same time take account of social interaction recognising that the two are part of the same communication phenomena.

Acknowledgement. The authors would like to thank CONACYT and UABC for the financial support given to this project. The students (Carelia Gaxiola and Dora-Luz Flores) were supported by a scholarship from CONACYT.

References

1. André, E., Klesen, M., Gebhard, P., Allen, A., Rist, T.: Exploiting Models of Personality and Emotions to Control the Behavior of Animated Interface Agents. In: Proceedings of the Workshop on Achieving Human-Like Behavior in Interactive Animated Agent in conjunction with the Fourth International Conference on Autonomous Agents, Barcelona, pp. 3–7 (2000)
2. André, E., Klesen, M., Gebhard, P., Allen, A., Rist, T.: Integrating Models of Personality and Emotions into Lifelike Characters. In: Proceedings of the Workshop on Affect in Interactions Towards a new Generation of Interfaces in conjunction with the 3rd i3 Annual Conference, Siena, Italy, pp. 136–149 (1999)
3. Beer, S.: Diagnosing the System for Organizations. John Wiley & Sons, London (1985)
4. Berne, E.: Principles of group treatment. Oxford University Press, New York (1964)
5. Boeree, C.G.: Teorias de la Personalidad (2001), `http://www.ship.edu/~cgboeree/introduccion.html`
6. Cheng, Z., Capretz, A.M., Osano, M.: A model for negotiation among agents based on the transaction analysis theory. In: Proceedings of Second International Symposium on Autonomous Decentralized Systems, ISADS 1995, April 25-27, pp. 427–433 (1995)
7. Cheng, Z., Capretz, A.M., Osano, M.: A Model for Negotiation among Agents based on the Transaction Analysis Theory. In: Proceedings of the Seventh International Workshop on Computer-Aided Software Engineering, Washington, p. 427 (1995)

8. Gmytrasiewicz, P.J., Lisetti, C.: Emotions and Personality in Agent Design and Modeling. In: Bauer, M., Gmytrasiewicz, P.J., Vassileva, J. (eds.) UM 2001. LNCS (LNAI), vol. 2109, p. 237. Springer, Heidelberg (2001)
9. Jones, S.J., Rodriguez-Diaz, A., Hall, L., Castanon-Puga, M., Flores-Gutierrez, D.L., Gaxiola-Pacheco, C.: A cybernetic approach to multi-agent system simulation in Tijuana-San Diego using the Viable Systems Model. In: IEEE International Conference on Systems, Man and Cybernetics, ISIC 2007, October 7-10, pp. 1648–1652 (2007)
10. Kopecek, I.: Personality and Emotions - Finite State Modeling by Dialogue Automata. In: Proceedings of UM 2001 Workshop on Attitudes, pp. 1–6. University of Bari (2001)
11. Nitta, T., Tanaka, T., Nishida, K., Inayoshi, H.: Modeling Human Mind. Systems, Man, and Cybernetics. In: 1999 IEEE International Conference on IEEE SMC 1999 Conference Proceedings, October 12-15, vol. 2, pp. 342–347 (1999)
12. Ovum Report Intelligent agents: the new revolution in software (1994)
13. Rodríguez-Díaz, A., Cristóbal-Salas, A., Castanón-Puga, M., Jáuregui, C., González, C.: Personality and Behaviour Modelling Based on Cathexis Flux. In: Proceedings of the Joint Symposium on Virtual Social Agents AISB 2005: Social Intelligence and Interaction in Animals, Robots and Agents, ED. AISB The Society for the Study of Artificial Intelligence and the Simulation of Behaviour UH, pp. 130–136. The University of Hertfordshire (2005)
14. Sargent, P.: Back to school for a brand new ABC, in The Guardian, p. 28 (1992)
15. Slogan, A.: What Sort of Control System is able to have a Personality. In: Proceedings Workshop on Designing Personalities for Synthetic Actors, Vienna (1995)
16. Velasquez, J.D., Maes, P.: Cathexis: A Computational Model of Emotions. In: Proceedings of the First International Conference on Autonomous Agents, pp. 518–519 (1997)
17. Virtual Person Project, Virtual Persons, Virtual Worlds, Sponsored by The PT-Project at Illinois State University (2003), `http://www.ptproject.ilstu.edu/vp/welcome.htm`
18. Transactional Analysis Journal Internet, Science and TA, `http://www.tajnet.org/articles/boyd-science-and-ta.html`

Part IV
Robotics and Hardware Implementations

Controlling Unstable Non-Minimum-Phase Systems with Fuzzy Logic: The Perturbed Case

Nohe R. Cazarez-Castro[1], Luis T. Aguilar[2], Oscar Castillo[3], and Antonio Rodríguez-Díaz[1]

[1] Universidad Autonoma de Baja California, Tijuana, BC, Mexico
nohe@ieee.org, ardiaz@uabc.mx
[2] Instituto Politectico Nacional, Mexico
luis.aguilar@ieee.org
[3] Instituto Teconlogico de Tijuana, Mexico
ocastillo@hafsamx.org

Abstract. In this paper Fuzzy Logic Systems (FLS) for controlling non-minimum-phase systems are proposed. A generalized Proportional Integral Derivative (PID) Fuzzy Logic Controller (FLC) for a benchmarking second order problem with an unstable zero is presented. The same Fuzzy Rule Base (FRB) of the PID FLC is used in a PD FLC to regulate a plant consisting in a non-minimum-phase servomechanism with nonlinear backlash. Simulations demonstrate that the proposed FLC can be used to handle the non-minimum-phase systems.

1 Introduction

There exist several papers dealing with the stabilization of non-minimum-phase systems. For example in [7], a simple design method by means of which it is possible to robustly stabilize, using output feedback, a significant class of uncertain nonlinear systems whose zero dynamics are unstable, the proposed procedure leads to the construction of a dynamic controller yielding robust, semi-global practical stability.

In [1] nonlinear H_∞ control synthesis is extended to an output regulation problem for a servomechanism with backlash. The problem in question is similar to one of the problems proposed in this paper, the design a feedback controller so as to obtain the closed-loop system in which all trajectories are bounded and the load of the driver is regulated to a desired position while also attenuating the influence of external disturbances.

On the other hand, Computational Intelligent (CI) strategies have been used to in order to solve the non-minimum-phase stabilization control problem. For example in [14] a fuzzy logic controller (FLC) is developed for the nonminimum phase system. The well-designed fuzzy rules are exploited to resolve the undershoot problem caused by the unstable zeros. A 3rd-order plant with two unstable zeros is used to verify the performance of the fuzzy controlled system.

O. Castillo et al. (Eds.): Evolutionary Design of Intelligent Systems, SCI 257, pp. 245–257.
springerlink.com

A fuzzy logic based approach is proposed in [2] for the control of non-minimum phase systems. In this approach, a variable structure controller can be designed using fuzzy logic and linguistic control rules implementation.

In [3] an adaptive fuzzy logic system is incorporated with the Variable Structure Control (VSC) system for the purpose of improving the performance of the control system. A sliding surface with an additional tunable parameter is defined as a new output based on the idea of output redefinition, as a result the overload system of missile with the characteristic of non-minimum phase can be transformed into minimum-phase system by tuning the parameters of the sliding surface, and a sliding-mode controller can be designed.

In [5] and [8], an hybridization of FLS and Genetic Algorithms (GA) [6] is used in order to control non-minimum-phase systems, by finding the optimal parameter, of each MF in a FLS.

This paper addresses the problem of designing FLS to control non-minimum-phase systems, first for non-minimum-phase Linear Time Invariant (LTI) [13] class of systems, and then for the design a feedback controller so as to obtain the closed-loop system in which all trajectories are bounded and the load of the driver is regulated to a desired position while also attenuating the influence of external disturbances, where the provided servomotor position is the only measurement available for feedback. Performance issues of the proposed FLC output regulator constructed are illustrated in a simulation study. In the rest of this paper a non-minimum-phase system will be considered as defined in [13]: a system that has at least one positive (unstable) zero.

The paper is organized as follows: Section 2 present FLS theoretical aspects, and the designing of a FLC for controlling non-minimum-phase systems, Section 3 presents the configuration as PID FLC of the FLC designed in Section 2 to control a non-minimum-phase LTI system with an unstable zero, Section 4 presents the configuration as PD FLC of the FLC designed in Section 2 for the output regulation of a servomechanism with non-minimum-phase backlash, and finally, in Section 5 we present our conclusions.

2 Fuzzy Logic Control

A FLS is a numerical system that makes a nonlinear mapping from input to output data. A FLS consists of four basic elements (see Fig.1): the ***fuzzifier***, the ***fuzzy rule-base***, the ***inference engine***, and the ***defuzzifier***. The ***fuzzy rule-base*** is a collection of rules in the form of (2), which are combined in the ***inference engine*** to produce a fuzzy output. The ***fuzzifier*** maps the crisp input into Fuzzy Sets (FS), which are subsequently used as inputs to the ***inference engine***, whereas the ***defuzzifier*** maps the FS produced by the ***inference engine*** into crisp numbers.

A FS can be interpreted as a MF U_x that associates with each element of x of the universe of discourse, U, a number $\mu_X(x)$ in the interval [0,1]:

$$\mu_x : U \to [0,1]. \tag{1}$$

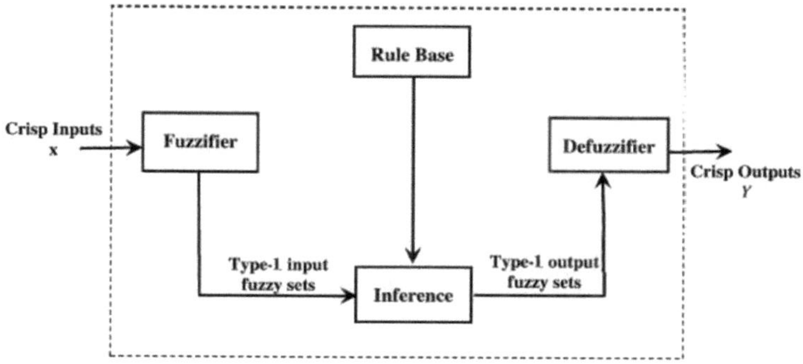

Fig. 1. Structure of Fuzzy Logic System.

2.1 FLC Design

To solve the regulation problem for non-minimum-phase systems, we propose two-input one-output rules in the formulation of the knowledge base. The IF-THEN rules are according with the Mamdani type of Fuzzy Inference Systems [10][9][11]:

$$\text{IF } y_1 \text{ is } A_1^l \text{ AND } y_2 \text{ is } A_2^l \text{ THEN } y_3 \text{ is } B^l \tag{2}$$

where $[y_1, y_2]^T = y \in U = U_1 \times U_2 \subset \mathbb{R}^2$, where y_1 is the fuzzified value of a measured variable, y_2 is the fuzzified value of the derivative of the measured variable and $y_3 \in V \subset \mathbb{R}$. For each input fuzzy set A_k^l in $y_k \subset U_k$ with $k = 1,2$; and output fuzzy set B^l in $y_3 \subset V$ exists an input membership function $\mu_{A_k^l}(y_k)$ whit $k = 1,2$, and output membership function $\mu_{B^l} \in y_3 \subset V$, respectively, with l being the number of membership functions associated to the input k.

The particular choice of each $\mu_{B^l}(y_3)$ will depend on the heuristic knowledge of the experts over the plant.

We select triangular membership functions for each input (*error* and *change of error*) and output (*control*) variables, granulating each one of these three variables

Table 1. Fuzzy rules

No.	***error***	***change of error***	***control***
1	n	n	p
2	n	z	p
3	n	p	z
4	z	n	z
5	z	z	z
6	z	p	z
7	p	n	z
8	p	z	n
9	p	p	n

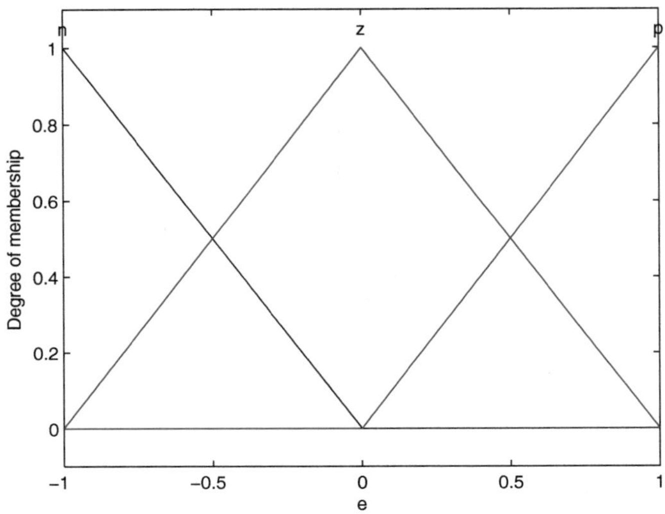

Fig. 2. Input variable *error*.

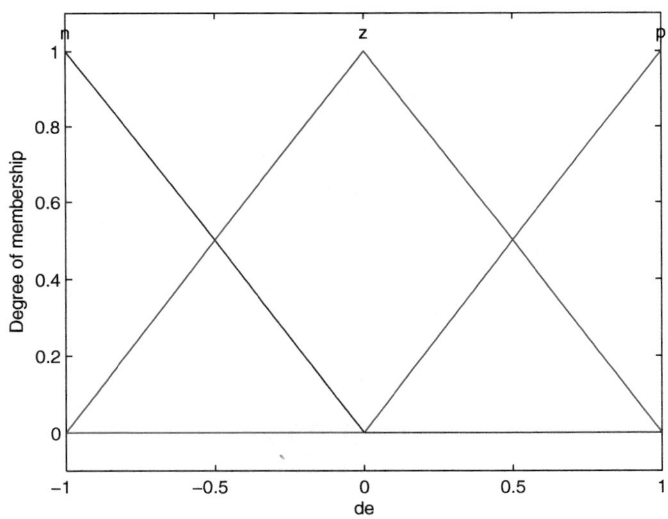

Fig. 3. Input variable *change of error*.

in three fuzzy sets: *negative* (**n**), *zero* (**z**) and *positive* (**p**). The shape of the variables can be seen in Fig. 2, 3 and 4 respectively.

These input and output variables are combined in a FRB in the form of (2), and we select the fuzzy rules shown in Table 1.

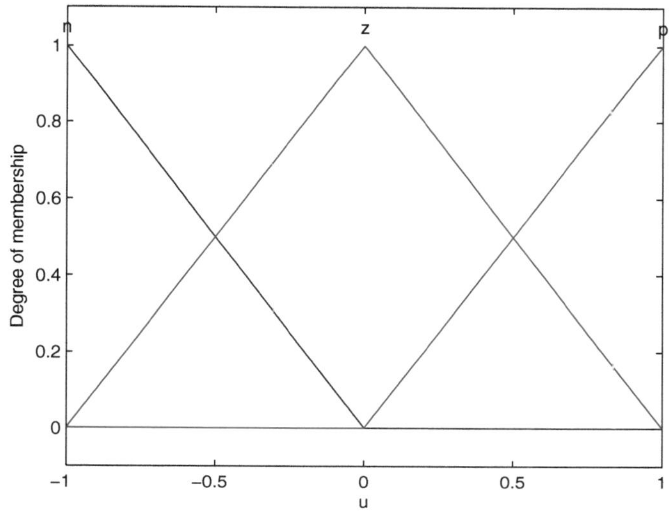

Fig. 4. Input variable *control*.

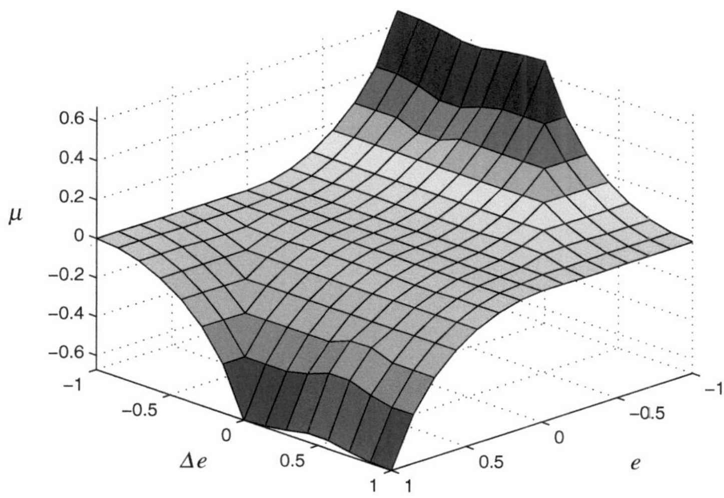

Fig. 5. Surface of control.

For the inference process, we implement the Mamdani [10][9] type of Fuzzy Inference, with *minimum* as disjunction operator, *maximum* as conjunction operator, *minimum* as implication operator, *maximum* as aggregation operator and Centroid (COA) [4] as our defuzzification method.

Defuzzification refers to the way a numeric value is extracted from a fuzzy set as a representative value. In general, the defuzzification methods take a fuzzy set B of a universe of discourse V, where B is usually represented by an aggregated output membership function.

The COA defuzzication method is expressed as

$$\tau_m = \frac{\int_{y_3} \mu_B(y_3) y_3 dy_3}{\int_{y_3} \mu_B(y_3) dy_3}, \tag{3}$$

where $\mu_B(y_3)$ is the aggregated output MF. This is the most widely adopted defuzzification strategy, which is reminiscent of the calculation of expected values in probability distributions.

The FLS is builded with the FRB of Table 1 and triangular MF of Figs. 2, 3 and 4, this FLS outputs the control surface of Fig. 5.

3 Controlling a LTI System with an Unstable Zero

The LTI [13] systems, are systems that can be represented by ordinary differential equations.

A LTI system that has a positive zero is called a non-minimum-phase system, the fact that a system has a positive zero means that the zero is unstable.

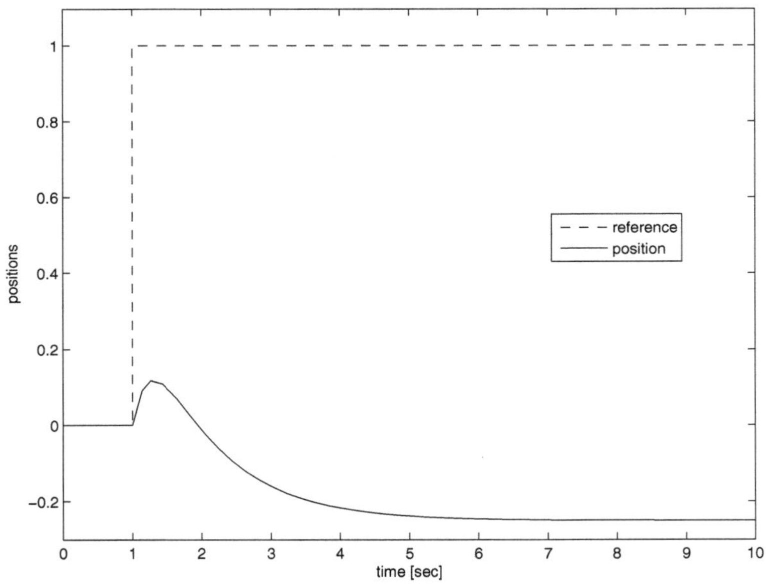

Fig. 6. System (4) response to the unitary step.

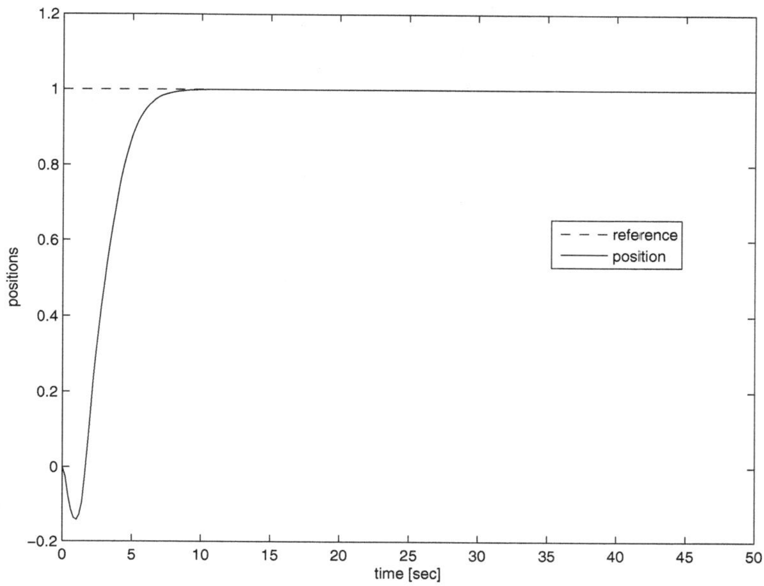

Fig. 7. Closed-loop system response for system (4).

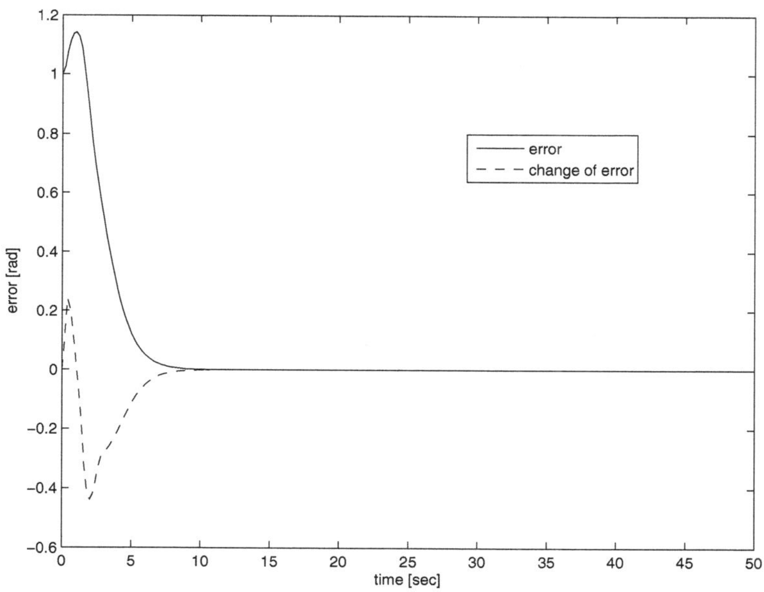

Fig. 8. Behavior of input variables *error* and *change of error* in the closed-loop system for system (4).

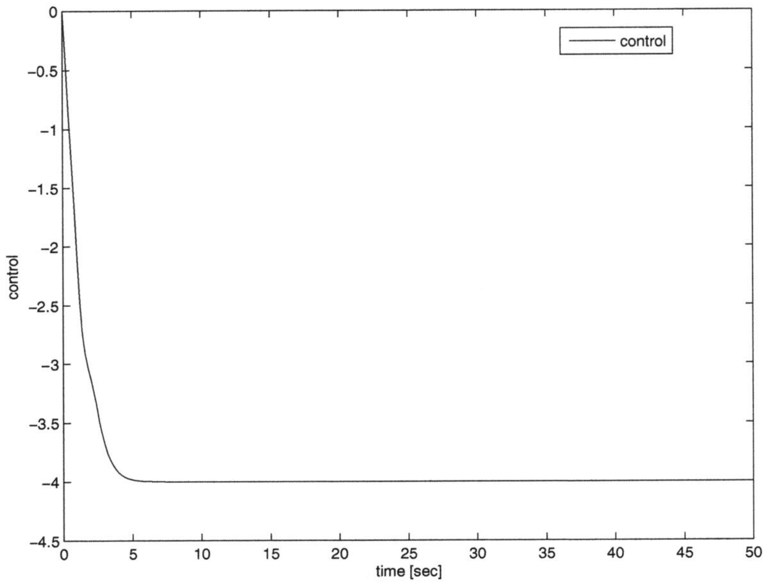

Fig. 9. Control signal applied to system (4) in the closed-loop.

Let us consider the following non-minimum-phase system:

$$H(s) = \frac{(s-1)}{s^2+5s+4}. \tag{4}$$

The open-loop system response of (4) to an unitary step is depicted in Fig. 6.

In order to regulate system (4) in a set-point of 1, we implement a Fuzzy PID Controller with the control law (3), with (2), FRB of Table. 1 and the MF parameters proposed in Section 2. The closed-loop systems' response is in Fig. 7, the behavior of the input variables *error* and *change of error* is in Fig. 8, and the control signal applied is in Fig. 9. Considering a settling criterion of $\pm 1\%$, the closed-loop system achieves the control objective in 5.4 seconds.

4 Controlling a Servomechanism with Non-Minimum-Phase Backlash

In order to prove the robustness of the proposed FLC a different problem is proposed: *the output regulation problem of a servomechanism with non-minimum-phase backlash.*

4.1 Dynamic Model

The dynamic models of the angular position $q_i(t)$ of the DC motor and the $q_0(t)$ of the load are given according to

$$\begin{aligned} J_0 N^{-1}\ddot{q}_0 + f_0 N^{-1}\dot{q}_0 &= T + w_0 \\ J_i\ddot{q}_i + f_i\dot{q}_i + T &= \tau_m + w_i \end{aligned} \tag{5}$$

hereafter, J_0, f_0, $\ddot{q}_0$ and $\dot{q}_0$ are, respectively, the inertia of the load and the reducer, the viscous output friction, the output acceleration, and the output velocity. The inertia of the motor, the viscous motor friction, the motor acceleration, and the motor velocity denoted by J_i, f_i, $\ddot{q}_i$ and $\dot{q}_i$, respectively. The input torque τ_m serves as a control action, and T stands for the transmitted torque. The external disturbances $w_i(t)$, $w_0(t)$ have been introduced into the driver equation (5) to account for destabilizing model discrepancies due to hard-to-model nonlinear phenomena, such as friction and backlash.

The transmitted torque T through a backlash with an amplitude j is typically modeled by a dead-zone characteristic [12, p.7]:

$$T(\Delta q) = \begin{cases} 0 & |\Delta q| \le j \\ K\Delta q - Kj\text{sign}(\Delta q) & \text{otherwise} \end{cases} \tag{6}$$

with

$$\Delta q = q_i - Nq_0, \tag{7}$$

where K is the stiffness, and N is the reducer ratio. Such a model is depicted in Fig. 10. Provided the servomotor position $q_i(t)$ is the only available measurement on the system, the above model (5)-(7) appears to be non-minimum-phase because along

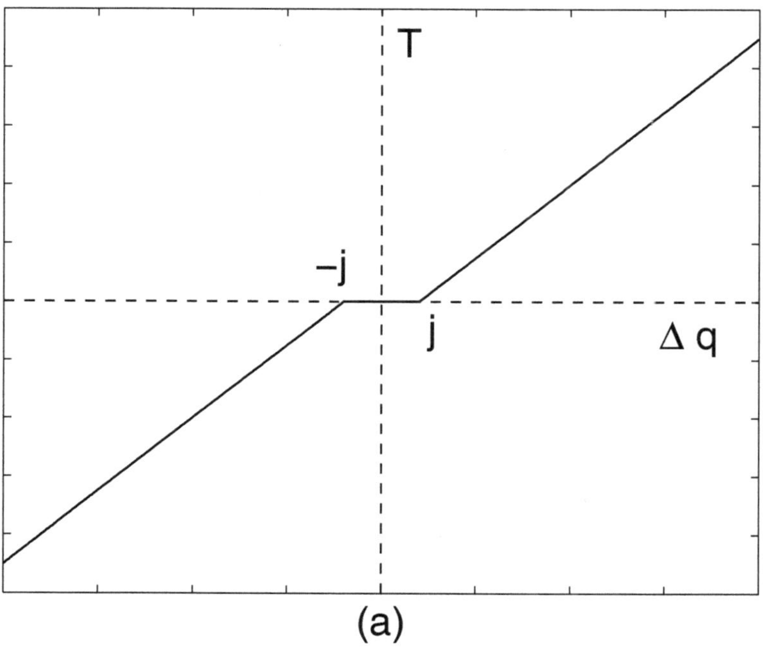

Fig. 10. The dead-zone model of backlash.

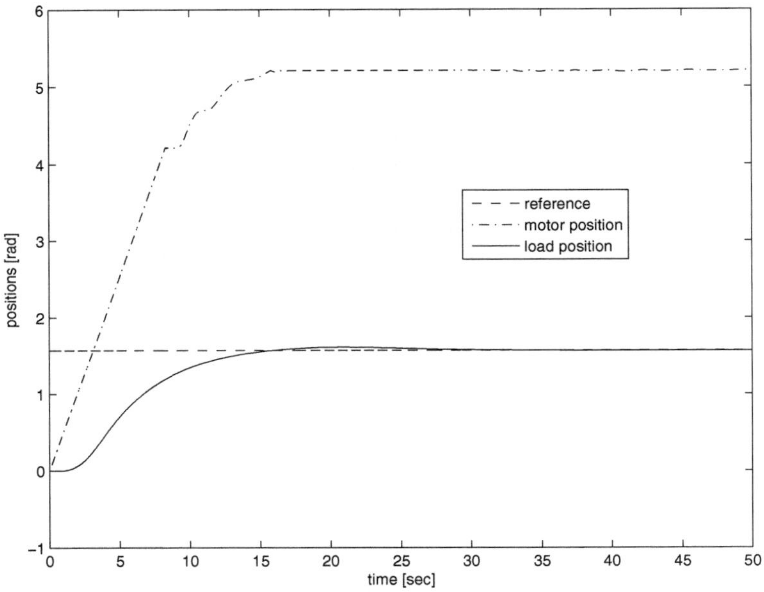

Fig. 11. Closed-loop system response for system (5)-(8).

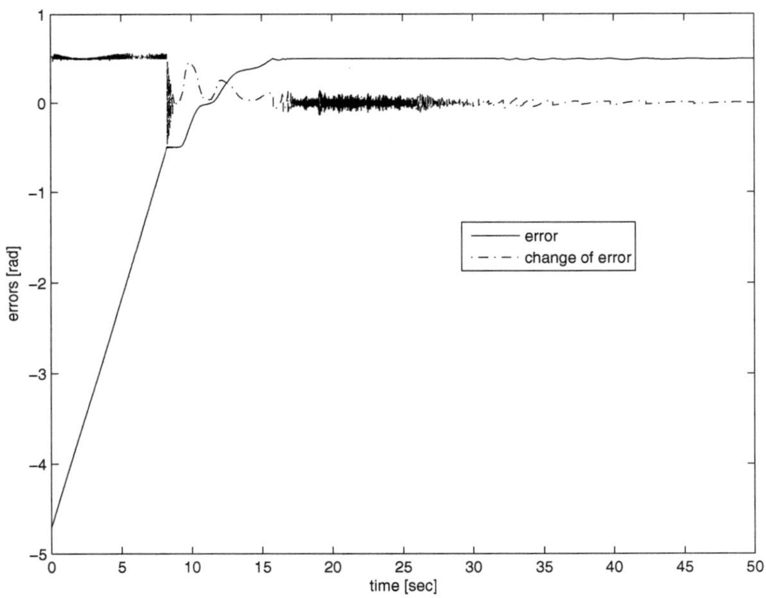

Fig. 12. Behavior of input variables *error* and *change of error* in the closed-loop system for system (5)-(8).

with the origin the unforced system possesses a multivalued set of equilibria (q_i, q_0) with $q_i = 0$ and $q_0 \in [-j, j]$.

4.2 Problem Statement

To formally state the problem, let us introduce the state deviation vector $x = [x_1, x_2, x_3, x_4,]^T$ with

$$\begin{aligned} x_1 &= q_0 - q_d \\ x_2 &= \dot{q}_0 \\ x_3 &= q_i - Nq_d \\ x_4 &= \dot{q}_i \end{aligned}$$

where x_1 is the load position error, x_2 is the load velocity, x_3 is the motor position deviation from its nominal value, and x_4 is the motor velocity. The nominal motor position Nq_d has been pre-specified in such a way to guarantee that $\Delta q = \Delta x$, where

$$\Delta x = x_3 - N x_1,$$

and the output is given by

$$y = x_3. \tag{8}$$

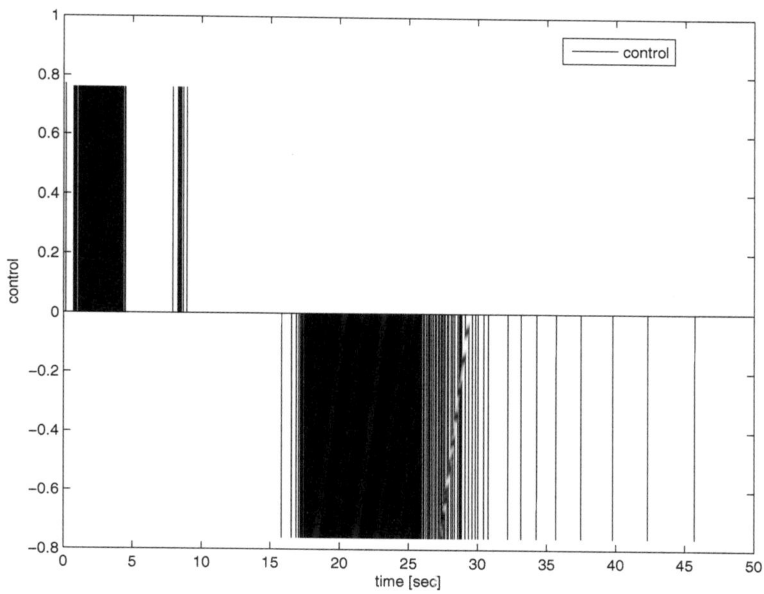

Fig. 13. Control signal applied to system (5)-(8) in the closed-loop.

The objective of the Fuzzy Control output regulation of the nonlinear driver system (5) with backlash (6), is thus to design a Fuzzy Controller so as to obtain the closed-loop system in which all these trajectories are bounded and the output $q_0(t)$ asymptotically decays to a desired position q_d as $t \to \infty$ while also attenuating the influence of the external disturbances $w_i(t)$ and $w_0(t)$.

4.3 Simulation Results

In order to regulate system (5)-(8) in a desired set-point of $q_d = \pi/2$, we implement a Fuzzy PD Controller with the control law (3), with (2), FRB of Table. 1 and MF parameters proposed in Section 2. Fig. 11 shows the closed-loop systems response, the behavior of the input variables *error* and *change of error* is in Fig. 12, and the control signal applied is in Fig. 13. Considering settling criterion of $\pm 1\%$, the closed-loop system achieves the control objective in 14.5 seconds.

5 Conclusions

The main goal of this paper was to show that a FLC is a good strategy to control non-minimum-phase systems. This goal was achieved designing the FLC of Section 2. This FLC was configured in a close-loop system in to different control problems:

- the regulation of a LTI non-minimum-phase system, and
- the output regulation of a servomechanism with non-minimum-phase backlash.

In the first problem, regulation of a LTI non-minimum-phase system, the proposed FLC was configured as a Fuzzy PID controller in a closed-loop system, achieving the control objective satisfactory.

In the second problem of the output regulation of a servomechanism with non-minimum-phase backlash, the proposed FLC was configured as a Fuzzy PD controller in a closed-loop system, achieving the control objective in a satisfactory manner.

The simulations presented in this paper show that the proposed FLC is robust, that is, the FRB and parameters of membership functions can be configured in different ways to achieve different control objectives, furthermore, this same FLC can be used in different control problems that do not include the non-minimum-phase phenomenon.

References

1. Aguilar, L.T., Orlov, Y., Cadiou, J.C., Merzouki, R.: Nonlinear H_∞-output regulation of a nonminimum phase servomechanism with backlash. Journal of Dynamic Systems, Measurement, and Control 129(4), 544–549 (2007), `http://link.aip.org/link/?JDS/129/544/1`
2. Antic, D., Dimitrijevic, S.: Non-minimum phase plant control using fuzzy sliding mode. Electronics Letters 34(11), 1156–1158 (1998)

3. Bao, Y., Du, W., Tang, D., Yang, X., Yu, J.: Adaptive Fuzzy Sliding-Mode Control for Non-minimum Phase Overload System of Missile. In: Intelligent Control and Automation, 1st edn. LNCIS, vol. 344, pp. 219–228. Springer, Heidelberg (2006)
4. Castillo, O., Melin, P.: Type-2 Fuzzy Logic: Theory and Applications, 1st edn., vol. 223. Springer, Heidelberg (2008)
5. Chen, P.C., Chiang, W.L., Chen, C.W., Tsai, C.H.: Adaptive fuzzy controller for nonlinear systems via genetic algorithm. In: AEE 2008: Proceedings of the 7th WSEAS International Conference on Application of Electrical Engineering, pp. 71–76. World Scientific and Engineering Academy and Society (WSEAS), Stevens Point (2008)
6. Holland, J.M.: Adaptation in Natural and Artificial Systems. University of Michigan Press, Ann Arbor (1975)
7. Isidori, A.: A tool for semi-global stabilization of uncertain non-minimum-phase nonlinear systems via output feedback. IEEE Transactions on Automatic Control 45(10), 1817–1827 (2000)
8. Li, T.-H.S., Shieh, M.-Y.: Design of a ga-based fuzzy pid controller for non-minimum phase systems. Fuzzy Sets Syst. 111(2), 183–197 (2000), `http://dx.doi.org/10.1016/S0165-0114(97)00404-1`
9. Mamdani, E.H.: Advances in the linguistic synthesis of fuzzy controllers. J. Man- Machine Studies 8, 669–678 (1976)
10. Mamdani, E.H., Assilian, S.: An experiment in linguistic synthesis with fuzzy logic controller. J. Man- Machine Studies 7(1), 1–13 (1975)
11. Mamdani, E.H., Assilian, S.: An experiment in linguistic synthesis with a fuzzy logic controller. Int. J. Hum.-Comput. Stud. 51(2), 135–147 (1999)
12. Nordin, M., Bodin, P., Gutman, P.O.: New Models and Identification Methods for Backlash and Gear Play. In: Adaptive Control of Nonsmooth Dynamic Systems, pp. 1–30. Springer, Berlin (2001)
13. Ogata, K.: Modern Control Engineering. Prentice Hall PTR, Upper Saddle River (2001)
14. Tsai, C.Y., Li, T.H.S.: Fuzzy logic control of non-minimum phase system. In: IEEE World Congress on Computational Intelligence, Proceedings of the Third IEEE Conference on Fuzzy Systems, 1994, vol. 1, pp. 199–204 (1994), doi:10.1109/FUZZY.1994.343686

Genetic Optimization for the Design of Walking Patterns of a Biped Robot

Selene L. Cardenas-Maciel[1], Oscar Castillo[2], Luis T. Aguilar[3], and Antonio Rodríguez-Díaz[1]

[1] Universidad Autonoma de Baja California, Tijuana, BC, Mexico
lilettecardenas@ieee.org, ardiaz@uabc.mx
[2] Instituto Tecnologico de Tijuana, Mexico
ocastillo@hafsamx.org
[3] Instituto Politecnico Nacional
Mexico luis.aguilar@ieee.org

Abstract. This paper presents an Genetic Algorithm (GA) for the design of walking patterns of a 3-DOF biped robot formulated as a system with impulse effects. The GA optimizes the coefficients of a polynomial which represents the desired behavior of the walking which is included into the dynamics of the biped robot to obtain periodic motions while fulfills a minimal energy consumption over a complete walking cycle assumed as single support and instantaneous double support phases. Optimization results are presented showing walking patterns with low energy consumption and periodic motions.

1 Introduction

In the last years several results has been provided for the stabilization of the biped locomotion, which usually are based on find a suitable reference trajectory i. e. walking pattern such that guarantee to perform stable steps while obey a certain criteria such as walking velocity, step length, step frequency, including energy minimization [16], [17]. Recent methods proposed for the walking pattern synthesis incorporate the zero moment point approach (ZMP) [18] as stability criteria. Qiang et. al. in [9] considering the ground condition and walking patterns with small torque and velocity of the joint actuators are obtained. Capi et. al. in [3] generate the joint trajectories for walking and going up-stairs based on consumed energy and torque change using evolutionary computation methods. Denk et. al. in [5] propose a systematic approach for the design of a walking pattern database using optimal control techniques and ZMP for prevention of the foot rotation and ensure the feasibility of the walking; Kajita et. al. in [11] introduce a method for a walking pattern generation by using a preview control of ZMP that uses reference future.

On the other hand the formulation for the walking pattern synthesis is done as a optimal control problem and optimization techniques are used. Bessonnet et al. in [2], by quadratic programming define the generalized coordinates as spline functions such that fulfill the minimization of the amount of the driving torques.

O. Castillo et al. (Eds.): Evolutionary Design of Intelligent Systems, SCI 257, pp. 259–271.
springerlink.com

A Genetic Algorithm is used as optimization method by Sang-Ho et. al. [4] to determine the shape of velocities and accelerations trajectories that minimizes the sum of deviation of each one through the search of suitable via points for obtain smooth and stable walking. Likewise, Park et. al. in [15] obtain an optimal locomotion pattern with minimal energy consumption and optimal locations of the center of mass of the links.

This paper is concerned in the energy optimal walking pattern design for a biped robot modeled as a system with impulse effects. The joint trajectories are parameterized as an polynomial (as are reported in [15]), which are included into the dynamics of the biped robot. A real coded Genetic Algorithm (GA) is used as optimization method to find the compatible polynomial coefficients which minimize the required input energy during the walking cycle while holding a stable walking motions. The optimization process is based on the evaluation of the biped robot in closed-loop, considering a set of constraints which guarantee no violation of the biped model assumptions and reach periodic walking motions. The difference with related publications is the generation of walking patterns whose motions are restricted to be described as periodic orbits and the design is not based on the ZMP approach, thus the reported walking patterns, allow more maneuvers than in this work is so far of our goal.

The paper is organized as follows: Section 2 presents the mathematical model of an 3-DOF biped robot. In Section 3, the problem formulation is summarized. Section 4, presents the GA implementation details used for the problem optimization established in previous section. Section 5, presents the results of numerical simulations. Section 6 provide the conclusions.

2 Biped Robot Model

In this section the dynamical model of a planar biped robot is introduced. The biped robot consists of a torso, hips, and two legs of equal length without ankles and knees. Two torques are applied between the legs and torso, so the system is under actuated. The definition of the angular coordinates and the disposition of the masses of the torso, hips and legs of the biped robot are shown in Figure 1, note that positive angles are computed in clockwise with respect to the indicated vertical lines and all masses of the links are lumped.

The walking cycle takes place in sagittal plane and on a level of surface, and it is assumed as successive phases where only one leg (stance leg) touching the walking surface (swing phase), with the transition from one leg to another taking place in a smaller length of time. During swing phase, the stance leg is modeled like a pivot and the swing leg is assumed move into the frontal plane [13] and to renter the plane of the motion when the angle of the stance leg attains a given desired value. The assumptions concerning to the walking cycle, define the biped robot model in two parts: the model that describe the swing phase and another one that describe the contact event of the swing leg with the walking surface; which are presented below.

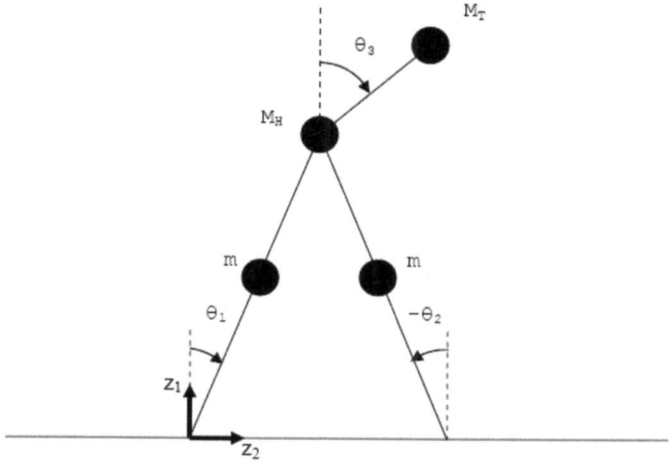

Fig. 1. 3-linkbiped robot.

2.1 Dynamic Model

The dynamical model of the robot during the swing phase is a second order system derived of Lagrange method [12]:

$$M(\theta)\ddot{\theta}+C(\theta,\dot{\theta})\dot{\theta}+G(\theta)=Bu, \tag{1}$$

where $\theta=[\theta_1,\theta_2,\theta_3]^T$ are the generalized coordinates, θ_1 parameterizes the stance leg, θ_2 the swing leg and θ_3 the torso; $u=[u_1,u_2]^T$ are the input; $M(\theta)$ is a positive-definite inertia matrix, $C(\theta,\dot{\theta})\dot{\theta}$ is the vector of the centripetal and Coriolis forces; and $G(\theta)$ is the vector of the conservative forces and B is the input matrix.

Transforming the second order equation (1) to state-space form by defining

$$\begin{aligned}\dot{x} &:= \tfrac{d}{dt}\begin{bmatrix}\theta\\ \dot{\theta}\end{bmatrix}=\begin{bmatrix}\dot{\theta}\\ M^{-1}(\theta)[-C(\theta,\dot{\theta})\dot{\theta}-G(\theta)+Bu]\end{bmatrix}\\ &:= f(x)+g(x)u.\end{aligned} \tag{2}$$

2.2 Impact Model

The impact between the swing leg and the ground is modeled like a contact between two rigid bodies. The main objective is to obtain the velocity of the generalized coordinates after the impact of the swing leg with the walking surface in terms of the velocity and position before the impact. The impact model for the biped robot is based in the rigid impact model of [10] and assume the case where the contact of the swing leg with de walking surface produce either no rebound nor slipping of the swing leg, and the stance leg lifting the walking surface without interaction. The conditions in [10] for these assumptions to be valid are that:

1. the impact takes place over an infinitesimally small period of time;
2. the external forces during impact can be represented by impulses;
3. impulsive forces may results in instantaneous change in the velocities of the generalized coordinates, but positions remain continuous;
4. the torques supplied by the actuators are not impulsional.

Taking into account the previous assumptions, the impact model is expressed with the equation [19]:

$$\begin{bmatrix} M_e & -E^T \\ E & 0 \end{bmatrix} \begin{bmatrix} \dot{\theta}_e^+ \\ F \end{bmatrix} = \begin{bmatrix} M_e \dot{\theta}_e^- \\ 0 \end{bmatrix}, \tag{3}$$

where $\theta_e = [\theta_1, \theta_2, \theta_3, z_1, z_2]^T$ are the generalized coordinates resulting of the add the Cartesian coordinates $[z_1, z_2]^T$ of the end of the stance leg as shown in Figure 1; M_e is the positive-definite inertia matrix of the biped model with the new generalized coordinates; $F = [F_T, F_N]^T$ is the tangential and normal forces applied at the end of the swing leg; $\dot{\theta}_e^+$ is the velocity after the impact; $\dot{\theta}_e^-$ is the velocity before the impact and

$$E := \frac{\partial Y}{\partial \theta_e} = \begin{bmatrix} r\cos(\theta_1) & -r\cos(\theta_2) & 0 & 1 & 0 \\ -r\sin(\theta_1) & r\sin(\theta_2) & 0 & 0 & 1 \end{bmatrix}, \tag{4}$$

obtaining $\dot{\theta}_e^+$ of the equation (3) must be done a change of coordinates and re-initialize the model (2). The change of coordinates is an expression for $x^+ := (\theta^+, \dot{\theta}^+)$ in terms of $x^- := (\theta^-, \dot{\theta}^-)$, which is given for the mapping function:

$$x^+ = \Delta(x^-) := \left[\, \theta_2^- \;\; \theta_1^- \;\; \theta_3^- \;\; \dot{\theta}_2^+(x^-) \;\; \dot{\theta}_1^+(x^-) \;\; \dot{\theta}_3^+(x^-) \,\right]. \tag{5}$$

Thus, the overall model of the biped walking is the combination of the swing phase model and impact model, as follows:

$$\Sigma : \begin{cases} \dot{x} = f(x) + g(x)u & x^- \notin S \\ x^{=}\Delta(x^-) & x^- \in S \end{cases} \tag{6}$$

where

$$S := \{(\theta, \dot{\theta}) | z_1 > 0, z_2 = 0, \theta_1 = \theta_1^d\} \tag{7}$$

3 Problem Formulation

This paper is based on the feedback controller for walking proposed by Grizzle et al. in [19] which designs a feedback for posture control and swing leg advancement of the biped robot (6), where the posture control means maintaining the torso angle at a constant value θ_3^d and swing leg advancement consists in driving the swing leg to behave as mirror image of the stance leg ($\theta_2 = \theta_1$). Thus, the behavior is encoded into the dynamics of the robot by defining the following output function:

$$y := \begin{bmatrix} y_1 \\ y_2 \end{bmatrix} := \begin{bmatrix} h_1(\theta, a) \\ h_2(\theta, a) \end{bmatrix} := \begin{bmatrix} \theta_3 - h_{d,1}(\theta_1, a) \\ \theta_2 - h_{d,2}(\theta_1, a) \end{bmatrix} \tag{8}$$

where

$$\begin{aligned} h_{d,1}(\theta_1,a) &:= \qquad\qquad a_1^0+\cdots+a_1^3(\theta_1)^3 \\ h_{d,2}(\theta_1,a) &:= -\theta_1+(a_2^0+\cdots+a_2^3(\theta_1)^3)\times(\theta_1+\theta_1^d)\times(\theta_1-\theta_1^d) \end{aligned} \tag{9}$$

with the control objective being to zeroing the outputs (8), they define the feedback

$$u(x) := (L_gL_fh(x))^{-1}(\Psi(h(x),L_fh(x))-L_f^2h(x)) \tag{10}$$

and

$$\Psi(y,\dot{y}) := \begin{bmatrix} \frac{1}{\varepsilon^2}\psi_\alpha(y_1,\varepsilon\dot{y}_1) \\ \frac{1}{\varepsilon^2}\psi_\alpha(y_2,\varepsilon\dot{y}_2) \end{bmatrix} \tag{11}$$

$$\psi_\alpha(x_1,x_2) := -sign(x_2)|x_2|^\alpha - sign(\phi_\alpha(x_1,x_2))|\phi(x_1,x_2)|^{\frac{\alpha}{2-\alpha}} \tag{12}$$

$$\phi_\alpha := x_1+\left(\frac{1}{2-\alpha}\right)sign(x_2)|x_2|^{2-\alpha} \tag{13}$$

The problem statement is described as follows: Given the biped robot model (6) and the feedback (10), find the parameters a_i^j of the function in (9) such that it obtains a walking pattern which produce periodic walking motions with low energy consumption.

4 Optimization Strategy

According to the problem, a Genetic Algorithm (GA) is proposed to solve the optimization problem. GA are search algorithms based on natural selection and natural genetic principles [8], apply them to optimization problems differ from traditional optimization methods that they use: payoff information (based on a objective function), no derivatives or other auxiliary knowledge; a coding of the parameters, not the parameters themselves and probabilistic rules as a tool to guide the search [14].

The implementation of the GA begins with the election of consistent decision variables which to solve the problem continuing with the coding of them into a string called individual, as well as an objective function which provide a numerical value that indicate the goodness degree or fitness of the individual to solve the problem. The GA operation consist of several stages, that to imitate the natural principles in which are based: The *initialization* stage where a set of individuals called population is created randomly; the *selection* stage in which each individual is evaluated through the objective function, the values of the evaluation are used by a selection algorithm that identifies and choose the fittest individuals that to serve as the parents for the creation of new individuals; The new individuals are created by applying the genetic operations to the individuals selected in the previous stage. This operations are: crossover, which with certain probability, combine the parameters encoded between pairs of individuals producing two individuals with information of both parents and the mutation, which is the alteration of the value of one parameter encoded in the individual, its function is to introduce diversity in the population, this stage is called *reproduction*; the last stage is the *replacement*, it consist in the creation of the new population with the new individuals and the parents. The overall stages is performed of iterative manner, each iteration is called generation. The GA operation explained above is expressed through an algorithm as shown below:

1. Begin
2. $i \leftarrow 0$
3. Create a population P_i of N individuals
4. Evaluate fitness of all individuals of the population: $P_i^e \leftarrow Fitness(P_i)$
5. Repeat the steps until the conditions are reached
   a) Select the parents for matting: $P_i^s \leftarrow Select(P_i, P_i^e)$
   b) Apply crossover and mutation: $P_i^g \leftarrow Genetic\_Operation(P_i^s)$
   c) $i \leftarrow i+1$
   d) Create the new population: $P_i \leftarrow New\_Pop(P_i^g, P_{i-1})$
   e) Evaluate fitness of new individuals: $P_i^e \leftarrow Fitness(P_i)$
6. Print out the best solution
7. End

Thus, the details of our implementation of the each stage are provided below.

4.1 Individual Representation

The individual representation selected is the real coding scheme, whose genotype is composed by eight real numbers such that represent to each parameter a_i^j (decision variables) of the function in (9).

4.2 Objective Function

The goal of the GA is the minimization of the cost function which represents the approximation to the average energy consumed over the walking cycle:

$$\hat{J}(a) = \int_0^T (u_1^2 + u_2^2)dt \tag{14}$$

where T is such that $\theta_1(T) - \theta_1^d = 0$, and $u(t)$ is obtained of applying (10) into the closed loop with the biped robot model (2) and (8).

This cost function is subject to the following constraints taking of [7]:

$$\begin{array}{ll} \hat{\theta}_1^- - 0.99\lambda(\hat{\theta}_1^-) & \leq 0 \\ \|y(\bar{t})\| & \leq 0 \\ |\frac{F_T}{F_N}| - \mu & \leq 0 \\ -\dot{z}_2^+ & \leq 0 \end{array} \tag{15}$$

The first constraint assures the existence of the one fixed point, the second one assures that the controller has converged before the impact and the last two constraints verify the validity of the impact model.

4.3 Genetic Operations

The genetic operations comprise the classical one point crossover and non-uniform mutation [6][14]. Non-uniform mutation is defined as follows: if $s_v^t = [v_1, ..., v_m]$ is a individual (with t, which means the generation number) and the element v_k was selected for this mutation, the result is a vector $s_v^t = [v_1, ..., v_k', ..., v_m]$ where:

$$v_k' = \begin{cases} v_k + \Delta(t, UB - v_k) & \text{if a random digits is 0} \\ v_k - \Delta(t, v_k - LB) & \text{if a random digit is 1} \end{cases} \tag{16}$$

and LB and UB are the lower an upper domains bounds of the variable v_k the function $\Delta(t,y) = y\left(1 - r\left(1 - \frac{t}{T}\right)^b\right)$ return a value in the range $[0,y]$; r is a random number, T is the maxima generation number and b is the dependency degree on the iteration number.

5 Simulation Results

The GA was developed in MatLab with the characteristics described in Section III, the individual evaluation is during one walking cycle and uses the implementation in software of 3-DOF model (6) with the controller (10) available in [1]. The physical parameters of the biped model are shown in the Table 1, pre-impact angular velocity is taken as $\dot{\theta}_1^- = 1.55$, the friction coefficient $\mu \geq \frac{2}{3}$, and the controller parameters for equations (11) and (12) are $\varepsilon = 0.1$ and $\alpha = 0.9$ respectively. Details of the biped model can be found in [19].

In the Table 3 presents the parameters for the function (9) and the energy consumption as results of five independent GA executions using the GA parameters

Table 1. Biped model parameters

Parameter	Value
Mass of Hips (M_H)	15 kg
Mass of Torso (M_T)	10 kg
Mass of Legs (m)	5 kg
Length of legs (r)	1 m
Length of torso (l)	0.5 m

Table 2. Genetic algorithm parameters

Parameter	Value
Number of individuals	100
Number Maximum of Generations	400
Crossover Rate	0.9
Mutation Rate	0.2
Selection Rate	0.4

Table 3. Parameter a_i^j and energy magnitude results with the GA

Number	Parameters								$\hat{J}(a)$
	a_1^0	a_1^1	a_1^2	a_1^3	a_2^0	a_2^1	a_2^2	a_2^3	
1	0.5893	0.4644	1.5558	-0.0367	-1.2656	1.0308	1.1086	3.7436	719.32
2*	0.5120	0.0730	0.0350	-0.8190	-2.2700	3.2600	3.1100	1.8900	761.00
3	0.7098	0.0581	0.0536	-1.0280	-1.9803	2.5752	3.6109	2.0000	1225.40
4	0.3992	0.7515	-2.5759	2.5365	1.8022	0.5344	0.2620	6.0033	1679.20
5	1.0730	-0.4986	-0.4127	1.6194	2.7029	-1.4978	1.6201	-0.8384	2541.70
6	0.2892	1.9435	-2.8760	0.4328	2.0252	4.5643	-0.5546	-4.1460	2980.50

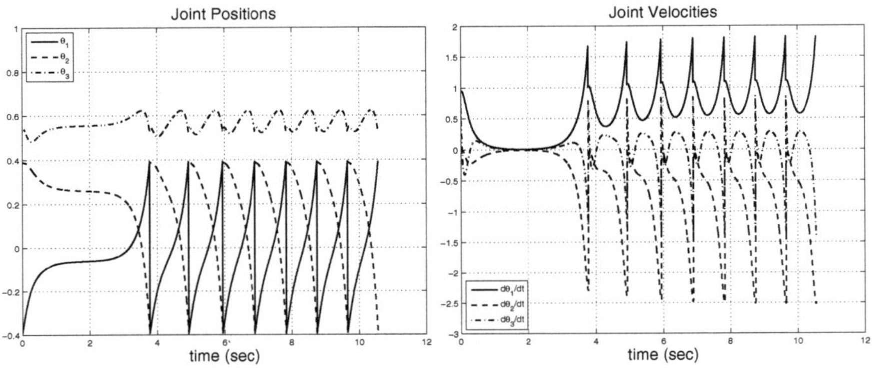

Fig. 2. Joint positions and velocities trajectories corresponding to closed-loop system execution with the parameters of the first entry of the Table 3

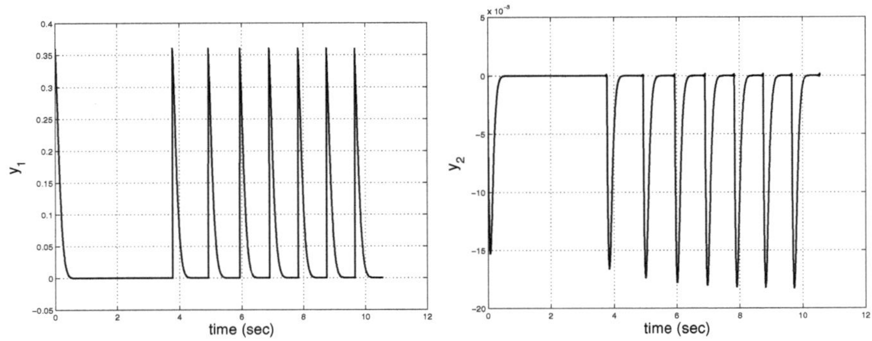

Fig. 3. Output function trajectories corresponding to closed-loop system execution with the parameters of the first entry of the Table 3

showed in the Table 2. The second entry on the table 3 include the parameters obtained by Grizzle et. al. in [19] with the same biped robot model using a traditional optimization method.

The results summarized in the Figures 2 to 5, corresponds to the closed-loop system simulations using the walking patterns defined by parameters of the first entry of the Table 3, and Figure 6 shows the fitness values of the best individual in each generation, finishing the optimization process about 140th generation. Overall results depict the execution of the closed-loop system performing eight steps. Note that the energy magnitude obtained with GA is slightly lower than the reported in [19]. The joint trajectories in the Figure 2 shows that the time needed to execute the first step is longer due that minimal change in the input torques succeeded, hence

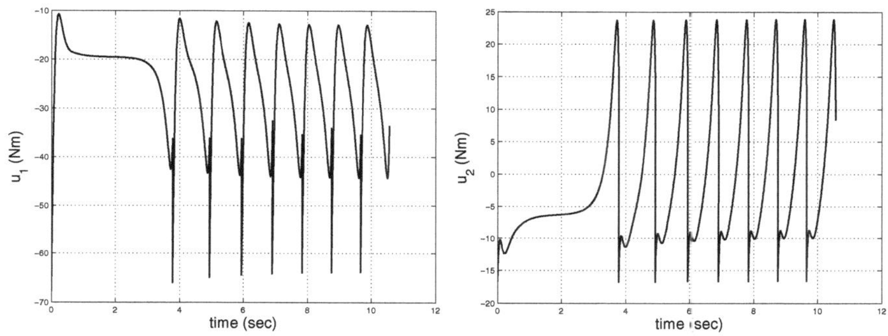

Fig. 4. Control signals corresponding to closed-loop system execution with the parameters of the first entry of the Table 3

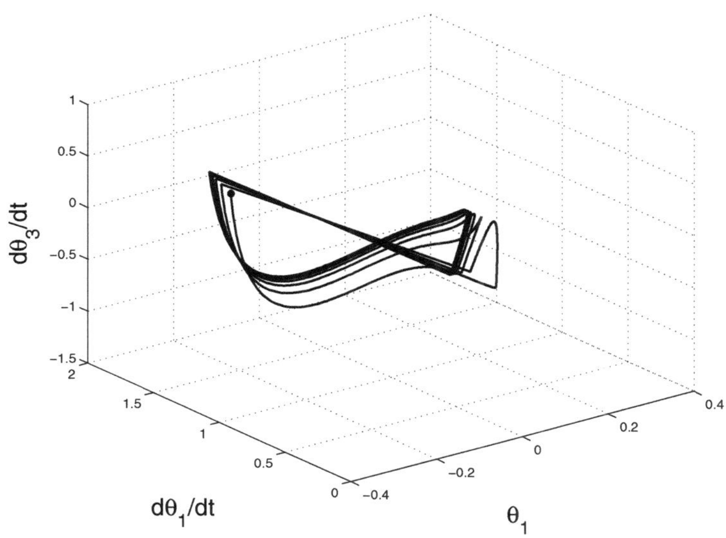

Fig. 5. Orbits obtained corresponding to closed-loop system execution with the parameters of the first entry of the Table 3

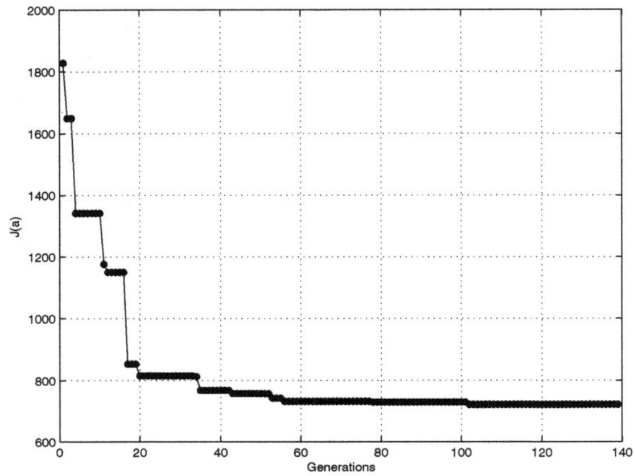

Fig. 6. Fitness values obtained through the optimization process for the first execution of the Table 3

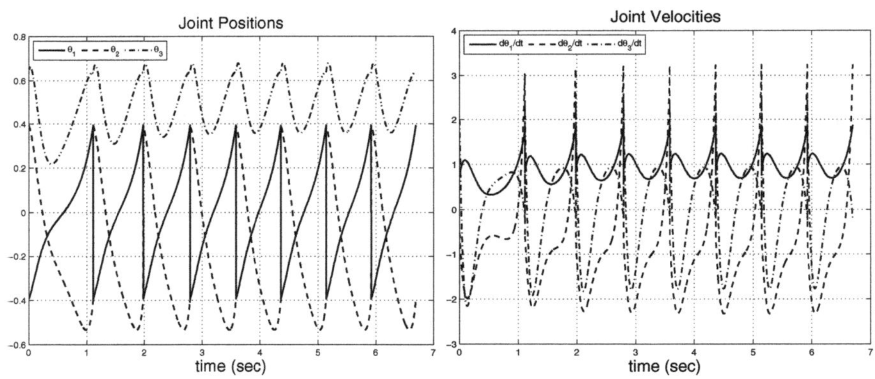

Fig. 7. Joint positions and velocities trajectories corresponding to closed-loop system execution with the parameters of the sixth entry of the Table 3

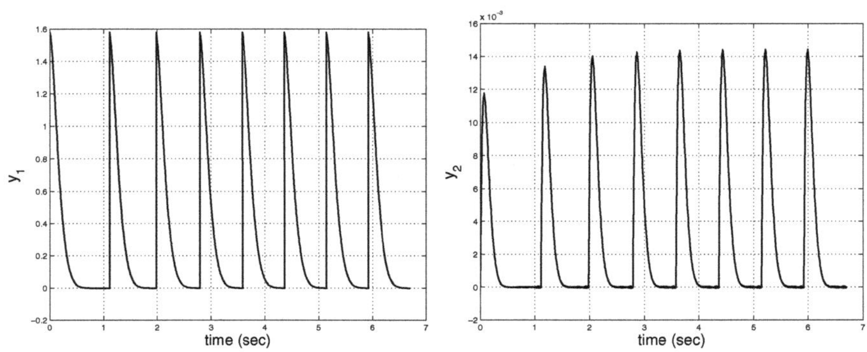

Fig. 8. Output function trajectories corresponding to closed-loop system execution with the parameters of the sixth entry of the Table 3

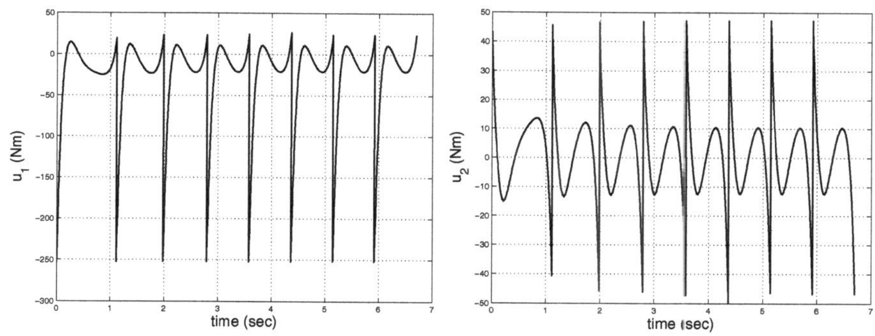

Fig. 9. Control signals corresponding to closed-loop system execution with the parameters of the sixth entry of the Table 3

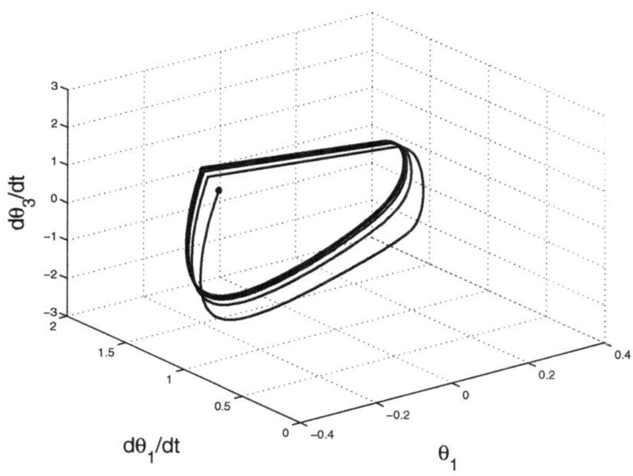

Fig. 10. Orbits corresponding to closed-loop system execution with the parameters of the sixth entry of the Table 3

the motion turning slow, after these, the step time tend to be the same through the following steps. In the Figures 7 to 10 demonstrate the results whose parameters are not optimal in energy magnitude, but still produce periodic walking motions. Moreover, in the Figures 5 and 10 present the orbits for both simulations cases, which shown the periodic nature of the walking.

6 Conclusions

In this paper a GA for the design of walking patterns of a biped robot was presented. The design approach consists in the search of suitable coefficients of a function, which describe the desired behavior of the joint trajectories such that produce periodic walking motion while keeping a minimal energy consumption. The numerical results with the function obtained by GA in the closed-loop system shows that this approach can provide a set of different walking patterns whose motions are described as periodic orbits and was able to obtain near optimal energy magnitudes.

References

1. Feedback control of dynamic bipedal robot locomotion book webpage (2007), `http://www.dynamicbipedcontrol.org`
2. Bessonnet, G., Seguin, P., Sardain, P.: A Parametric Optimization Approach to Walking Pattern Synthesis. The International Journal of Robotics Research 24(7), 523–536 (2005)
3. Capi, G.: Application of genetic algorithms for biped robot gait synthesis optimization during walking and going upstairs. Advanced Robotics 15(20), 675–694 (2001)
4. Choi, S.H., Choi, Y.H., Kim, J.G.: Optimal walking trajectory generation for a biped robot using genetic algorithm. In: Proceedings of IROS 1999 IEEE/RSJ International Conference on Intelligent Robots and Systems 1999, vol. 3, pp. 1456–1461 (1999)
5. Denk, J., Schmidt, G.: Synthesis of Walking Primitive Databases for Biped Robots in 3D-Environments. In: Proceedings of the IEEE International Conference on Robotics and Automation, ICRA 2003, Taipei, Taiwan (2003)
6. Goldberg, D.E.: Genetic Algorithms in Search. In: Optimization and Machine Learning. Addison-Wesley Longman Publishing Co., Boston (1989)
7. Grizzle, J., Abba, G., Plestan, F.: Proving asymptotic stability of a walking cycle for a 5 DOF biped robot model. In: International Conference on Climbing and Walking Robots. Angleterre, Portsmouth (1999)
8. Holland, J.M.: Adaptation in Natural and Artificial Systems. University of Michigan Press, Ann Arbor (1975)
9. Huang, Q., Yokoi, K., Kajita, S., Kaneko, K., Arai, H., Koyachi, N., Tanie, K.: Planning walking patterns for a biped robot. IEEE Transactions on Robotics and Automation 17(3), 280–289 (2001)
10. Hurmuzlu, Y., Marghitu, D.: Rigid body collisions of planar kinematic chains with multiple contact points. Int. J. Robot Res. 13(1), 82–92 (1994)
11. Kajita, S., Kanehiro, F., Kaneko, K., Fujiwara, K., Harada, K., Yokoi, K., Hirukawa, H.: Biped walking pattern generation by using preview control of zero-moment point. In: Proceedings of ICRA 2003. IEEE International Conference on Robotics and Automation 2003, vol. 2, pp. 1620–1626 (2003)
12. Kelly, R.: Control de Movimiento de Robots Manipuladores. Prentice Hall, Madrid (2003)
13. McGeer, T.: Passive dynamic walking. Int. J. Robot Res. 9(2), 62–82 (1990)
14. Michalewicz, Z.: Genetic Algorithms + Data Structures = Evolution Programs. Springer-Verlag New York, Inc., Secaucus (1994)
15. Park, J.H., Choi, M.: Generation of an optimal gait trajectory for biped robots using a genetic algorithm. JSME International Journal Series C Mechanical Systems, Machine Elements and Manufacturing 47(2), 715–721 (2004)
16. Roussel, L., de Wit, C.C., Goswami, A.: Generation of energy optimal complete gait cycles for biped robots. In: Proceedings of the IEEE International Conference on Robotics and Automation, pp. 2036–2041 (1998)

17. Shiriaev, A.S., Freidovich, L.B., Manchester, I.R.: Can we make a robot ballerina perform a pirouette? orbital stabilization of periodic motions of underactuated mechanical systems. Annual Reviews in Control 32(2), 200–211 (2008)
18. Vukobratovic, M., Juricic, D.: Contribution to the synthesis of biped gait. IEEE Transactions on Biomedical Engineering 16(1), 1–6 (1969)
19. Westervelt, E.R., Grizzle, J.W., Chevallereau, C., Choi, J.H., Morris, B.: Systematic Design of Within-Stride Feedback Controllers for Walking. In: Feedback Control of Dynamic Bipedal Robot Locomotion, pp. 137–11891. Taylor & Francis/CRC, Boca Raton (2007)

Design and Simulation of the Type-2 Fuzzification Stage: Using Active Membership Functions

Oscar Montiel[1], Roberto Sepúlveda[1], Yazmín Maldonado[2], and Oscar Castillo[3]

[1] Centro de Investigación y Desarrollo de Tecnología Digital del Instituto Politécnico Nacional (CITEDI-IPN), Av. del Parque No.1310, Mesa de Otay, 22510, Tijuana, B.C., México
o.montiel@ieee.org,r.sepulveda@ieee.org
[2] M.S. Student at CITEDI-IPN
maldonado@citedi.mx
[3] Division of Graduate Studies and Research, Calzada Tecnológico S/N, Tijuana, B.C., México
ocastillo@hafsamx.org

Abstract. This paper describes the design and implementation of the fuzzification stage for type-1 and type-2 fuzzy inference systems (FIS). A versatile method to calculate the membership values was used, it handles real numbers using decimal floating point binary encoding to calculate the slopes of triangular and trapezoidal membership functions. The designs were developed using VHDL code for FPGA implementation. The type-1 implementation is shown to give the basis of the type-2 implementation, which is based on the average method that consists in substituting an interval type-2 FIS by two type-1 FISs to cope with uncertainty. The functionality of the designs were evaluated by the analysis of the control surface plots of a speed controller for a DC motor. The plots were obtained from Simulink models that includes the VHDL designs developed in the Xilinx ISE. They were imported to the Simulink environment through the Xilinx System Generator.

1 Introduction

Recently, there has been an increasing interest in the research and implementations of type-2 fuzzy systems because they offer bigger advantages in handling uncertainty with respect to type-1 fuzzy systems [4, 7, 9, 10, 11]; so, many researchers have found this characteristic very interesting to be applied in the design of new digital controllers. This statement is supported by several research reports that cover from theoretical studies to physical implementations [20, 21]. However, it is well known that the implementation of type-2 fuzzy systems demand higher computational resources, this requirement is very difficult to satisfy for low cost microcontroller systems for industrial or consumer use.

With the rapid development of the Very Large Scale Integrated Circuits (VLSI), such as the Field Programmable Gate Array (FPGA) and the Applied Specific Integrated Circuit (ASIC), as well as the development of new programming tools that allow to take complex digital designs to practice in a very small time, now, it is

O. Castillo et al. (Eds.): Evolutionary Design of Intelligent Systems, SCI 257, pp. 273–293.
springerlink.com

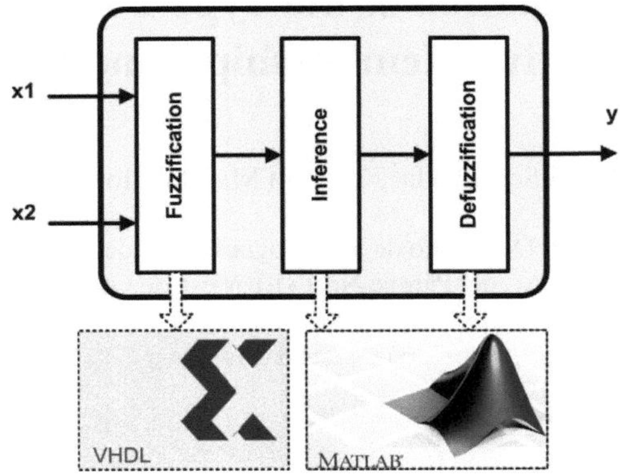

Fig. 1. Overview of the computer architecture used to test and validate the VHDL code of the fuzzification stage.

possible to design and implement high performance systems embedded into an integrated circuit. Nowadays, the study and proposals to implement a fuzzy system into an FPGA is growing up [23].

The main objective of this work is to show how to implement the fuzzification stage of a type-2 fuzzy system based on the average of two type-1 fuzzy systems using the method proposed in [20]. The proposed algorithms were tested using the Xilinx System Generator (XSG) from Xilinx, and the Matlab/Simulink from Mathworks. In [26], the details to implement and validate through the Xilinx System Generator (XSG) the type-1 fuzzification stage were published. Fig. 1 shows the setting up of the software architecture that was used to test and validate the type-1 and type-2 fuzzification stages. Note in this figure, that there is a main block containing three sub-blocks: Fuzzification, inference, and defuzzification. The fuzzification block has two inputs, and one output, this block is related with another block labeled as VHDL (Very High Description Language) to indicate that this stage was programmed using this high level hardware description language; the other two stages (inference and defuzzification) are related with one block labeled as MATLAB to indicate that these stages were programmed using Matlab/Simulink. Regarding the fuzzification stage, two different architectures were proposed, they are: Fuzzification using only the active MFs, and fuzzification using all the MFs. In the first one, only the affected MFs by the inputs were considered. In the second architecture, all the MFs were considered. This work is focused on the design of the fuzzification stage for active MFs.

There are other works related to this one; for example, for type-1 fuzzy systems [13, 15, 18, 19, 22, 24, 25], in the book [17] the authors give an academic example of the VHDL coding of a type-1 fuzzy controller. With respect to type-2 fuzzy systems there are some works like [14, 15, 16].

This work has been organized as follows: Section 2 is devoted to give an introduction from crisp sets to type-1 and type-2 fuzzy sets (FSs), some useful definitions about generalized type-2 fuzzy sets are given with special emphasis in the MFs. In section 3 the mathematics for interval type-2 fuzzy sets is given. In section 4 the average type-2 fuzzy inference system is explained, emphasizing on the fuzzification stage. Section 4.1 was dedicated to explain the most common methods to perform digital fuzzification, we explain in this section the method used in this work, two important piece of VHDL code to fuzzify using triangular and trapezoidal MFs are given; also, we explain the arithmetic method used to handle real and integers numbers to improve the performance, a numeric example is given. Section 5 was dedicated to explain the general idea that was followed in testing and validating this stage. First, using a type-1 FS, an explanation of the input and output MFs, and their respective linguistic variables and terms is given, we show the design entity for this system. Next, it is explained how to implement the type-2 fuzzification stage for the average method, by taking advantage of VHDL code reusability; i.e., using two times the same VHDL code previously developed for the type-1 FS. Moreover, a Simulink model that allows to test the developed type-2 VHDL code for the fuzzification stage is shown. Finally, in section 5.4 some conclusions about this work are given.

2 Sets

This section is dedicated to provide definitions of set theory, starting with "crisp sets", in order to introduce the basic concepts of type-1 and type-2 fuzzy set theory.

2.1 Crisp Sets

In classical set theory (George Cantor, 1845-1918) [2], a set A is comprised of elements x with membership into a universe of discourse X; i.e., $x \in X$. The membership of an element can be expressed using a membership function (MF) that is also known as characteristic function or discrimination function [12]. The definition 1 is valid to represent a MF in the classical bivalent set (crisp set) theory. One of the most common method to represent the crisp set A is,

$$A = \{x | x \in X\} \tag{1}$$

Definition 1. For a crisp set A, the membership function μ_A is defined as

$$\forall x \in X,\ \mu_A(x) = \begin{cases} 1 \text{ if } x \in A \\ 0 \text{ if } x \notin A \end{cases} \tag{2}$$

in other words, the function μ_A maps the elements in the universal set X to the set $\{0,1\}$,

$$\mu_A(x) : X \to \{0,1\}. \tag{3}$$

□

2.2 Fuzzy Sets

Ordinary fuzzy sets were developed by Lotfi Zadeh in 1965 [28], they are an extension of classical set theory where the concept of membership was extended to have various grades of membership on the real continuous interval [0,1]. The original idea was to use a fuzzy set (FS); i.e, a linguistic term to model a word; however, after almost ten years, Zadeh introduced the concept of type-n FS as an extension of an ordinary FS (type-1 FS) with the idea of blurring the degrees of membership values [27].

Type-1 fuzzy sets have demonstrated to work efficiently in many applications, most of them use the mathematics of fuzzy sets but losing the focus on words, that are mainly used in the context to represent a function that is more mathematics than linguistic [8]. Membership functions are used to characterize type-1 and type-2 fuzzy sets.

2.2.1 Type-1 Fuzzy Sets

Previously, we mentioned that a FS and its MF is an extension of a crisp set defined in (1) and in (3), respectively. Therefore, that definition has been modified for the FS and the MF. A type-1 FS is a set of ordered pairs expressed by (4) [6], and its corresponding MF is given by definition 2,

$$A = \{(x, \mu_A(x)) | x \in X\} \tag{4}$$

Definition 2. For a type-1 fuzzy set A, each element is mapped to [0,1] by its MF μ_A, where [0,1] means real numbers between 0 and 1, including the values 0 and 1,

$$\mu_A(x) : X \to [0,1]. \tag{5}$$

□

2.2.2 Type-2 Fuzzy Sets

In a type-1 FS the word fuzzy has the connotation of uncertainty, the problem is that once the MF parameters have been specified the type-1 FS is completely certain; as a direct consequence of limitations of type-1 fuzzy sets to handle appropriately linguistic uncertainties type-n fuzzy sets were proposed [28], being type-2 fuzzy sets an special case, the advent of this new type of fuzzy sets came with the fact that uncertainty is always encountered in real life problems and in different ways; so, type-2 fuzzy sets allow to handle appropriately linguistic and random uncertainties [1, 10].

A pointwise definition of a type-2 FS is given as follows,

Definition 3. For a type-2 FS $\widetilde{A}$ characterized by a type-2 MF $\mu_{\widetilde{A}}(x,u)$ were $x \in X$ and $u \in J_x \subseteq [0,1]$, i.e.,

$$\widetilde{A} = \left\{(x,u),\ \mu_{\widetilde{A}}(x,u) | \ \forall x \in X, \forall u \in J_x \subseteq [0,1]\right\} \tag{6}$$

where $0 \leq \mu_{\widetilde{A}}(x,u) \leq 1$. □

Another way to express $\widetilde{A}$ is,

$$\widetilde{A} = \int_{x\in X} \int_{u\in J_x} \mu_{\widetilde{A}}(x,u)/(x,u) \quad J_x \subseteq [0,1] \tag{7}$$

Similar to type-1 FS, the symbol $\int$ stands for union of continuous functions, and the corresponding symbol for discrete functions is $\sum$; then in (7), $\int\int$ denotes union over all admissible x and u.

Definition 4. The 2D plane with axes u and $\mu_{\widetilde{A}}(x',u)$ formed with $x = x'$ is called a *vertical slice of* $\mu_{\widetilde{A}}(x,u)$. □

Definition 5. A secondary MF is a *vertical slice* of $\mu_{\widetilde{A}}(x,u)$; in particular, $\mu_{\widetilde{A}}(x = x',u)$ for $x' \in X$ y $\forall u \in J_{x'} \subseteq [0,1]$, i.e.,

$$\mu_{\widetilde{A}}(x = x',u) \equiv \mu_{\widetilde{A}}(x') = \int_{u\in J_{x'}} f_{x'}(u)/u \quad J_{x'} \subseteq [0,1]. \tag{8}$$

where $0 \leq f_{x'} \leq 1$. □

We can drop the prime notation on $\mu_{\widetilde{A}}(x')$ $\forall x' \in X$ and refer to $\mu_{\widetilde{A}}(x)$ as a secondary membership function also known as the secondary set, which is a type-1 FS.

Using the concept of secondary sets, a type-2 FS can be reinterpreted as the union of all the vertical slices,

$$\widetilde{A} = \{(x,\mu_{\widetilde{A}}) | \forall x \in X\} \tag{9}$$

or, as

$$\widetilde{A} = \int_{x\in X} \mu_{\widetilde{A}}(x)/x = \int_{x\in X} [\int_{u\in J_x} f_x(u)/u]/x \qquad J_x \subset [0,1]. \tag{10}$$

Definition 6. The *primary membership* of x, J_x ($J_x \subseteq [0,1]$ $\forall x \in X$), is the domain of a secondary MF. □

Definition 7. A *secondary grade* is the amplitude of a secondary MF. For example, in (10) $f_x(u)$ is a secondary grade. □

Definition 8. A *main membership function* of a type-2 FS is the union of all the secondary grades equal to 1 (considering that each secondary MF has only one secondary grade equals to 1); i.e.,

$$\mu_{main}(x) = \int_{x\in X} u/x \quad \text{where} \quad f_x(u) = 1 \tag{11}$$

□

Generalized type-2 fuzzy sets are those where secondary membership grades are in the range of [0,1]; but there is the case when secondary grades are always 1, then we have the next definition,

Definition 9. A type-2 FS where its secondary MF is always 1 ($f_x(u) = 1, \ \forall u \in J_x \subseteq [0,1]$), is called an *interval type-2 FS*.

Definition 10. The *footprint of uncertainty* (FOU) is the union of all primary memberships, it is represented as a bounded region that contains uncertainties in the primary memberships of a type-2 FS, $\widetilde{A}$, i.e.,

$$FOU(\widetilde{A}) = \bigcup_{x \in X} J_x \tag{12}$$

□

The concept of FOU is very useful because it allows to focus the attention only on the inherent uncertainties of a type-2 MF. For convenience, the FOU can be described in terms of upper and lower MFs, so we have the next definition.

Definition 11. There are two type-1 membership functions that are the bounds of the FOU, they are the upper and the lower membership functions. The upper bound of $FOU(\widetilde{A})$ is associated with the upper MF, which is denoted by $\bar{\mu}_{\widetilde{A}}(x)$, $\forall x \in X$. The lower bound of $FOU(\widetilde{A})$ is associated with the lower MF denoted by $\underline{\mu}_{\widetilde{A}}(x)$, $\forall x \in X$, i.e.,

$$\bar{\mu}_{\tilde{A}}(x) = \overline{FOU(\tilde{A})} \ \ \forall x \in X. \tag{13}$$

$$\underline{\mu}_{\tilde{A}}(x) = \underline{FOU(\tilde{A})} \ \ \forall x \in X. \tag{14}$$

□

Figure 2 shows an interval type-2 MF, the shadow region is the FOU. At the points x_1 and x_2 are the primary MFs J_{x_1} and J_{x_2}, the corresponding secondary MFs $\mu_{\widetilde{A}}(x_1)$ and $\mu_{\widetilde{A}}(x_2)$ are shown in Figure 3.

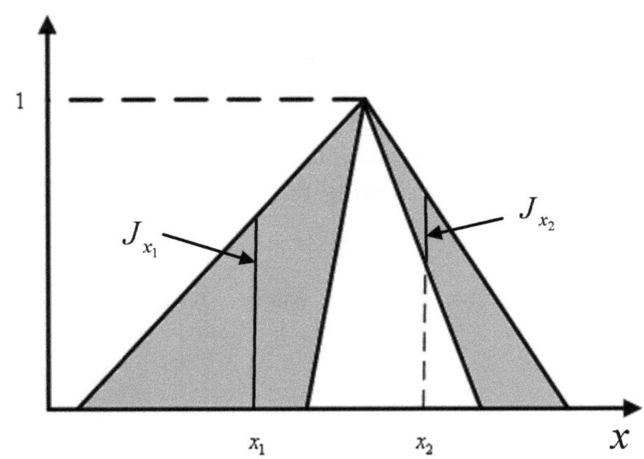

Fig. 2. Type-2 membership function with FOU showing the primary memberships.

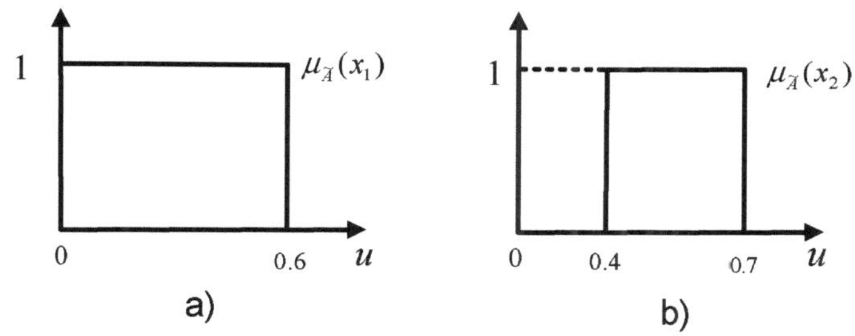

Fig. 3. Secondary membership values a) J_{x_1}, b) J_{x_2}.

3 Interval Type-2 Fuzzification Stage

In section 2.2.2 was mentioned that type-2 FSs were proposed to cope with the uncertainty that might come from several sources; and it was used the term linguistic uncertainties when the FS is used to model a word, this to embrace uncertainties from fuzziness (or vagueness) that result from imprecise boundaries of FSs, non-specificity which is connected with sizes (information based-imprecision), and strife (or discord) that expressed conflicts among the various sets or alternatives [8]. In this section, a summary of type-2 FIS is provided, emphasizing on the fuzzification stage, since it is the aim of this chapter. Figure 4 shows a schematic of a type-2 FIS, it has been divided in four stages: fuzzification, inference engine, type reducer and defuzzification. The task of the fuzzifier is to map a crisp value $x = (x_1, \ldots, x_p)^T \in X_1 \times X_2 \times \ldots \times X_p \equiv \mathbf{X}$ into a type-2 FS $\widetilde{A}_x$ in $\mathbf{X}$. $\widetilde{A}_x$ is a type-2 fuzzy singleton if $\mu_{\tilde{A}_x} = 1/1$ for $x = x'$ and $\mu_{\tilde{A}_x} = 1/0$ for all other $x \neq x'$ [10, 9].

Figure 5 shows an example of the fuzzification process for an interval type-2 FIS, the universe of discourse X is in the range of $[-10, 10]$. Considering that the input x =-4 cuts the type-2 MFs "Negative" and "Zero", for the upper bound of the first MF the membership value is $\bar{\mu}_N$=0.6, and for the lower bound A is $\underline{\mu}_N$=0.5 a set of

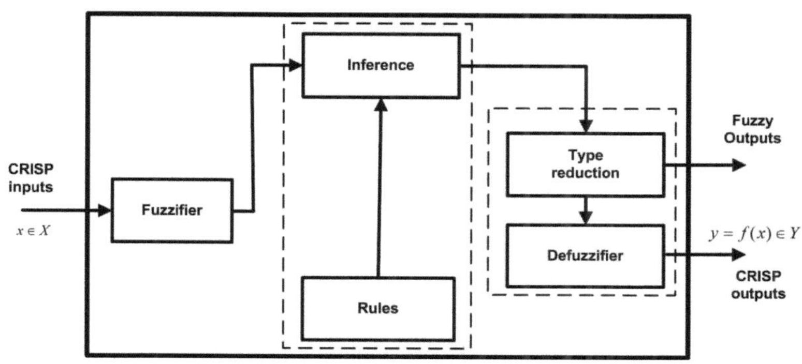

Fig. 4. Type-2 fuzzy system.

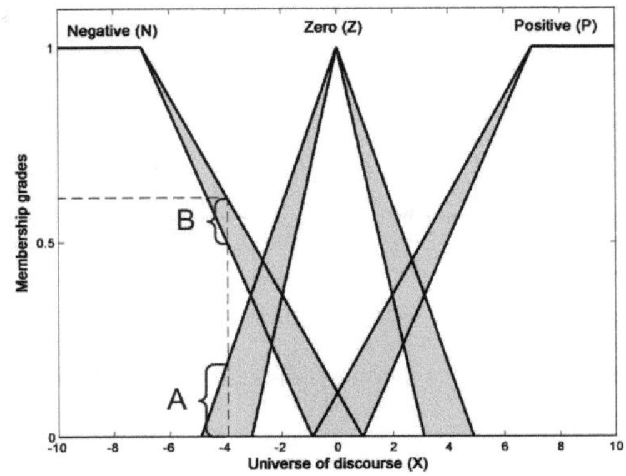

Fig. 5. Interval type-2 MFs.

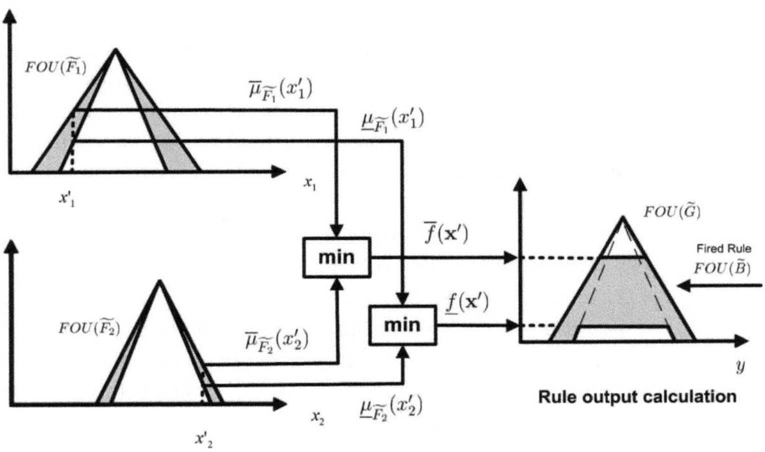

Fig. 6. Pictorial description of input and antecedent operations for an interval type-2 singleton, the process uses the minimum t-norm. The average type-2 method is compatible with this idea; however, instead of using type-2 MFs in the antecedents, the method splits each type-2 MF in two type-1 MF placed at the uncertainty boundaries; i.e., $FIS_u(\widetilde{F_1})$ is used for the $\overline{FOU(\widetilde{F_1})}$, and $FIS_l(\widetilde{F_1})$ for $\underline{FOU(\widetilde{F_1})}$.

primary membership values in the range [0,0.25], the secondary membership values are equal to one for both fuzzy terms. A type-2 fuzzy inference engine combines the if-then rules and gives a mapping from input type-2 fuzzy sets to output type-2 fuzzy sets. Figure 6 shows a pictorial description for the inputs x_1 and x_2, and the minimum t-norm. According to Liang and Mendel [5], for an interval type-2 FIS the result of the input and antecedent operations are contained in the interval type-1 firing set $[\underline{f}^l, \bar{f}^l]$, where

$$\underline{f}^l = \underline{\mu}_{\tilde{F}_1^l}(x_l') \bigstar \cdots \bigstar \underline{\mu}_{\tilde{F}_P^l}(x_p') \tag{15}$$

$$\bar{f}^l = \bar{\mu}_{\tilde{F}_1^l}(x_l') \bigstar \cdots \bigstar \bar{\mu}_{\tilde{F}_P^l}(x_p') \tag{16}$$

Depending on the desired output, it is possible to use the "output processing" block to obtain a type-reduced set (type-1), or a crisp output.

4 Average Type-2 FIS

An average type-2 FIS consists in substituting an interval type-2 FIS (Figure 5) with two type-1 FIS (Figure 7). The main idea of this method is to embrace the uncertainty of the interval type-2 FSs using type-1 FSs. One FIS is used for the upper bound uncertainty, and the second FIS for the lower bound uncertainty. Considering the type-2 FS "A", the interval of uncertainty can be embraced by two type-1 FS. The MFs of the first type-1 FIS are placed such as they match with the upper bound of $\overline{FOU(\tilde{A})}$, i.e., $(\bar{\mu}_{\tilde{A}}(x))$; similarly, the MFs of the second type-1 FIS are chosen to match with the lower bound of $\underline{FOU(\tilde{A})}$, i.e., $\underline{\mu}_{\tilde{A}}(x)$. Each type-1 FIS handles an identical set of rules, however the fuzzy output will be different because the MFs of the inputs are not the same. The average type-2 FIS is consistent with the nomenclature of interval type-2 FIS; so, Fig. 6 remains being useful to explain the fuzzification method that was used in this work. Basically, in this first stage, the difference lies in the fact that an average type-2 FIS uses two type-1 FS in the fuzzification stage to substitute each type-2 FS. A crisp value is obtained by averaging the defuzzified output or the two type-1 FIS.

The first step to split the IT2 FIS is to begin defining that one of the FIS is going to be used for the lower bound of the uncertainty, FIS_l; and the second one for the upper bound, FIS_u. Hence, for the linguistic variables X_1 and X_2 of an IT2 FIS, we will write $FIS_l(X_1)$ for the lower bound of uncertainty, and $FIS_u(X_2)$ for the upper bound. Following a similar idea of nomenclature, the linguistic terms of the IT2 FIS of Figure 6, $FOU(\widetilde{F}_1)$ and $FOU(\widetilde{F}_2)$, will be written as $FIS_l(\widetilde{F}_1,\widetilde{F}_2)$, and $FIS_u(\widetilde{F}_1,\widetilde{F}_2)$ to indicate that the $FIS_l(\widetilde{F}_1,\widetilde{F}_2)$ was developed considering the lower bound of the type-2 linguistic terms $\widetilde{F}_1$ and $\widetilde{F}_2$; similarly, $FIS_u(\widetilde{F}_1,\widetilde{F}_2)$ indicates the use of the upper bounds of uncertainty of the mentioned terms.

4.1 Type-2 Fuzzification Stage Architecure

In Figure 1 the proposal to test and validate the type-2 fuzzification stage coded in VHDL was illustrated. This figure depicts that the inference and defuzzification stages were programmed in Matalb/Simulink using the existing Fuzzy Simulink modules of the Fuzzy toolbox, whereas the fuzzification stage was coded in VHDL, this was achieved to evaluate only the VHDL codification.

As was mentioned before, the type-2 FIS explained in this work is based on calculating the average of two type-1 FIS (for the upper and lower bounds).

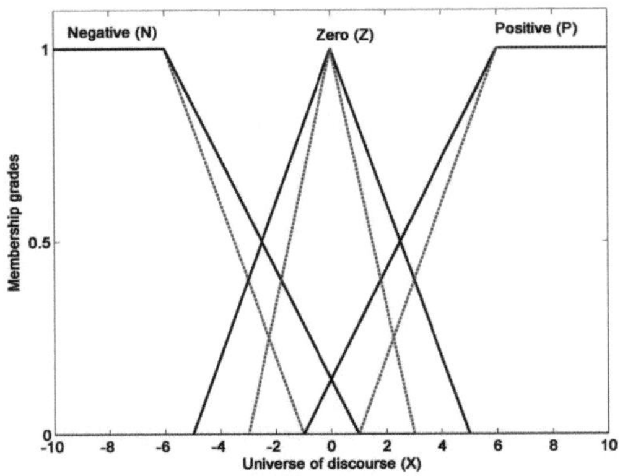

Fig. 7. The interval type-2 MFs of Fig. 5 were substituted with type-1 MFs. The MFs plotted with solid lines correspond to the fuzzification stage of one FIS, so we have for the upper bounds of uncertainty $FIS_u(\widetilde{N},\widetilde{Z},\widetilde{P})$; and for the lower bounds, $FIS_l(\widetilde{N},\widetilde{Z},\widetilde{P})$.

Microelectronic implementations of the fuzzification stage can be achieved in two ways: Storage memory and arithmetic calculation. Next, a brief review of each method is given.

4.1.1 Storage Memory Method

This method consists in sampling the universe of discourse and saving into memory the corresponding membership values and linguistic terms. This alternative allow us to define any shape of membership function, since the only requirement is to sample and save the membership values that are addressed by their corresponding binary input in the universe of discourse. One limitation of this method is regarding precision, because it depends on the number of samples and bits needed for an application, high precision applications require higher memory resources. In [25] can be found more information about the amount of memory required to achieve this implementation.

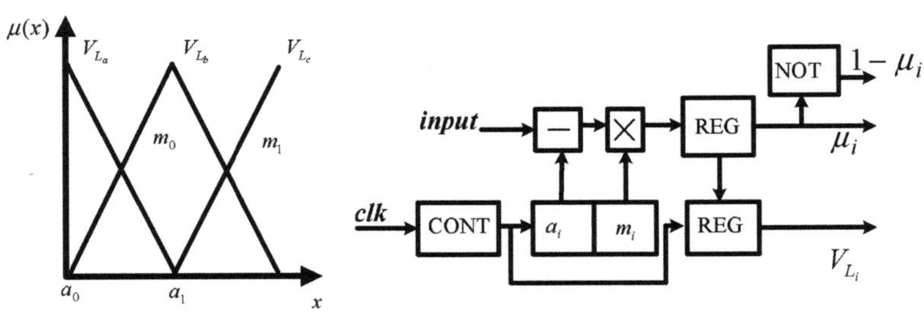

Fig. 8. Block diagram to illustrate the arithmetic calculation method.

4.1.2 Arithmetic Calculation Method

The arithmetic method consists in performing a progressive calculation of the degree of membership of the antecedents. This method is restricted to use only symmetric triangular MFs. The method uses two memory allocations for each straight line, where their slopes values m_i, and interception points a_i are saved into memory. An arithmetic circuit for each input shown in Fig. 8(a), solves the corresponding straight-line equation to obtain the membership value, as can be seen in Fig. 8(b). A counter driven by the system clock generates the memory addresses, the breakpoint value is subtracted from the input and the result is multiplied by the slope; if $x > xi$, the product is stores in a latch an the counter is increased in the next clock pulse. The process continues until $x < xi$. This circuit requires as many clock cycles to perform its operation as fuzzy sets are defined.

4.1.3 Novel Arithmetic Fuzzification Method

The arithmetic fuzzification method (AFM) was originally proposed by the authors in [26], in this work some improvements to the original proposal are presented. The AFM method is more versatile than the storage memory and the arithmetic calculation methods that were previously explained.

In comparison with the storage memory method explained in section 4.1.1, this method does not need to have a preloaded singleton table values nor to use high memory resources to obtain high resolution values.

Comparing with the arithmetic calculation method of section 4.1.2, the AFM allows to use symmetric and non-symmetric triangular and trapezoidal MFs. The intersection and slopes values can be calculated online and updated, which is a nice capability to implement adaptive fuzzy systems. In addition, the AFM only needs one clock cycle to achieve the fuzzification, this is a very attractive feature when

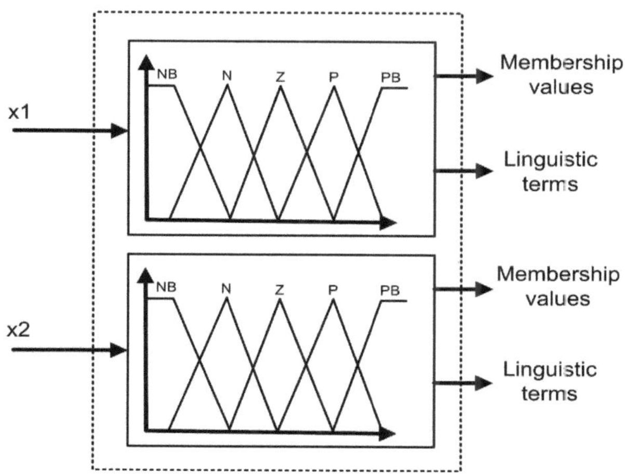

Fig. 9. Fuzzification stage with two inputs. For each input there are two outputs: The membership value, and the linguistic term.

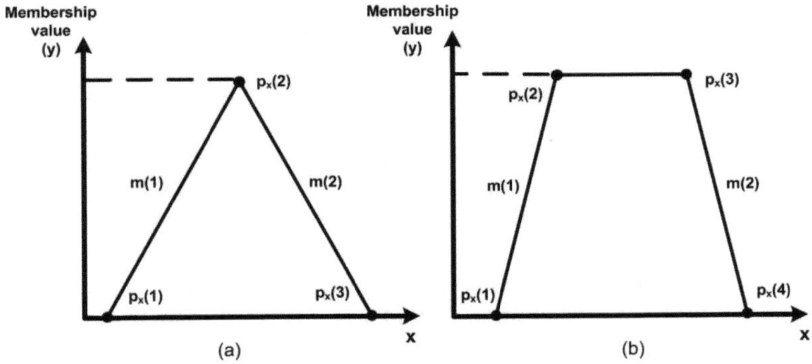

Fig. 10. Triangular and trapezoidal MFs. The slopes and intercept points are shown.

we compare with the arithmetic calculation method of section 4.1.2 that requires several clock cycles to perform this task.

The AFM code is divided in two parts. The first part is parallel code devoted to update the intersection points, calculate slopes values, and assign the corresponding linguistic tag to each MF. It uses one status bit (status flag) related with each slope to inform that the corresponding slope value has been updated. Once the calculations have been achieved, all the status flags are set. If any change occurs, the implied status flags should be reset. The second part of the code checks, whether all the status flags are set; if so, then the crisp input is fuzzyfied. Fig. 9 is a big picture of how the fuzzification is achieved for a two input system, the outputs are the membership grades and linguistic terms of the active MFs. The AFM for a given input (crisp value) can be summarized into three steps:

1. Calculate the slope values.
2. Calculate the membership values.
3. Assign to signals the membership values and linguistic terms.

Figures 10(a) and 10(b) show the slopes and intercepts of the straight lines of a triangular MF, and a trapezoidal MF, respectively. Both, use the basic straight line equation given by the slope-intercept form $y = mx + b$ where "m" is the slope and "b' is the intercept. The MFs have two slopes, the positive slope "$m(1)$", and the negative slope "$m(2)$"; although they have different signs, both are handled as positive to avoid signed operations; later, the sign is handled by the software. Figure 11 and 12 show the VHDL code to implement the triangular and trapezoidal MF, respectively. These MFs can be implemented using fixed point arithmetic to represent integers and real numbers, as well as floating point arithmetic. The decision of the best numeric representation most of the time depends on how the problem will be handled in later stages. This decision must be done carefully at an early project phase, in a broad sense, fixed point implementations provide higher speeds and lower cost implementations, on the other hand, floating point offers higher dynamic range and it is not necessary to perform scaling task that is a very attractive feature for complicated algorithms. Its main virtue is that it allows easy conversion to

```
function trian (x: std_logic_vector; y:triangular; m: pendiente)return std_logic_vector is
          variable sal: std_logic_vector(n downto 1):=(others =>'0');
          variable sal1: std_logic_vector(n*2 downto 1):=(others =>'0');
  begin
    if x<=y(1)  then sal:="00000000";
      elsif x> y(1) and x<y(2)
          then
          sal1:=m(1)*(x-y(1));
          sal := sal1(8 downto 1);
      elsif x= y(2) then sal:=tope;
      elsif x> y(2) and x< y(3)
          then
            sal1:= m(2)*(y(3)-x);
            sal := sal1(8 downto 1);
      else   sal:="00000000";
    end if;
return sal;
end function trian;
```

Fig. 11. VHDL codification to implement the triangular MF.

```
function trape (x: std_logic_vector; y:trapezoidal; m: pendiente)return std_logic_vector is
          variable sal: std_logic_vector(n downto 1):=(others =>'0');
          variable sal1: std_logic_vector(n*2 downto 1):=(others =>'0');
  begin
    if (x<=y(1))
    then
          if (y(1) = y(2)) then
            sal := tope;
    else
          sal:="00000000";
    end if;
    elsif (x> y(1) and x<y(2)) then
          sal1:=m(1)*(x-y(1));
          sal := sal1(8 downto 1);
    elsif x>= y(2) and x<= y(3) then
          sal:=tope;
    elsif x> y(3) and x<= y(4)  then
           if (y(3) = y(4))then
           sal:= tope;
     else
          sal1:= m(2)*(y(4)-x);
          sal := sal1(8 downto 1);
     end if;
     else
          sal:="00000000";
      end if;
  return sal;
end function trape;
```

Fig. 12. VHDL codification to implement the trapezoidal MF.

decimal digits for printing or display and faster decimal calculations, its drawbacks are the increased complexity of circuits needed to implement mathematical operations and a relatively inefficient encoding that occupies more space than a pure binary representation.

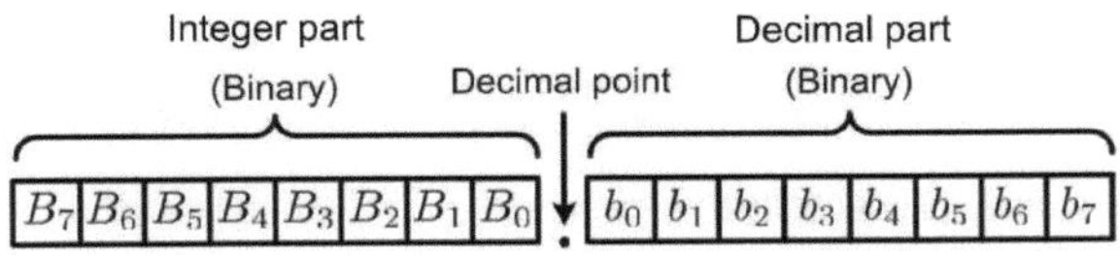

Fig. 13. VHDL code for the triangular MF.

4.1.4 Fuzzification Using Real Numbers

The arithmetic used in this work is based on the encoding method called Decimal Floating Point Binary Encoding (DFPBE), but differently to the common way to implement this encoding method, we used a fixed point format with no exponents. Figure 13 shows the general idea of encoding a real number using two integers, and to calculate the slope "$m(1)$" of the triangular MF showed in Figure 10(a), it is necessary to achieve a method with several steps; so, Fig. 16 is going to be used in order to explain the algorithm. Considering that we are going to fuzzify the crisp binary value 48 (i.e, x1=48), to obtain $\mu(48) = 0.3325$. First, it is necessary to calculate the positive slope of the Negative (N) MF, using the slope formula $m = \frac{y}{x}$, i.e., $m = \frac{255-0}{80-32} = 5.3125 \approx 5.3$ since the binary equivalent for an eight-bits MF is 255, hence the membership value is $\frac{5.3\times(48-32)}{255}$.

1. Obtain the divisor of the slope formula: Considering that the x value is given by the points $p_x(1)$=32 and $p_x(2)$=80, the divider is obtained using $p_x(2) - p_x(1) = 80 - 32 = 48$. Since the codification method DFPBE is being used, it is necessary for convenience to divide by ten the divider, instead of multiplying by ten the dividend as is common; therefore, we have $\frac{p_x(2)-p_x(1)}{10} = 4.8$, then the divider for the integer numbers in the range of (0 to 255) is 48; $Quotient_1(4)$, $residue_1(8)$; hence, then the divider for the decimal part is 4.
2. Obtain the slope: The slope is given as the number "$B_m.b_m$"; where "B_m" is an integer number, and "b_m" is a decimal number.
For the integer part: $Bm = \frac{255}{p_x(2)-p_x(1)} = 5.3$: $Quotient_2(5)$, $residue_2(15)$. For the decimal part, $bm = \frac{residue_2}{Quotient_1} = \frac{15}{4} = 3.7$: $Quotient_3(3), residue_3(3)$. The slope is a two bytes binary numbers B_m=5, b_m=3; i.e. slope "m(1)=5.3".
3. The membership value for a given crisp input, for example "x=48", is obtained as follows: Adjust the crisp input to the corresponding slope, in this case calculate $x_{48} = x - p_x(1) = 48 - 32 = 16$. Using the slope formula $y = mx$, which is equivalent to $y = B_m x + \frac{b_m}{10} x$. This applied to our example considering scaling, $y = \frac{5(16)+0.3(16)}{255} = 0.3325$; so, the membership value is 0.3325 for a crisp input "x=48".

Note that DFPBE method was used only to calculate the slopes, the membership values are real numbers coded in one binary word.

5 Experiment Design and Validation

In Fig. 1 the general idea followed in testing and validating the fuzzification stage for type-1 and type-2 FIS was presented. More specifically, the FIS was developed to work as a PD (Proportional Derivative) controller into a PD+I (PD+ Incremental) controller configuration to regulate the speed of a DC motor. The next two sections of this chapter are dedicated to explain how to achieve the codification of the type-1 and type-2 fuzzification stages.

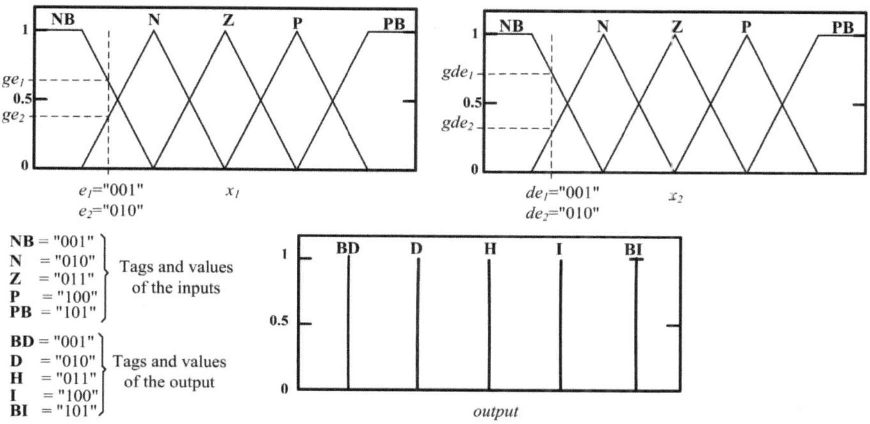

Fig. 14. Fuzzy inputs and outputs. Each fuzzy input has five MFs called Negative Big (NB), Negative (N), Zero (Z), Positive (P), Positive Big (PB). The output has five singletons called Big Decrease (BD), Decrease (D), Hold (H), Increase (I), Big Increase (BI).

5.1 Type-1 Fuzzification Stage: Active MFs

Fig 14 shows the linguistic variables for the inputs and for the output, these are: *error*, *change of error*, and *output*. Each input has five MFs, Negative Big (**NB**), Negative (**N**), Zero (**Z**), Positive (**P**), and Positive Big (**PB**); and each label an associated binary value, for example **NB** is associated with the value **"001"**, etc. The output has five singletons values called Big Decrease (**BD**), Decrease (**D**), Hold (**H**), Increase (**I**), and Big Increase (**BI**); hence, **BD** is associated with the binary value **"001"**, etc. [19, 18]. Additionally, it is shown that each input produces two fuzzified values; i.e., in input 1 (x_1) the dashed line indicates a crisp value that is being fuzzified, so it will produce the variables e_1 and e_2, associated with their corresponding degree of membership stored in variables ge_1 and ge_2. The variable content value for e_1 is "010", and "001" for e_2. Input x_2 works in a similar way. Figure 15 shows the obtained design entity for the aforementioned design, at the output are the active grades values and their corresponding linguistic terms. The variable values for e_1, e_2, ge_1, and ge_2 are assigned to output signals. After debugging, only some minor modifications have to be performed to the VHDL code to take the tested code to the

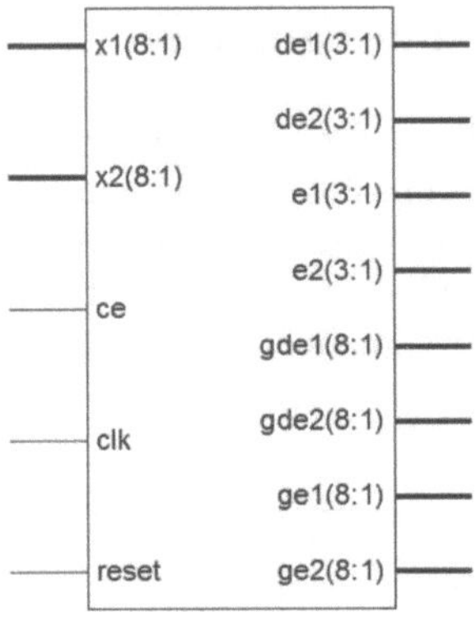

Fig. 15. Design entity of the fuzzification stage. The "clk" input is 50 Mhz for the Spartan 3 (XC3S1000). The "ce" (clock enable input) enables the algorithm in the Simulink. The "reset" input restart the variables of the algorithm.

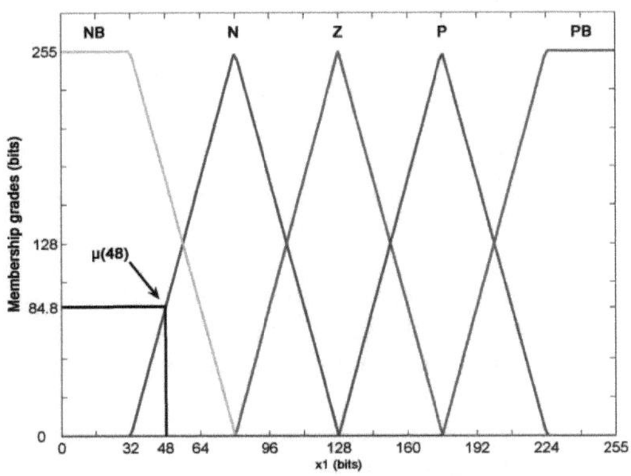

Fig. 16. MFS for the input x_1. An input $x1$=48 is fuzzified; i.e., $\mu(48) = 84.8$, because we are using an 8-bits binary codification system, rounding towards towards minus infinity, $\mu(48) \approx 84$.

physical implementation into the FPGA. In Fig. 16 we have a crisp input x_1=48 that is fuzzified using type-1 MFs.

5.2 Type-2 Fuzzification Stage: Active MFs

The implementation of the type-2 fuzzification stage for the average method [20, 21] is achieved using two type-1 fuzzification stages; in other words, two instances like the one shown in Figure 15 are used, the only difference between them is the parameters values of the MFs, with the purpose to represent the existing FOU in an interval type-2 MF, the rules and defuzzification stage are similar, although each

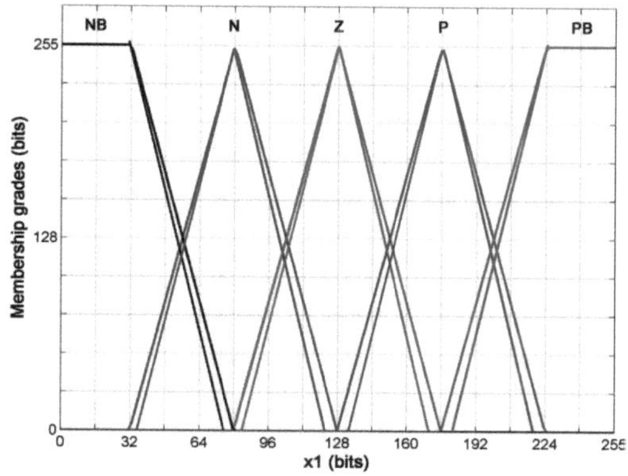

Fig. 17. MFs for the input x_1 of the type-2 fuzzification stage.

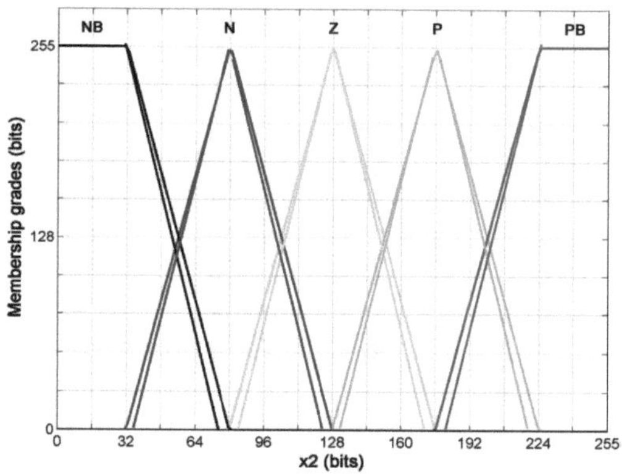

Fig. 18. MFs for the input x_2 of the type-2 fuzzification stage.

system performs independently, and only after each FIS has calculated its respective output the averaging operation is achieved. Figures 17 and 18 show the type-2 MFs used for the "error" and "change_of_error" inputs, each variable has five linguistic terms; however, to take this approach to practice we will need ten MFs for each linguistic variable (two MFs for each linguistic term); hence two FIS are used, the $FIS_u(X_1, X_2)$ for the $\overline{FOU}$ of X_1 ("error" input) and X_2 ("change of error" input); and a second FIS, the $FIS_l(X_1, X_2)$ for the $\underline{FOU}$ of X_1 and X_2. Note, that the signal "error" is connected to X_1, and the signal "change of error" to X_2 of the fuzzification entity in Fig. 15.

5.3 Simulink Model for the Type-2 Fuzzification Stage

Fig. 19 shows the Simulink model that was used to test the type-2 FIS. The type-1 FIS identified by the circled numbers 1, 2, and 3 represent the upper bound FOU;

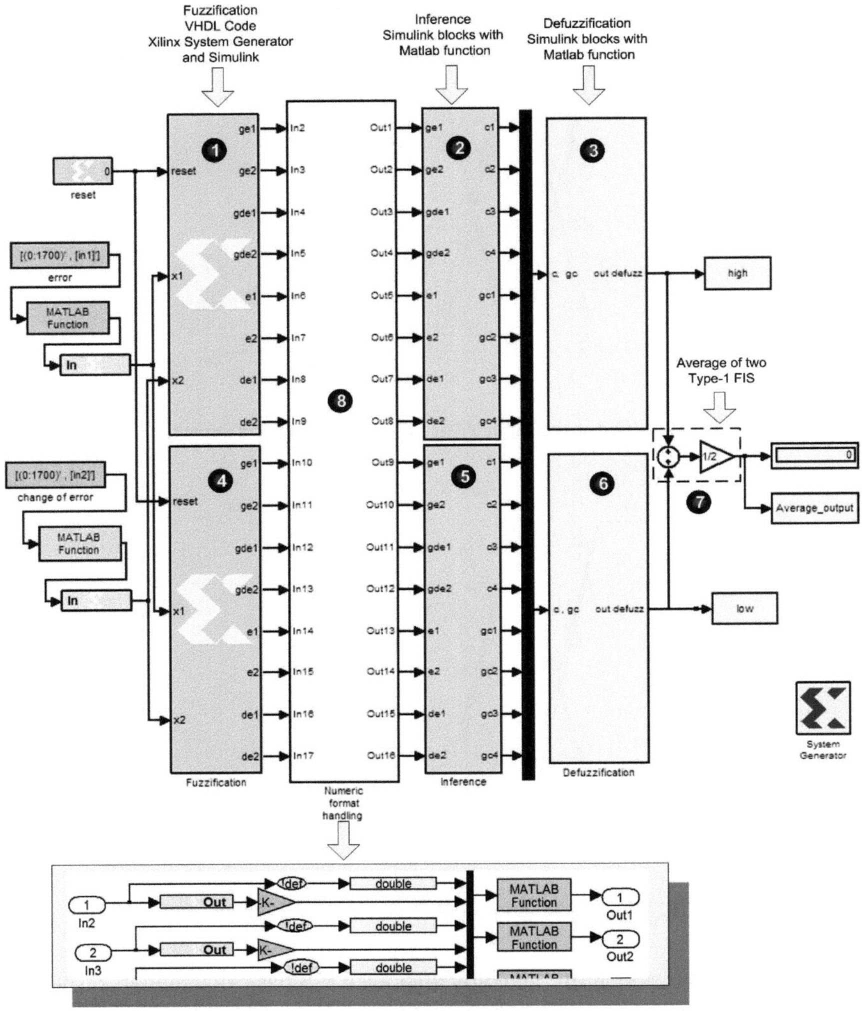

Fig. 19. Simulink model of the type-2 FIS (Average method) used to plot the control surface.

and the numbers 4, 5, and 6 identify the second type-1 FIS for the lower bound FOU. The average of the output of both FIS is achieved in the section marked with number 7. In the blocks 1 and 4 are the fuzzification stages, these blocks belongs to the XSG, they are used to import the VHDL code to Simulink. The inference engine of each type-1 FIS are in the blocks 2 and 5. The defuzzification stage of each type-1 FIS are identify by the numbers 3 and 6. Blocks 2, 3, 5, and 6 were programmed using Simulink blocks from the toolbox as well as Matlab functions, they were not programmed in VHDL because in this part the goal is only to test fuzzification stage. The subsystem block number 8 is to handling numeric format between the code of

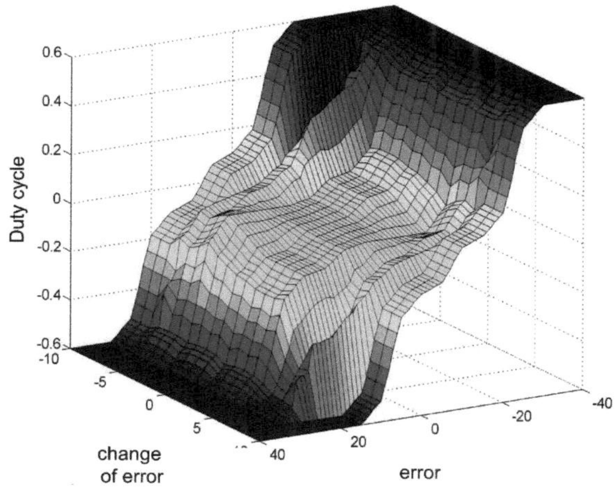

Fig. 20. Control surface obtained with the Simulink model (Average method) showed in Fig. 19.

the VHDL fuzzification stages and the Matlab inference engines, below this block there is a shadowed box showing a small section of the content of this subsystem, the rest is quite similar. The other small Simulink blocks that are at the input of both fuzzification blocks give two vectors of 1700 values each, for the "error" and "change of error" inputs in order to obtain the plot shown in Fig. 20, the small Simulink blocks that are shown between the the fuzzification and inference engine stages are used to achieve numeric compatibility from the numeric codification of real numbers used in the fuzzification stages coded in VHDL, to the floating point numeric representation used in Matlab.

5.4 Conclusions

One of the most stressing task in the development of new methods to improve a system is related with the use of new software and/or hardware technologies. This is specially true when the target of the developed software is code for a VLSI circuit, for example an FPGA. The use of type-2 FIS for real world applications has been plenty demonstrated in many works; so, the idea of taking some existing fuzzy controllers from type-1 to type-2 technology is very appealing. However, we have to expect that the programming cost in terms of developing time grew, as well as the size of code, and performance, this depending on the selected algorithm. There are several proposals to implement an interval type-2 FIS [3, 4, 5], but there is no way to go with little effort from interval type-1 to interval type-2 FIS. The average method provides an option to handle uncertainty similarly than an interval type-2 FIS does; moreover, the method makes easier the transition from type-1 to type-2 FIS, basically the only additional thing that designers will need do is to duplicate and modify the MFs values of the original type-1 FIS. This advantage is more emphatic when the type-1 FIS was coded using VHDL, because this language uses the concept of design entity that can contain a whole design that can be duplicated as many times as needed. Hence, the cost of programming time is reduced to the minimal, the code is practically twice the size of the type-1 FIS, and the performance of the hardware implementation of a type-2 is almost the same than the type-1 FIS, the small difference is due to the time that the electronics takes to perform an averaging operation, i.e; one addition and one divide operation. The Xilinx System Generator takes VHDL code to the Simulink environment offering advantages, since it is possible to analyze the system's behavior using the control surface, or achieving the control of the physical plant by interfacing the Simulink model with real world.

Acknowledgements. The authors thank Comisión de Operación y Fomento de Actividades Académicas del Instituto Politécnico Nacional, Instituto Tecnológico de Tijuana, and CONACYT for supporting our research activities.

References

1. Klir, G.J., Wierman, M.J.: Uncertainty-Based Information: Elements of Generalized Information Theory. Physica-Verlag (1999)
2. Cantor, G.: Contributions to the Founding of the Theory of Transfinite Numbers. Dover Publications Inc. (1915)
3. Wu, H., Mendel, J.M.: Uncertanty Bounds and Their use in the Design of Interval type-2 Fuzzy Logic Systems. IEEE Transactions on Fuzzy Systems 10, 1–16 (2002)
4. Karnik, N.N., Mendel, J.M., Liang, Q.: Type-2 Fuzzy Logic Systems. IEEE Transactions on Fuzzy Systems 7, 643–658 (1999)
5. Liang, Q., Mendel, J.M.: Interval Type-2 Fuzzy Logic Systems: Theory and Design. IEEE Trans. on Fuzzy Systems 8, 535–550 (2000)
6. Jang, J.S.R., Sun, C.T., Mizutani, E.: Neuro-Fuzzy and Soft Computing. Prentice-Hall, Englewood Cliffs (1997)
7. Mendel, J.M.: Type-2 Fuzzy Sets and Systems: an Overview. IEEE Computational Intelligence Magazine 2, 20–29 (2007)
8. Mendel, J.M.: Type-2 Fuzzy Sets: Some Questions and Answers. IEEE Neural Networks Society, United States (2003)
9. Mendel, J.M., John, R.I.B.: Type-2 Fuzzy Sets Made Simple. IEEE Transactions on Fuzzy Systems 10, 117–127 (2002)
10. Mendel, J.M.: Uncertain Rule-Based Fuzzy Logic Systems: Introduction and New Directions. Prentice-Hall, Upper-Saddle River (2001)
11. Karnik, N., Mendel, J.M.: Centroid of a Type-2 Fuzzy Set. Information Sciences 132, 195–220 (2001)
12. Lee, K.H.: Advances in Soft Computing: First Course on Fuzzy Theory and Applications. Springer, Germany (2005)
13. Lago, E., Jiménez, C.J., López, D.R., Sánchez-Solano, S., Barriga, A.: XFVHDL: A Tool for the Synthesis of Fuzzy Logic Controllers. Design Automation and Test in Europe, 102–107 (1998)
14. Melgarejo, M., Peña-Reyes, C.A.: Implementing Interval type-2 Fuzzy Processors. IEEE Computational Intelligence Magazine 2, 63–71 (2007)
15. Miguel, A., Melgarejo, R., Peña-Reyes, C.A.: Hardware architecture and FPGA implementation of a type-2 fuzzy system. In: Proceedings of the 14th ACM Great Lakes symposium on VLSI, pp. 458–461 (2004)

16. Melgarejo, M.A., Garcia, R.A., Peña-Reyes, C.A.: Pro-Two: a hardware based platform for real time type-2 fuzzy inference. In: Proceedings of 2004 IEEE International Conference on Publication Fuzzy Systems, vol. 2, pp. 977–982 (2004)
17. Cirstea, M.N., Dinu, A., Khor, J.G., McCormick, M.: Neural and Fuzzy Logic Control of Drives and Power System, Newnes (2002)
18. Montiel, O., Maldonado, Y., Sepúlveda, R., Castillo, O.: Simple tuned fuzzy controller embedded into an FPGA. In: 2008 NAFIPS Conference Proceedings, pp. 1–6 (2008)
19. Montiel, O., Olivas, J., Sepúlveda, R., Castillo, O.: Development of an Embedded Simple Tuned Fuzzy Controller. In: 2008 IEEE World Congress on Computational Intelligence (WCCI 2008), pp. 555–561 (2008)
20. Sepúlveda, R., Castillo, O., Melín, P., Díaz, A.R., Montiel, O.: Experimental Study of Intelligent Controllers under Uncertainty using type-1 and type-2 Fuzzy Logic. Information Sciences 177, 2023–2048 (2007)
21. Sepúlveda, R., Castillo, O., Melín, P., Montiel, O.: An Efficient Computational Method to implement Type-2 Fuzzy Logic in Control Applications. In: Analysis and Design of Intelligent Systems Using Soft Computing Techniques. Advances in soft computing, vol. 41, pp. 45–52. Springer, Heidelberg (2007)
22. Vuong, P.T., Madni, A.M., Voung, J.B.: VHDL Implementation for a Fuzzy Logic Controller. In: World Automation Congress, WAC apos;2006, pp. 1–8 (2006)
23. Iregui, S., Linares, D., Melgarejo, M.: Performance Evaluation of Fuzzy Operators for FPGA Technology. In: NAFIPS 2008, New York (2008)
24. Sánchez-Solano, S., Senhadji, R., Cabrera, A., Baturone, I., Jiménez, C.J., Barriga, A.: Prototyping of Fuzzy Logic Based Controllers Using Standard FPGA Development Boards. In: 13th IEEE International Workshop on Rapid System Prototyping on Volume, pp. 25–32 (2002)
25. Sánchez Solano, S., Barriga, A., Jiménez, C.J., Huertas, J.L.: Design and Application of Digital Fuzzy Controllers. In: Sixth IEEE International Conference on Fuzzy Systems (FUZZ-IEED 1997), Barcelona, Spain, vol. 2, pp. 869–874 (1997)
26. Maldonado, Y., Montiel, O., Sepuúlveda, R., Castillo, O.: Design and simulation of the fuzzification stage through the Xilinx System Generator. In: Soft Computing for Hybrid Intelligent Systems, vol. 154, pp. 297–305. Springer, Heidelberg (2008)
27. Zadeh, L.A.: The concept of a linguistic variable and its application to approximate reasoning. Information Sciences 8, 199–249 (1975)
28. Zadeh, L.A.: Fuzzy sets. Information and Control 8, 338–353 (1965)

Methodology to Test and Validate a VHDL Inference Engine of a Type-2 FIS, through the Xilinx System Generator

Roberto Sepúlveda[1], Oscar Montiel[1], José Olivas[2], and Oscar Castillo[3]

[1] Centro de Investigación y Desarrollo de Tecnología Digital del Instituto Politécnico Nacional (CITEDI- IPN), Av. del Parque No.1310, Mesa de Otay, 22510, Tijuana, B.C., México
r.sepulveda@ieee.org, o.montiel@ieee.org
[2] M.S. Student at CITEDI-IPN
olivas@citedi.mx
[3] Division of Graduate Studies and Research, Calzada Tecnológico S/N, Tijuana, B.C., México
ocastillo@hafsamx.org

Abstract. In this paper an improved high performance type-1 inference engine (IE) is proposed that can be applied with no modifications to the implementation of a type-2 FIS using the average method. The performance of the type-2 FIS will not be diminish for the use of this stage since it is achieved in parallel. The proposals are focused to be implemented into an FPGA. Simulink models to test the type-1 and type-2 inference engines are presented. The type-2 IE was tested in a speed controller for a DC motor.

1 Introduction

There have been published several papers about the implementation of type-1 and type-2 fuzzy inference systems (FIS) in [1, 4, 6, 8, 9]. Good references for the required fuzzy mathematics to implement type-1 and type-2 FIS are [3, 7, 10, 12]. In [25] was explained how to make a hardware implementation of the fuzzification stage for type-1 FIS, considering integers and real numbers, the results can be easily extended for the type-2 case using the [17, 18] method. Other proposals in the same line of type-1 FIS are [11, 19, 20, 21, 22, 23, 24]; and for type-2 FIS [13, 14, 15]. The use of the software Xilinx System Generator (XSG) to import the FIS VHDL code to the Simulink environment for testing the entities was explained in [25], also a model that allows to plot the control surface to test the fuzzification stage for an application was given.

Considering a FIS with crisp inputs and crisp outputs, the fuzzification is the first stage of any FIS. In this paper, a method to achieve the hardware implementation of the second stage, the inference engine for type-1 and type-2 FIS is presented. The designs are oriented to achieve high performance computations and with a slightly

O. Castillo et al. (Eds.): Evolutionary Design of Intelligent Systems, SCI 257, pp. 295–308.
springerlink.com

different architecture from other proposal [16]. In order to accomplish this goal several strategies has been followed, some of them are:

- The use of active rules. The idea is to consider only the rules affected by the antecedents. This has pros and cons, for example, it offers an easier implementation and provides a faster system, however, it lacks of flexibility.
- The use of integers and real numbers. It uses the encoding method called Decimal Floating Point Binary Encoding (DFPBE).
- The use of the average method. This method is based in handling the uncertainty of the type-2 fuzzy set (FS) of an IT2 FIS by two type-1 FIS located at the uncertainty boundaries of the fuzzy sets.
- The average method allows to go from a type-1 FIS to a type-2 FIS with a minimal of effort, since the only thing that is necessary to do is to split the type-2 FIS in two type-1 FIS; then, the required fuzzy mathematics is the same used for type-1 FIS.
- An improved inference engine is presented, the previous proposal presented in [26] was optimized for faster computations.

This paper is organized as follows: Section 2 is devoted to briefly comment on the origins of Fuzzy Logic (FL), the foundations of IT2 FIS are presented with special emphasis on the inference engine. Section 3 explains the advantages of using the average method to achieve type-2 fuzzy inferences, moreover specific nomenclature for the average method is proposed. In Section 4, the improved architecture to achieve hardware implementation of the type-1 and type-2 inference engines is presented. In Section 5, three Simulink models to test the VHDL code are given: The first model is a type-1 inference engine, which purpose is to verify the accuracy of the results considering that the inputs were individual values given by hand. The second model is a type-2 inference engine that was tested in a similar way than the first model. The third model is a complete type-2 FIS where the source of the inference engine is VHDL code imported to Simulink through the Xilinx System Generator (XSG), the fuzzification and defuzzification stages are models from the Simulink, the aim of this model is to test the type-2 inference engine behavior in a controller application, for this purpose each input linguistic variable was fed with a vector value that covers the whole universe of discourse, the result is the surface control of the system. Finally, in Section 6 the conclusions of this work are given.

2 Type-2 Fuzzy Logic Systems

After almost 40 years that fuzzy set theory was proposed by Zadeh [28], and hundreds of applications working successfully; ten years later, the concept of type-2 fuzzy set was introduced by Zadeh in [27], to cope with uncertainty from words, because "words mean different things for different people, and are therefore uncertain" [5]. Type-1 FISs are unable to handle uncertainties directly, because they use type-1 fuzzy sets that are certain [7]. On the other hand, type-2 FLSs, are very useful in circumstances where it is difficult to determine an exact membership function because there is uncertainty from many sources. Type-2 fuzzy sets are more difficult

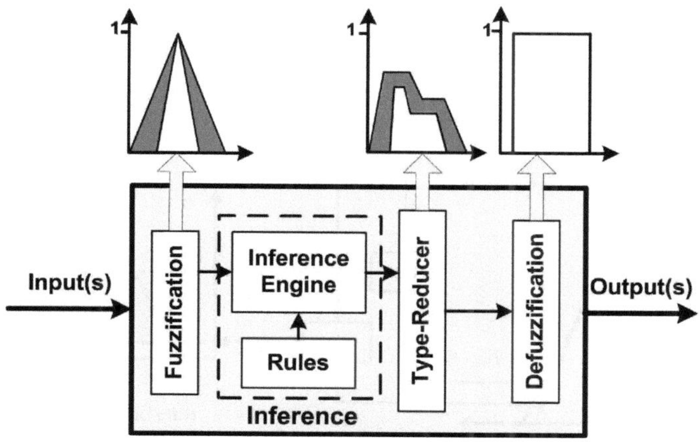

Fig. 1. Typical schematic representation of a type-2 FIS.

to use and understand than type-1 fuzzy sets; hence, their use is not widespread yet. We can say that a FIS is type-2 if there is at least one type-2 FS. An interval type-2 Fuzzy Inference System (IT2 FIS), contains four components: fuzzification, inference engine, type-reducer, and defuzzification, which are connected as it is shown in Fig. 1. The IT2 FIS can be seen as a mapping from the inputs to the output and it can be interpreted quantitatively as $Y = f(X)$, where $X = \{x_1, x_2, \cdots, x_n\}$ are the inputs to the IT2 FIS f, and $Y = \{y_1, y_2, \cdots, y_n\}$ are the defuzzified outputs. Note that, in Fig. 1 that the fuzzification is achieved using type-2 MFs, the output of the inference engine are type-2 FSs; therefore, a type reducer to go from type-2 to type-1 FS, and a defuzzifier to obtain crisp outputs are needed.

The rules are the core of a FIS, they can be obtained from human experts (Models for words approach), or extracted without the intervention of humans by the use of a numerical method (Abstract Mathematics approach) [5]. It is usual to express rules using the **if-then** nomenclature; so, for a IT2 FIS with p inputs $x_1, \ldots, x_p \in X_p$, and one output $y \in Y$, with M rules, the ith rule is given by,

$$R^i : \textbf{If } x_1 \text{ is } \tilde{F}_1^i \text{ and} \cdots \text{and } x_p \text{ is } \tilde{F}_p^i, \textbf{ then } y \text{ is } \tilde{G}^i \qquad i = 1, \ldots\ldots, M.$$

where $\tilde{F}_k^i$ are the antecedents, the MFs are given by $\mu_{\tilde{F}_1^i}(x_k)(k = 1, \ldots, p)$, and $\tilde{G}^i$(y) are the consequents.

This rule is a type-2 relation from the input space $X_1 \times \ldots \times X_p$ to the output space Y of the IT2 FIS.

2.1 Inference Engine of an Interval Type-2 FIS

Figure 2 depicts the input and antecedent operations for a two-antecedent single-consequent rule. First, the fuzzification is achieved; for the first crisp input $x_1 = x_1'$, the dashed vertical line at x_1' can intersect $FOU(\bar{F}_1)$ at any place in the interval

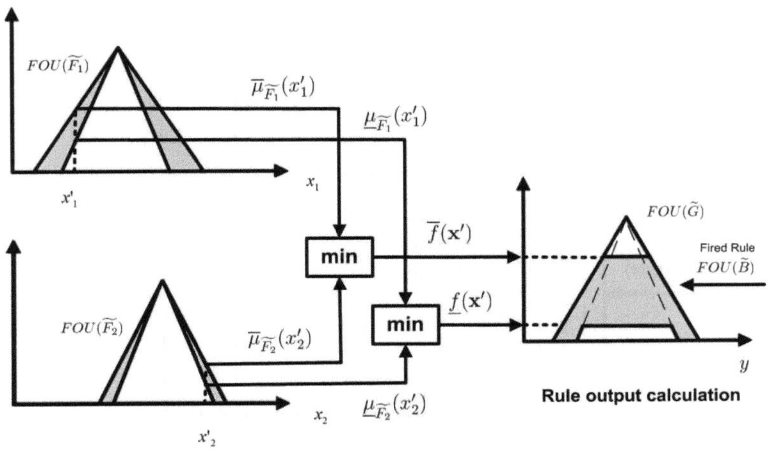

Fig. 2. Pictorial description for type-2 inference.

$[\underline{\mu}_{F_1}(x'_1), \overline{\mu}_{F_1}(x'_1)]$; and, for the second crisp input $x_2 = x'_2$, the dashed vertical line at x'_2 can intersect $FOU(\tilde{F}_2)$ everywhere in the interval $[\underline{\mu}_{F_2}(x'_2), \overline{\mu}_{F_2}(x'_2)]$. Then the lower firing level $\underline{f}(\mathbf{x}')$, and the upper firing level $\overline{f}(\mathbf{x}')$ are computed using the minimum t-norm, as follows:

Lower firing level,

$$\underline{f}(\mathbf{x}') = min[\underline{\mu}_{F_1}(x'_1), \underline{\mu}_{F_2}(x'_2)]$$

Upper firing level,

$$\overline{f}(\mathbf{x}') = min[\overline{\mu}_{F_1}(x'_1), \overline{\mu}_{F_2}(x'_2)]$$

For the purpose of this work, it is very important to note that the firing interval $F(\mathbf{x}') = [\underline{f}(\mathbf{x}'), \overline{f}(\mathbf{x}')]$ was obtained as the result of the input and antecedent operations. To obtain the output rule calculation, the firing interval $F(\mathbf{x}')$ is t-normed with the consequent $FOU(\widetilde{G})$; i.e., $\underline{f}(\mathbf{x}')$ is t-normed with the lower bound of $FOU(\widetilde{G})$, and $\overline{f}(\mathbf{x}')$ is t-normed with the upper bound of $FOU(\widetilde{G})$. The lower and upper bounds of $FOU(\widetilde{G})$ are usually written as $LMF(\widetilde{G})$ and $UMF(\widetilde{G})$, respectively. Figure 2 shows the resulting FS, $FOU(\widetilde{B})$, considering a triangular $FOU(\widetilde{G})$, and the minimum for the t-norm.

3 The Average Type-2 FIS

It is known that the type reduction stage is a bottle neck in a type-2 FIS and an option to avoid it is [2], also the average method [17, 18] is a proposal to handle interval type-2 FIS avoiding type reduction, which is achieved by substituting the IT2 FIS with two type-1 FIS located adequately at the uncertainty boundaries of the type-2 MFs, the rules and the defuzzification at each FIS remain being identical, the output is different because the MFs of each FIS are different. The type-2 crisp output is calculated taken the average of the crisp output of the type-1 FISs, we used the arithmetic average.

Average type-2 FIS is consistent with the nomenclature of interval type-2 FIS; so, Figure 2 is still being useful to explain the inference engine of the average method. The first step to split the IT2 FIS is to begin defining that one of the FIS is going to be used for the lower bound of the uncertainty, FIS_l; and the second one for the upper bound, FIS_u. Hence, for the linguistic variables X_1 and X_2 of an IT2 FIS, we will write $FIS_l(X_1)$ for the lower bound of uncertainty, and $FIS_u(X_2)$ for the upper bound, hence for a IT2 FIS with two linguistic variables as inputs, X_1 and X_2, valid nomenclature is $FIS_l(X_1,X_2)$, and $FIS_u(X_1,X_2)$. Following a similar idea of nomenclature, the linguistic terms of the IT2 FIS of Figure 2, $FOU(\widetilde{F}_1)$ and $FOU(\widetilde{F}_2)$, will be written as $FIS_l(\widetilde{F}_1,\widetilde{F}_2)$, and $FIS_u(\widetilde{F}_1,\widetilde{F}_2)$ to indicate that the $FIS_l(\widetilde{F}_1,\widetilde{F}_2)$ was developed considering the lower bound of the type-2 linguistic terms $\widetilde{F}_1$ and $\widetilde{F}_2$; similarly, $FIS_u(\widetilde{F}_1,\widetilde{F}_2)$ indicates the use of the upper bounds of uncertainty of the mentioned terms. The approximated type-2 FIS output is calculated by taking the arithmetic average of the outputs of the defuzzifier processors of each FIS; i.e., the FIS_l and FIS_u. Note that the type reduction stage was avoided.

4 Inference Engine for Type-1 and Type-2 Proposals

In Section 3 the advantages of using the Average method to handle a type-2 FIS were explained; in short, the method works with the uncertainty boundaries using type-1 fuzzy mathematics to carry out the whole process, therefore the inference engine of a type-1 system is also used for the average type-2 proposal.

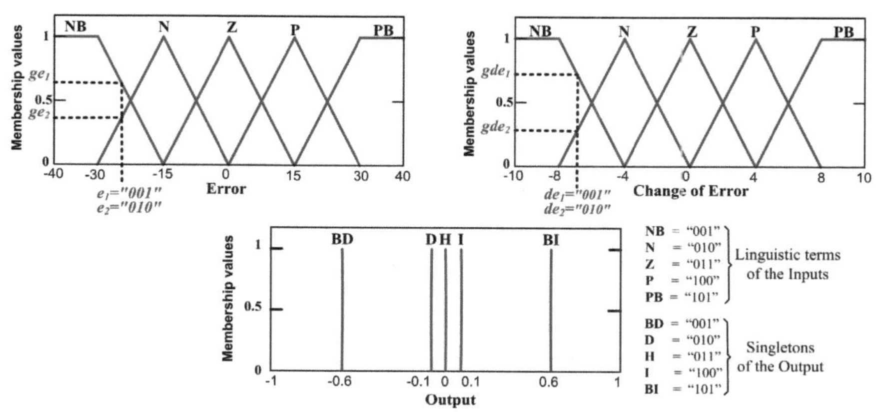

Fig. 3. Linguistic Variables for the inputs and output: Error, change of error, and output with their linguistic terms

4.1 Hardware Implementation of the Type-1 Inference Engine

Fig. 4 shows the interaction of the fuzzification stage, and the inference engine for a type-1 FIS. The main block of this picture labeled as "Inference Engine" is a VHDL entity, and it gives a general idea of its internal architecture. The Inference Engine Entity (IEE) has nine inputs and eight outputs, eight of inputs came from the fuzzification stage, the ninth input is a clock signal; with respect to the eight outputs, four are for the linguistic terms, the rest correspond to their firing strengths.

The fuzzification stage has two input variables: "Error" and "Change of Error". Each linguistic variable has five terms identified by linguistic tags, and each tag is related with three binary digits as follows, "Negative Big" (NB) with the value of "001", "Negative" (N) with "010", Zero (Z) with '011", "Positive" with "100", and Positive Big (PB) with "101". For this architecture, the maximal number of active membership function for each input is two, so in Fig. 3 at the input "Error" there is a crisp value of ≈ -25 that is activating the linguistic terms NB ("001") and N ("010"), and the corresponding membership values are ≈ 0.66 and ≈ 0.33, the active membership functions (linguistic terms) and membership values are assigned to VHDL variables, in the figure, $e_1 =$"001" (NB), $e_2 =$"010" (N), for the former, the membership value is $ge_1 = 0.66$, and for the latter, $ge_2 = 0.33$. Similarly, the second input "Change of Error" is handled, the linguistic terms are assigned to the VHDL variables de_1 and de_2, and the membership values are gde_1 and gde_2. Finally, the four VHDL variables of each input are assigned to VHDL output signals that are the inputs to the next stage; i.e. the IEE. The fuzzification Stage Entity (FSE) was designed using the behavioral style of VHDL codification, this style use sequential code, and it only needs one clock pulse to perform the fuzzification.

At the input of the IEE of Fig. 4 are the linguistic terms (e_1, e_2, de_1, and de_2), their membership values (ge_1, ge_2, gde_1, and gde_2), and the clock signal (clk). All the

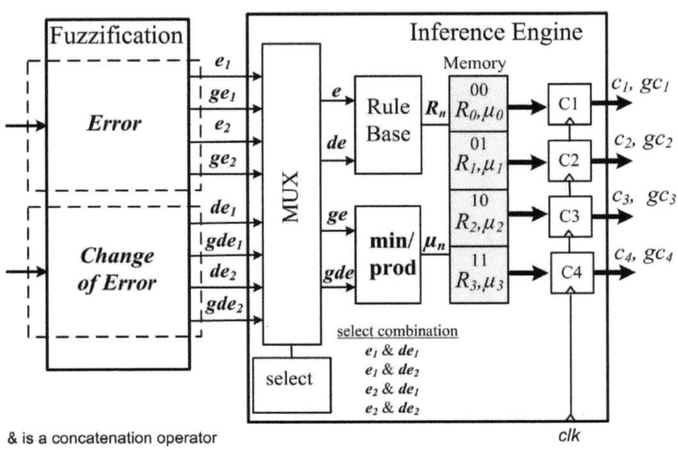

Fig. 4. Overview of the inference engine architecture.

```
case ante is
    when "001001" => c_n <= BI;
    when "001010" => c_n <= BI;
    when "001011" => c_n <= BI;
    when "001100" => c_n <= BI;
    when "001101" => c_n <= BI;
    when "010001" => c_n <= BI;
    when "010010" => c_n <= I;
    when "010011" => c_n <= I;
    when "010100" => c_n <= H;
    when "010101" => c_n <= I;
    when "011001" => c_n <= I;
    when "011010" => c_n <= H;
    when "011011" => c_n <= H;
    when "011100" => c_n <= H;
    when "011101" => c_n <= D;
    when "100001" => c_n <= D;
    when "100010" => c_n <= H;
    when "100011" => c_n <= D;
    when "100100" => c_n <= D;
    when "100101" => c_n <= BD;
    when "101001" => c_n <= BD;
    when "101010" => c_n <= BD;
    when "101011" => c_n <= BD;
    when "101100" => c_n <= BD;
    when "101101" => c_n <= BD;
    when others => c_n <= "000";
end case;
```

Fig. 5. VHDL code of the rule base using the Mamdani method

inputs, except *clk*, enter into a parallel selection process through a multiplexer; note that although the IEE uses sequential code, the circuits into the process are placed in parallel; the degree of parallelism can be tailored by an adequate codification style, in our case all the rules are processed in parallel, the eight outputs are obtained at the same time because the "clk" signal synchronize the process, hence this stage needs only one clock cycle to provide the output.

In the IEE, the required circuits to implement the MUX and Select blocks are placed in parallel, and they produce four outputs; by the concatenation of two of them, *e* and *de* the active rules are identified, then using the piece of code of Fig. 5 the corresponding consequent is ascertained. For example, for the combination of e_1 ="010" and e_2 ="010" the value of the variable "ante" is "010010" and the linked consequent is "I" with a binary value of "100", as it can be seen in Fig. 3, the common rule notation is shown next as rule number 4.

1. If e_1 is "001" and de_1 is "001" then C_1 is BI
2. If e_1 is "001" and de_2 is "010" then C_2 is BI
3. If e_2 is "010" and de_1 is "001" then C_3 is BI
4. If e_2 is "010" and de_2 is "010" then C_4 is I
5. $\vdots$

The other two outputs of the MUX block are the membership values "ge" and "gde", the minimal t-norm of the corresponding rule is calculated to obtain the firing strength. The e_1 and de_1 combination and its linked consequent are saved into VHDL variables, this has been illustrated in Fig. 4 as the memory position "00" that contains the consequent stored in R_0 and its firing strength stored in μ_0; the rule combination e_1 and de_2 and its consequent are stored in R_1 and in μ_1 in the memory position "01", etc.

For example, if the memory position "00" has the consequent of the active rule 1, i.e. the combination '001001", then the consequent is BI ("101"); if the memory position "01" has the consequent of the active rule 2, the combination '001010", then the consequent is BI; if the memory position "10" has the consequent of the active rule 3, the consequent of the rule combination "010001" is also BI; and so forth.

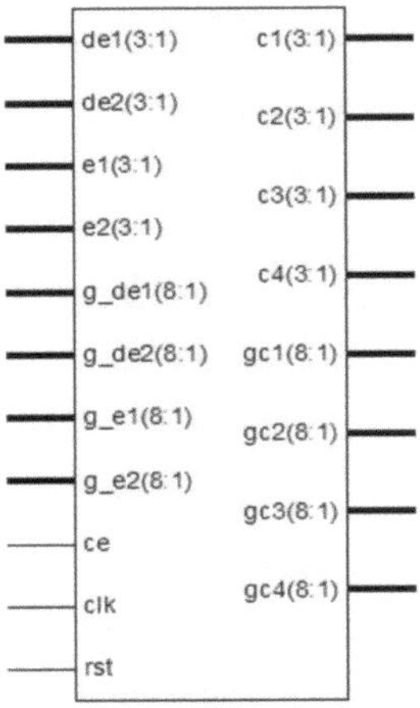


```
entity inference is
Port (
rst      : in std_logic;
clk      : in std_logic;
ce       : in std_logic;
e1       : in std_logic_vector (m downto 1);
e2       : in std_logic_vector (m downto 1);
de1      : in std_logic_vector (m downto 1);
de2      : in std_logic_vector (m downto 1);
g_e1     : in std_logic_vector (n downto 1);
g_e2     : in std_logic_vector (n downto 1);
g_de1    : in std_logic_vector (n downto 1);
g_de2    : in std_logic_vector (n downto 1);
c1       : out std_logic_vector (m downto 1);
c2       : out std_logic_vector (m downto 1);
c3       : out std_logic_vector (m downto 1);
c4       : out std_logic_vector (m downto 1);
gc1      : out std_logic_vector (n downto 1);
gc2      : out std_logic_vector (n downto 1);
gc3      : out std_logic_vector (n downto 1);
gc4      : out std_logic_vector (n downto 1));
end inference;
```

Fig. 6. Entity of the type-1 inference engine.

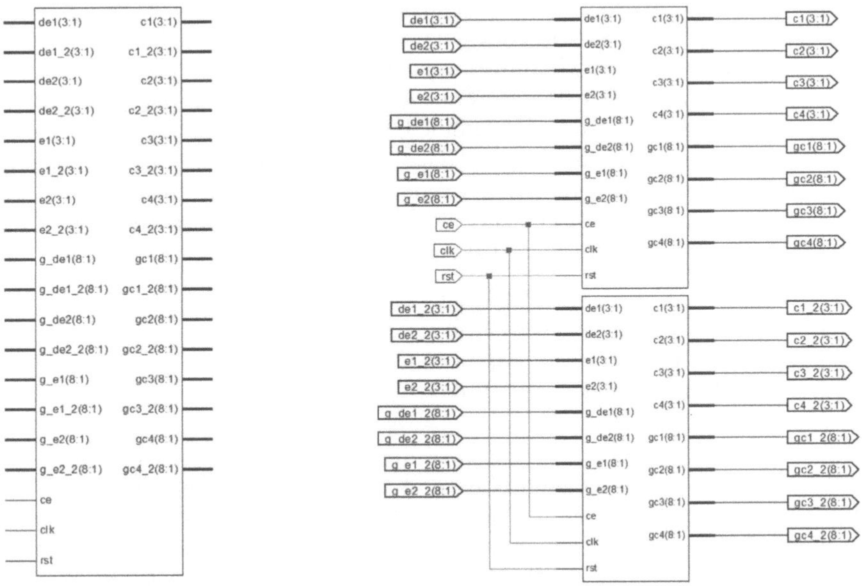

(a) (b)

Fig. 7. Entity of the type-2 inference engine; (a) General entity, (b) Internal entity.

The "Memory" block in Fig. 4 at the output is connected to four blocks named C_1 to C_4, each C_x block represents two VHDL signals where the triggered consequent c_x and its membership value gc_x are saved and sent to the entity output at the next clock cycle. Note that this method only needs four C_x blocks because the maximal number of active rules is four, and that the maximum required time to perform an inference is only one clock cycle.

Fig. 6 shows the IEE that was produced by the Xilinx ISE software as result of the synthesis process. This entity shows two extra inputs, "Chip Enable" (ce) and "Reset" (rst), they were added only to be compatible with the Xilinx System Generator (XSG) and Simulink.

4.2 Hardware Implementation of the Type-2 Inference Engine

In Section 3 was explained the advantages of using the average type-2 method to compute IT2 FIS. Here, we are going to focus just in the implementation. Fig. 7(a) shows the top level design of the type-2 VHDL entity of the Inference Engine. Fig. 7(b) shows one step into the type-2 Inference Engine, there are two type-1 Inference Engines connected in parallel, following the idea of representing FIS_u and FIS_l of the whole system. To go from type-1 to type-2 FIS using the average method, this stage do not need any modification because the uncertainty is handled at the Membership Function level and not at the inference level.

5 Models to Test the Type-1 and Type-2 Inference Engines

To test the VHDL inference engines the code generated with the Xilinx ISE was imported to Simulink using the XSG. We choose the application of the speed control of a DC motor to design the FIS, for the type-1 FIS we used the membership functions showed in Fig. 3, the corresponding Simulink Model is shown in Fig. 8. The

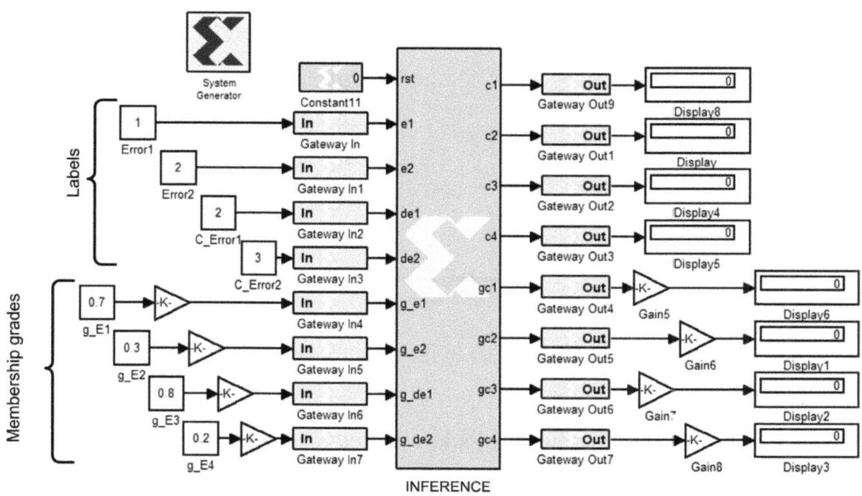

Fig. 8. Simulink model to test the type-1 Inference Engine.

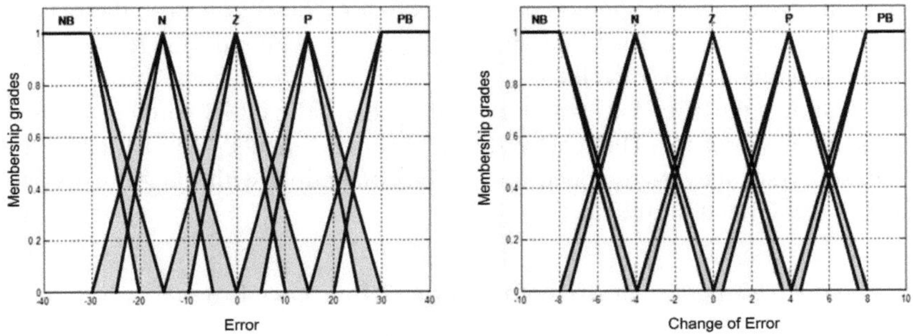

Fig. 9. Interval type 2 membership functions, for the inputs.

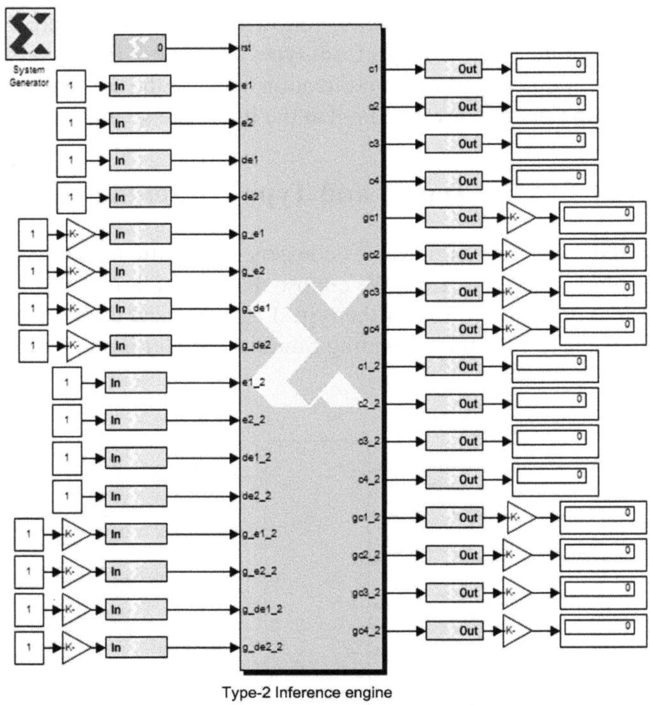

Fig. 10. Simulink model of the type-2 inference engine.

blocks with "K" and the "Gateway" blocks are used to adapt the inputs from floating to binary representation (for integers or real numbers), and the outputs from binary representation (integer or real numbers) to floating point, this is necessary because the inputs in this model are floating point numbers and the inputs and outputs handle real numbers using the special codification explained in the chapter "Design and

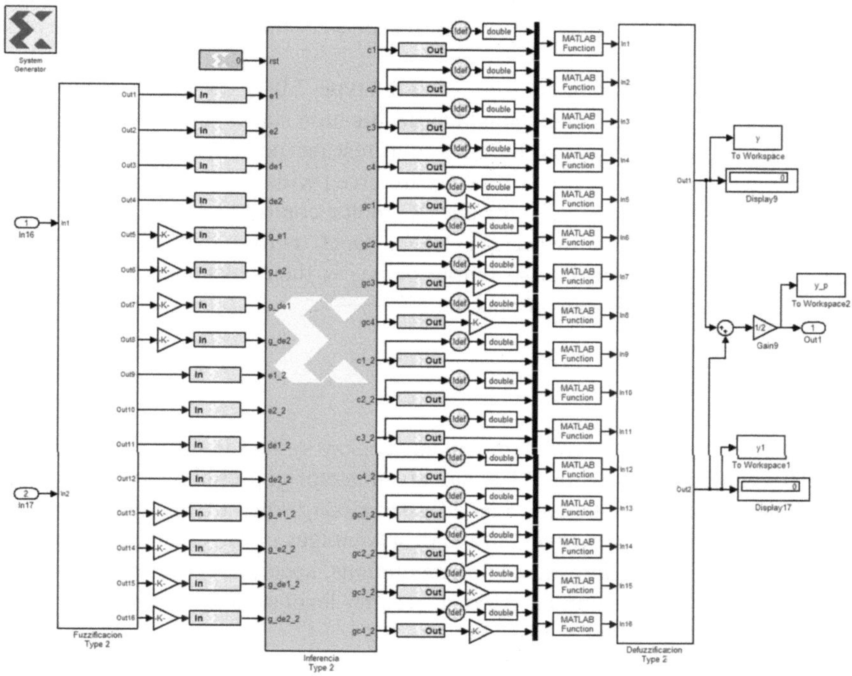

Fig. 11. Simulink model of a type-2 FIS to test the type-2 Inference Engine.

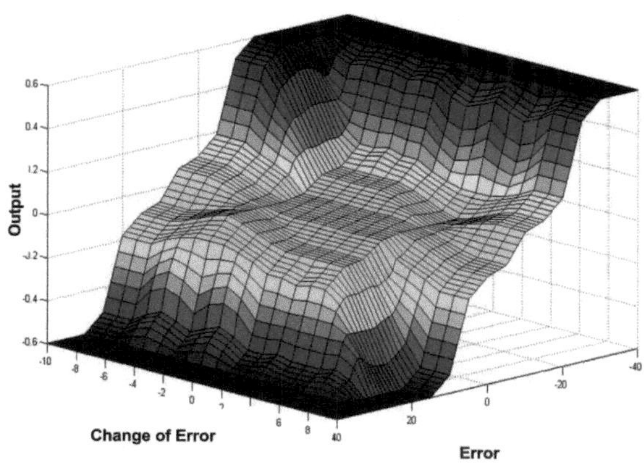

Fig. 12. Surface control of the type-2 FIS.

Simulation of the Type-2 Fuzzification Stage: Using Active Membership Functions" of this book.

For the type-2 FIS, the type-2 MF of Fig. 9 were implemented, Fig. 10 shows the type-2 test version of the type-1 showed in Fig. 8. The idea of this type-2 model is to provide values by hand and review the outputs that are sent to displays to analyze them.

Fig. 11 shows a second model for testing the type-2 IEE, the idea is to complete the FIS with the Inference Engine and Defuzzification stages, these two last stages were not coded in VHDL since the idea is to test the performance of the IEE, at model's inputs complete vectors of values are given with the aim of generating the control surface shown in Fig. 12 for the DC motor control application. At the defuzzification stage there are two crisp outputs, "Out1" is from the FIS_u, and "Out2" is from the FIS_l, the type-2 crisp output values is then calculated by taking the arithmetic average of this two outputs; i.e., $\frac{Out_1+Out_2}{2}$.

6 Conclusions

The most widely adopted way to implement a fuzzy system is software into a non-dedicated computer, this method has the advantage of giving high level support to the designer since there are many friendly specializing software that helps to obtain results in short time, however one of the disadvantages of the commented solution is the processing time. Hardware implementations, specifically those designs that have been carry out to be implemented at circuitry level offer the bigger advantages to designers to provide high speed solutions.

In this paper, an improved type-1 inference engine was proposed for hardware implementation, oriented to be implemented into an FPGA. The type-1 engine can be applied with no modifications to implement a type-2 FIS using the average method. The type-1 engine process all the rules in parallel providing high speed computations, the processing time of the whole inference engine is just one clock cycle, approximately 0.02 microseconds for the Spartan 3. The processing time of a type-2 system implemented with the type-1 inference engine will not grow up since both inference engine are connected in parallel, hence the processing time remains being the same for this stage.

Three Simulink models were presented for testing the inference engine for type-1 and type-2 FIS. One of the Simulink models was achieved to test the type-2 inference engine focused to control the speed of a DC motor, the resulting control surface is shown. We conclude that the inference engine worked fine.

Acknowledgement. The authors thank Comisión de Operación y Fomento de Actividades Académicas del Instituto Politécnico Nacional, Instituto Tecnológico de Tijuana, and CONACYT for supporting our research activities.

References

1. Hagras, H.: A hierarchical type-2 fuzzy logic control architecture for autonomous mobile robots. IEEE Transactions on Fuzzy Systems (2004)
2. Wu, H., Mendel, J.M.: Uncertanty Bounds and Their use in the Design of Interval type-2 Fuzzy Logic Systems. IEEE Transactions on Fuzzy Systems 10, 1–16 (2002)

3. Jang, J.S.R., Sun, C.T., Mizutani, E.: Neuro-Fuzzy and Soft Computing. Prentice-Hall, Englewood Cliffs (1997)
4. Mendel, J.M.: Type-2 Fuzzy Sets and Systems: an Overview. IEEE Computational Intelligence Magazine 2, 20–29 (2007)
5. Mendel, J.M.: Fuzzy Sets for Words: a New Beginning. In: Proc. of IEEE International Conf. on Fuzzy Systems, St. Louis, MO, pp. 37–42 (2003)
6. Mendel, J.M., Bob John, R.I.: Type-2 Fuzzy Sets Made Simple. IEEE Transactions on Fuzzy Systems 10, 117–127 (2002)
7. Mendel, J.M.: Uncertain Rule-Based Fuzzy Logic Systems: Introduction and New Directions. Prentice-Hall, Upper-Saddle River (2001)
8. Karnik, N.N., Mendel, J.M.: Qilian Liang: Type-2 Fuzzy Logic Systems. IEEE Transactions on Fuzzy Systems 7, 643–658 (1999)
9. Karnik, N., Mendel, J.M.: Centroid of a Type-2 Fuzzy Set. Information Sciences 132, 195–220 (2001)
10. Lee, K.H.: Advances in Soft Computing: First Course on Fuzzy Theory and Applications. Springer, Germany (2005)
11. Lago, E., Jiménez, C.J., López, D.R., Sánchez-Solano, S., Barriga, A.: XFVHDL: A Tool for the Synthesis of Fuzzy Logic Controllers. Design Automation and Test in Europe, 102–107 (1998)
12. Liang, Q., Mendel, J.M.: Interval Type-2 Fuzzy Logic Systems: Theory and Design. IEEE Trans. on Fuzzy Systems 8, 535–550 (2000)
13. Melgarejo, M., Peña-Reyes, C.A.: Implementing Interval type-2 Fuzzy Processors. IEEE Computational Intelligence Magazine 2, 63–71 (2007)
14. Melgarejo, M.A.R., Peña-Reyes, C.A.: Hardware architecture and FPGA implementation of a type-2 fuzzy system. In: Proceedings of the 14th ACM Great Lakes symposium on VLSI, pp. 458–461 (2004)
15. Melgarejo, M.A., Garcia, R.A., Peña-Reyes, C.A.: Pro-Two: a hardware based platform for real time type-2 fuzzy inference. In: Proceedings of 2004 IEEE International Conference on Publication Fuzzy Systems, vol. 2, pp. 977–982 (2004)
16. Cirstea, M.N., Dinu, A., Khor, J.G., McCormick, M.: Neural and Fuzzy Logic Control of Drives and Power System, Newnes (2002)
17. Sepúlveda, R., Castillo, O., Melín, P., Díaz, A.R., Montiel, O.: Experimental Study of Intelligent Controllers under Uncertainty using type-1 and type-2 Fuzzy Logic. Information Sciences 177, 2023–2048 (2007)
18. Sepúlveda, R., Castillo, O., Melín, P., Montiel, O.: An Efficient Computational Method to implement Type-2 Fuzzy Logic in Control Applications. In: Analysis and Design of Intelligent Systems Using Soft Computing Techniques. Advances in soft computing, vol. 41, pp. 45–52. Springer, Heidelberg (2007)
19. Montiel, O., Maldonado, Y., Sepúlveda, R., Castillo, O.: Simple tuned fuzzy controller embedded into an FPGA. In: 2008 NAFIPS Conference Proceedings, pp. 1–6 (2008)
20. Montiel, O., Olivas, J., Sepúlveda, R., Castillo, O.: Development of an Embedded Simple Tuned Fuzzy Controller. In: Zurada, J.M., Yen, G.G., Wang, J. (eds.) Computational Intelligence: Research Frontiers. LNCS, vol. 5050, pp. 555–561. Springer, Heidelberg (2008)
21. Vuong, P.T., Madni, A.M., Voung, J.B.: VHDL Implementation for a Fuzzy Logic Controller. In: World Automation Congress WAC apos; 2006, pp. 1–8 (2006)
22. Iregui, S., Linares, D., Melgarejo, M.: Performance Evaluation of Fuzzy Operators for FPGA Technology. In: NAFIPS 2008, New York (2008)

23. Sánchez-Solano, S., Senhadji, R., Cabrera, A., Baturone, I., Jiménez, C.J., Barriga, A.: Prototyping of Fuzzy Logic Based Controllers Using Standard FPGA Development Boards. In: 13th IEEE International Workshop on Rapid System Prototyping on Volume, pp. 25–32 (2002)
24. Sánchez Solano, S., Barriga, A., Jiménez, C.J., Huertas, J.L.: Design and Application of Digital Fuzzy Controllers. In: Sixth IEEE International Conference on Fuzzy Systems (FUZZ-IEED 1997), vol. 2, Barcelona, Spain, pp. 869–874 (1997)
25. Maldonado, Y., Montiel, O., Sepúlveda, R., Castillo, O.: Design and simulation of the fuzzification stage through the Xilinx System Generator. Soft Computing for Hybrid Intelligent Systems, vol. 154, pp. 297–305. Springer, Heidelberg (2008)
26. Olivas, J., Sepuúlveda, R., Montiel, O., Castillo, O.: Methodology to Test and Validate a VHDL Inference through the Xilinx System Generator. Soft Computing for Hybrid Intelligent Systems, vol. 154, pp. 325–331. Springer, Heidelberg (2008)
27. Zadeh, L.A.: The concept of a linguistic variable and its application to approximate reasoning. Information Sciences 8, 199–249 (1975)
28. Zadeh, L.A.: Fuzzy sets. Information and Control 8, 338–353 (1965)

Modeling and Simulation of the Defuzzification Stage of a Type-2 Fuzzy Controller Using the Xilinx System Generator and Simulink

Roberto Sepúlveda[1], Oscar Montiel[1], Gabriel Lizárraga[2], and Oscar Castillo[3]

[1] Centro de Investigación y Desarrollo de Tecnología Digital del Instituto Politécnico Nacional, Av. del Parque No.1310, Mesa de Otay, 22510, Tijuana, B.C., México
r.sepulveda@ieee.org, o.montiel@ieee.org
[2] M.S. Student at CITEDI-IPN
lizarraga@citedi.mx
[3] Division of Graduate Studies and Research, Calzada Tecnológico S/N, Tijuana, B.C., México
ocastillo@hafsamx.org

Abstract. This paper is focused on the study, analysis and development of code for the defuzzification stage of type-2 fuzzy systems, through the average of two type-1 fuzzy systems. This proposal is based on the average method for systems where the type-2 membership functions of the inputs and output, have no uncertainty in the mean or center. The codification is done using the hardware description language VHDL, and it was exported to Simulink through the Xilinx System Generator (XSG). Comparative tests were conducted between the type-2 fuzzy systems for different number of bits and noise levels.

1 Introduction

Most of the existing implementations of type-1 or type-2 fuzzy systems have been achieved in software on general-purpose computers, the main drawback of these implementations is the speed limitation due to the sequential computer program execution [26]. Thus the other option for the implementation of fuzzy systems that require high processing speed to operate in real-time is based on dedicated hardware. In this line of development, in order to achieve high performance systems, the use of VLSI devices is increasing [21]. The FPGA is a VLSI device highly flexible that allows us to implement digital circuitry through the use of specialized software, which is an appealing characteristic for designers. Other good features are that they consume low power and can be reprogrammables in field. Hence, there is an increasing interest in using FPGA devices to design digital controllers, and a growing interest in control systems based on fuzzy logic [13, 27, 29]. In fact it is possible to find some works where type-1 fuzzy inference systems (FIS)have been implemented in FPGA, such as in [26] for electrical vehicles, a Mamdani FIS based on FPGA using distributed arithmetic to calculate the defuzzification stage [17], or diverse designs and implementations on FPGA of fuzzy inference systems [1, 16, 22, 23, 24, 30].

O. Castillo et al. (Eds.): Evolutionary Design of Intelligent Systems, SCI 257, pp. 309–325.
springerlink.com

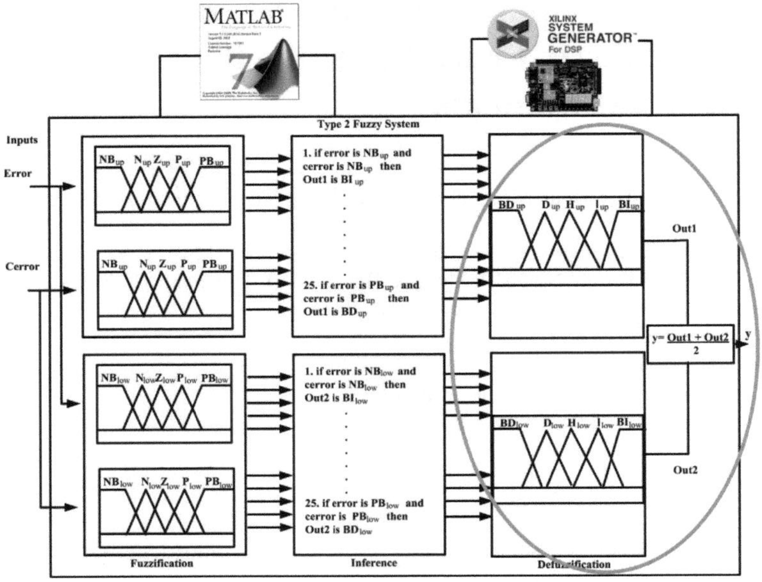

Fig. 1. Proposal for the solution of the defuzzification stage in a type-2 fuzzy system .

With regards to type-2 FIS in FPGA, in [19] is presented the implementation of the Wu-Mendel method in a XC2V3000ff1152-4 FPGA type of the Virtex family, taking advantages of the intrinsic parallelism that these devices offer; also we can find some others works related to this topic in [18, 20].

This work is about the design, test and implementation of the defuzzification stage of a type-2 FIS. The proposal is based on the use of two type-1 FIS, i.e. the average method, to emulate a type-2 FIS, so that the membership functions (MFs) were organized in such a way to emulate the footprint of uncertainty (FOU) of the type-2 MFs, and the final result is obtained as the average of the crisp values obtained in each type-1 FIS; it was used the height method for the type-2 defuzzification stage. In Figure 1 the scheme of the proposed method is shown.

A comparison was made between the control surface of the type-2 average method where the type-1 defuzzification stage is embedded, and the control surface of the type-2 FIS with the Wu-Mendel method.

This paper is organized as follows, Section 2 presents in a general context the type-2 FIS; in Section 3 it is presented the average of two type-1 FIS, as a proposal to avoid the type reduction; Section 4 is dedicated to explain the type-2 defuzzification stage method using VHDL code; Section 5 explains two experiments designed to test the type-2 Defuzzification stage and to verify the accuracy of the results considering that the inputs to this stage were individual values given by hand; Section 6 discusses two experiments and results with a complete type-2 FIS, where the source of the defuzzification stage is VHDL code imported to Simulink through the Xilinx System Generator (XSG) [15] , the fuzzification and the inference stages are models

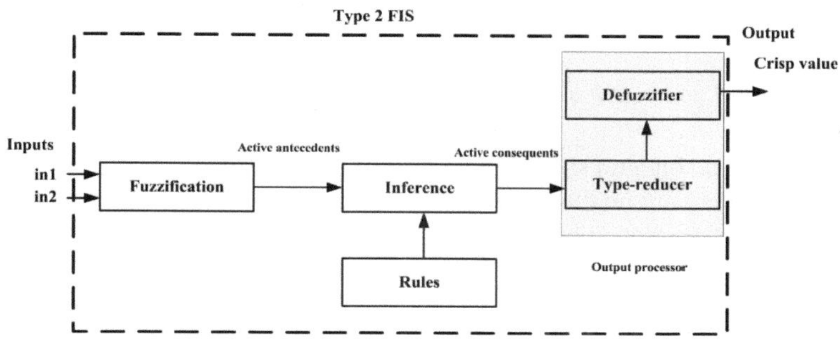

Fig. 2. Type-2 fuzzy system.

from the Simulink. The rules were used for the speed control of a DC motor for the system validation. Finally, Section 6 presents the conclusions of this work.

2 General Contexts of Type-2 FIS

The concept of a type-2 fuzzy set was introduced by Zadeh in 1975 [11], as an extension of the concept of an ordinary fuzzy set in 1965 [12], the membership grade of each of its elements is indeed a fuzzy set in [0, 1], unlike a type-1 set whose membership is a crisp number in [0, 1].

The basics and principles of fuzzy logic do not change from type-1 to type-2 fuzzy sets [4, 5, 6], they are independent of the nature of membership functions, and in general, will not change for any type-n. When a FIS uses at least one type-2 fuzzy set is a type-2 FIS, which is shown in Figure 2 with its components. The structure of the type-2 fuzzy rules, is the same as for the type-1 case because the distinction between them is associated with the nature of the membership functions. Hence, the only difference is that now some or all the sets involved in the rules are of type-2, and as long as any of its antecedents or consequents sets are interval type-2 fuzzy sets, it is called an interval type-2 FIS (IT2FIS) [7, 14] which is the most used. In a type-1 FIS, where the output sets are type-1 fuzzy sets, the defuzzification is performed to get a number, which is in some sense a crisp representation of the combined output sets. In the type-2 case, the output sets are type-2, so it is necessary the extended defuzzification operation to get type-1 fuzzy set at the output. Since this operation converts type-2 output sets to a type-1 fuzzy set, it is called type reduction, and the type-1 fuzzy set obtained is called a type-reduced set, the type-reduced fuzzy set may then be defuzzified to obtain a single crisp number.

2.1 Type-Reduction and Defuzzification in an Interval Type-2 FIS

Five different type-reduction (TR) methods are described in [7, 8]. Center-of-sets, centroid, center-of-sums, and height type-reduction can all be expressed as:

$$Y_{TR}(\mathbf{x'}) = [y_l(\mathbf{x'}), y_r(\mathbf{x'})] \equiv [y_l, y_r]$$

$$[y_l, y_r] = \int_{y^1 \in [y_l^1, y_r^1]} \cdots \int_{y^M \in [y_l^M, y_r^M]} \int_{f^1 \in [\underline{f}^1, \bar{f}^1]} \cdots \int_{f^M \in [\underline{f}^M, \bar{f}^M]} 1 \Big/ \frac{\sum_{i=1}^{M} f^i y^i}{\sum_{i=1}^{M} f^i}, \tag{1}$$

where the multiple integral signs denote the union operation [10].

In general, there are no closed-form formula for y_l and y_r; however, Karnik and Mendel [9] have developed two iterative algorithms (known as the Karnik-Mendel or KM Algorithms) for computing these end-points exactly, and they can be run in parallel [10]. Unfortunately this method is computationally costly so it is not well suited for hardware implementation as is the Wu-Mendel method [19].

Wu and Mendel [3] introduced a method to approximate the TR set by minimax uncertainty bounds. These uncertainty bounds are $\underline{y}_l(\mathbf{x}') \leq y_l(\mathbf{x}') \leq \bar{y}_l(\mathbf{x}')$ and $\underline{y}_r(\mathbf{x}') \leq y_r(\mathbf{x}') \leq \bar{y}_r(\mathbf{x}')$, where:

$$\bar{y}_l(\mathbf{x}') = min\left\{ \frac{\sum_{i=1}^{M} \underline{f}^i y_l^i}{\sum_{i=1}^{M} \underline{f}^i}, \frac{\sum_{i=1}^{M} \bar{f}^i y_l^i}{\sum_{i=1}^{M} \bar{f}^i} \right\} \tag{2}$$

$$\underline{y}_r(\mathbf{x}') = max\left\{ \frac{\sum_{i=1}^{M} \bar{f}^i y_r^i}{\sum_{i=1}^{M} \bar{f}^i}, \frac{\sum_{i=1}^{M} \underline{f}^i y_r^i}{\sum_{i=1}^{M} \underline{f}^i} \right\} \tag{3}$$

$$\underline{y}_l(\mathbf{x}') = \bar{y}_l(\mathbf{x}') - \left[\frac{\sum_{i=1}^{M}(\bar{f}^i - \underline{f}^i)}{\sum_{i=1}^{M} \bar{f}^i \sum_{i=1}^{M} \underline{f}^i} \times \frac{\sum_{i=1}^{M} \underline{f}^i (y_l^i - y_l^1) \sum_{i=1}^{M} \bar{f}^i (y_l^M - y_l^i)}{\sum_{i=1}^{M} \underline{f}^i (y_l^i - y_l^1) + \sum_{i=1}^{M} \bar{f}^i (y_l^M - y_l^i)} \right] \tag{4}$$

$$\bar{y}_r(\mathbf{x}') = \underline{y}_r(\mathbf{x}') - \left[\frac{\sum_{i=1}^{M}(\bar{f}^i - \underline{f}^i)}{\sum_{i=1}^{M} \bar{f}^i \sum_{i=1}^{M} \underline{f}^i} \times \frac{\sum_{i=1}^{M} \underline{f}^i (y_r^i - y_r^1) \sum_{i=1}^{M} \bar{f}^i (y_r^M - y_r^i)}{\sum_{i=1}^{M} \underline{f}^i (y_r^i - y_r^1) + \sum_{i=1}^{M} \bar{f}^i (y_r^M - y_r^i)} \right] \tag{5}$$

Observe that the four bounds in (2)-(5) can be computed without having to perform TR. Wu and Mendel [3] then approximate the TR set, as $[y_l(\mathbf{x}'), y_r(\mathbf{x}')] \approx [(\underline{y}_l(\mathbf{x}') + \bar{y}_l(\mathbf{x}'))/2, (\underline{y}_r(\mathbf{x}') + \bar{y}_r(\mathbf{x}'))/2]$ and compute the output of the IT2 FIS as

$$y(\mathbf{x}') = \frac{1}{2}\left[\frac{\underline{y}_l(\mathbf{x}') + \bar{y}_l(\mathbf{x}')}{2} + \frac{\underline{y}_r(\mathbf{x}') + \bar{y}_r(\mathbf{x}')}{2} \right] \tag{6}$$

So, by using the uncertainty bounds, they obtain both an approximate TR set as well as a defuzzified output [10].

3 Average Type-2 FIS

In cases where the performance of an IT2FIS is important, especially in real time applications, an option to avoid the computational delay of type-reduction, is the Wu-Mendel method [3], which is based on the computation of inner and outer bound

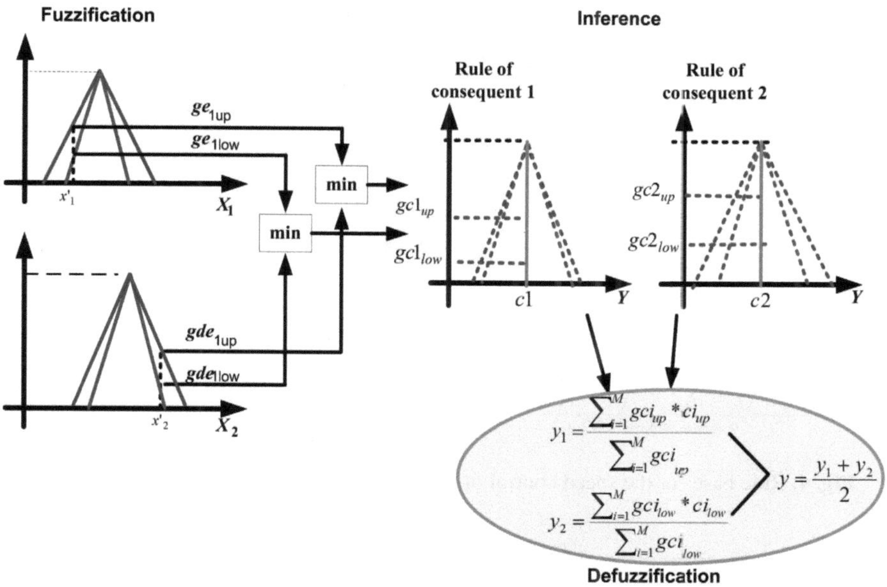

Fig. 3. The fuzzification, the inference and the defuzzification stages in the Average method uses two type-1 FIS.

sets. Another option to improve computing speed in an IT2FIS, is the average of two type-1 FIS method, which was proposed for systems where the type-2 MFs of the inputs and output, have no uncertainty in the mean or center; it is achieved by substituting the IT2FIS with two type-1 FIS, located adequately at the upper and lower footprint of uncertainty (FOU)of the type-2 MFs [25].

For the average method the fuzzification, the inference and the defuzzification stages at each FIS remain identical, the difference is at the output because the crisp value is calculated by taking the arithmetic average of the crisp output of each type-1 FIS, as it is shown in Figure 3, using the height method to calculate the defuzzified crisp output.

4 Type-2 Defuzzification Stage Method Using VHDL Code

In the average method, to achieve the defuzzification, one type-1 FIS is used for the upper bound of uncertainty, and the second FIS for the lower bound of uncertainty. So, as it was explained in Section 3, the defuzzification of a type-1 FIS is used in the average method and it will be explained next.

4.1 Defuzzification in Type-1 FIS

For the explanation of the type-1 defuzzification stage [2], the rule base of a speed control of a DC motor was used, which is presented in Figure 4, along with the

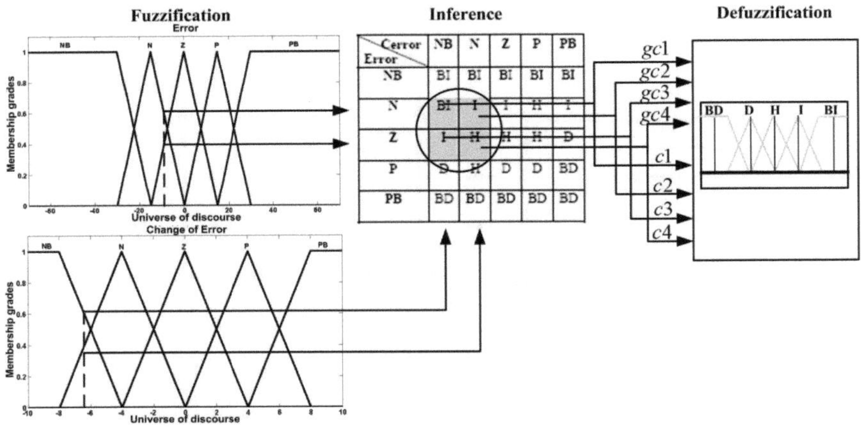

Fig. 4. Rule base for the speed control of a DC motor used for the system validation.

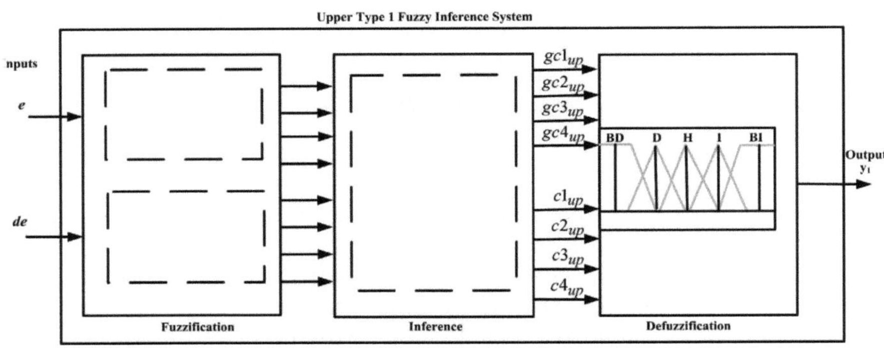

Fig. 5. Defuzzification stage proposal for a FIS.

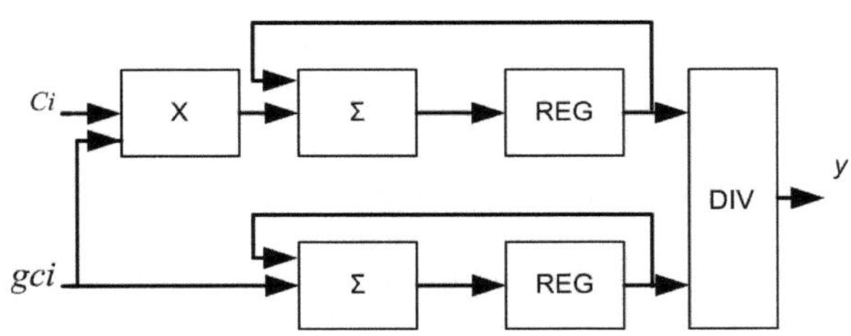

Fig. 6. Block diagram of the height method.

fuzzification and inference stages. In Figure 5 is shown the upper type-1 FIS with two inputs, *e* (error) and *de* (change of error), and one output *y*. Thus the inputs are connected to the fuzzification stage, and considering, for this case, that the method

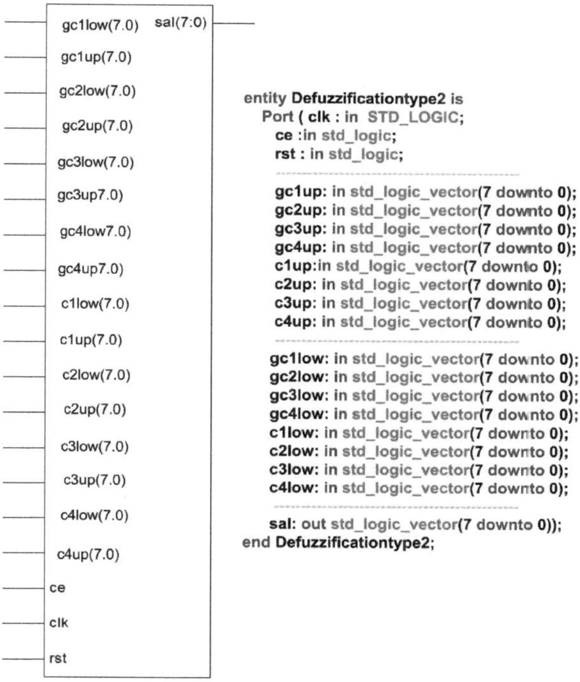

Fig. 7. Entity scheme of the type-2 defuzzification stage.

allows to have a maximum of four active MFs, the fuzzifier delivers two signal for each input, i.e, the membership grade and a tag that identifies the active MF.

The inference engine has eight outputs(considering the four maximum active rules), four are for the firing strengths, the rest correspond to the corresponding linguistic terms of the active consequents. So the output values of the inference engine named as $gc1_{up}$, $gc2_{up}$, $gc3_{up}$, $gc4_{up}$, are the four possible upper firing degrees of the active rules, meanwhile $c1_{up}$, $c2_{up}$, $c3_{up}$ and $c4_{up}$ are the respective center value of the active consequents. The upper part of the defuzzification stage using the gci_{up}, values and ci_{up}'s tags produces a crisp value $y1$ using the height defuzzification method. In the same manner as it is described for the upper defuzzification stage, it is obtained a crisp value $y2$ for the lower part of the defuzzification stage.

The upper defuzzification process using the height method, can be expressed by

$$y1 = \frac{\sum_{i=1}^{M} gci_{up} * ci_{up}}{\sum_{i=1}^{M} gci_{up}}, \tag{7}$$

where:

$y1$ is the crisp output value.
ci_{up} is the center of the active consequent.
gci_{up} is the upper firing degree of the active rule.

The height method was developed in VHDL based on the block diagram shown in Figure 6, where ci is the point of symmetry of the active consequent, gci is the firing degree of the active rule. These two inputs are connected to a block multiplier which corresponds to the part of the numerator in (7); only the addition of the gci, will produce the denominator. Both, numerator and denominator are connected to the block divider. The result of that division is $y1$ [28].

4.2 Implementation of the Type-2 Defuzzification Stage

The design entity of the type-2 defuzzification stage is shown in Figure 7. Such entity was programmed in VHDL to be implemented in a FPGA, but it can be used to simulate the Defuzzification stage without the necessity of designing and implementing any test bench. This entity has 19 inputs and one output. The first eight inputs ($gc1_{low}$, $gc1_{up}$, $gc2_{low}$, $gc2_{up}$, $gc3_{low}$, $gc3_{up}$, $gc4_{low}$, $gc4_{up}$) correspond to the lower and upper firing degrees of the active rules; the next eight inputs ($c1_{low}$, $c1_{up}$, $c2_{low}$, $c2_{up}$, $c3_{low}$, $c3_{up}$, $c4_{low}$ and $c4_{up}$) correspond to the centers of the active consequents. The remaining three input signals are the clock enable, clock, and reset

```
----Initialization----------------
        begin
        if (rst='1') then
        conta:=0;
        sal<="00000000";
        den:="0000000010";
        elsif (clk='1' and ce='1')
        then
        den:="0000000010";

---Data Acquisition ---------
        if(conta=0) then
        gra1up:="00"&gc1up;
        et1up:="00"&c1up;
        conta:=conta+1;
        end if;
        if(conta=1) then
        gra1low:="00"&gc1low;
        et1low:="00"&c1low;
        conta:=conta+1;
        end if;
        if(conta=2) then
        gra2up:="00"&gc2up;
        et2up:="00"&c2up;
        conta:=conta+1;
        end if;
        .
        .
        .
```

```
-------Defuzzificaton calculus----------------------

        if (conta=8) then
        divAup:=(gra1up*et1up)+(gra2up*et2up)..
        +(gra3up*et3up)+(gra4up*et4up);
        divBup:=gra1up+gra2up+gra3up+gra4up;
         resul:=divide(divAup,divBup);
        resulup:="000000000000"&resul(1)(7 downto 0);
        conta:=conta+1;
        end if;
        if(conta=9)then
        divAlow:=(gra1low*et1low)+(gra2low*et2low)..
        +(gra3low*et3low)+(gra4low*et4low);
        divBlow:=gra1low+gra2low+gra3low+gra4low;
        resul:=divide(divAlow,divBlow);
        resullow:="000000000000"&resul(1)(7 downto 0);
        conta:=conta+1;
        end if;

---------Average---------------------------------------------------

        if (conta=10) then
        resultemp:=resulup+resullow;
        resul:=divide(resultemp,den);
        sal<=resul(1)(7 downto 0);
        conta:=0;
        end if;
```

Fig. 8. VHDL code for the type-2 Defuzzification stage.

(ce, clk, rst) that allow the simulation of VHDL code in Simulink environment. The output port "sal", delivers the crisp value of the type-2 defuzzification stage.

In Figure 8 the VHDL code of the defuzzification stage for a type-2 FIS is presented, it is divided in four sections: Initialization, data acquisition, defuzzification computation, and the average calculus. In the first section the output port is initialized, the internal count register, which allows to carry out the data acquisition, besides determines the vector of the denominator of the division that will realize the average of the obtained results of each type-1 defuzzification stage. In the second section, the input data is obtained, i.e. the four possible firing strengths with the central values of the active consequents for each type-1 defuzzification stage. The input values are stored in registers, starting with the couple ($gc1_{up}$, $c1_{up}$ and $gc1_{low}$, $c1_{low}$) until the couple ($gc4_{up}$, $c4_{up}$ and $gc4_{low}$, $c4_{low}$) which correspond to the upper and lower firing strength and their respective activated consequent. The defuzzification calculus component, takes each input data couple and using aritmethic operations like addition, multiplication and division, computes the average of the two type-1 defuzzification stages with the height method, and sent it to the output port "sal". After that, the count register is initialized again, ready to the acquisition of new input data, and carry on the new defuzzification calculus in a cyclical process.

5 Test of the Type-2 Defuzzification Stage

To test the type-2 defuzzification stage, two experiments were conducted. The comparison was made between the defuzzification stage developed in VHDL code for several bits of resolution, and the Wu-Mendel method [3] programmed in Matlab code and whose block diagram is shown in Figure 9.

In order to achieve the experiments to test the defuzzification stage three main different software tools were used: The Simulink from Mathwork which is a practical high-level design and simulation tool because it provides a flexible design and simulation platform to test and correct designs at high level; the Xilinx Integrated Software Environment (Xilinx ISE) which is a Hardware Description Language (HDL) design software suite that allows taking designs through several steps in the ISE design flow finishing with final verified modules that can be implemented in a hardware target such a Field Programmable Gate Array (FPGA); and the Xilinx System Generator (XSG), which is a DSP design tool that enables the use of Simulink for FPGA design. This tool is very important because it allows generating VHDL code from the System Generator Simulink modules; and visceversa, VHDL modules can be included in the Simulink design platform by importing the VHDL code in a System Generator "Black box", the VHDL code in converted to Matlab functions.

As a first step the development of the stage was carried out using VHDL programming in the ISE Xilinx v8.2i [15], simulated and tested in Simulink/Matlab using the XSG. The experiments were based on the proposed type-2 MFs for the output variable, shown in Figure 10. This variable has five MFs, two trapezoidal and three triangular, on the universe of discourse proposed. As the height method was used as a defuzzification method, in fact there were used singletons to represent the central value of each MF.

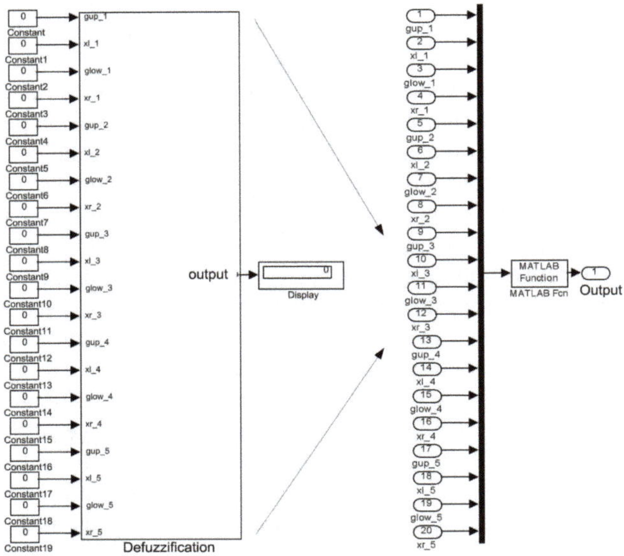

Fig. 9. Output using the Wu-Mendel method.

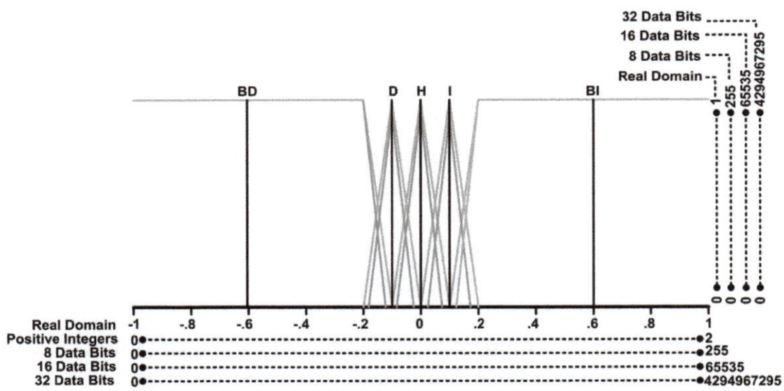

Fig. 10. Output variable Membership Functions.

The experimental defuzzification stage coded in VHDL using the average of two type-1 fuzzy systems is presented in Figure 11. The stage was coded in VHDL and simulated in Simulink/Matlab through the XSG tool. The stage was tested for 8, 16, and 32 bits resolution.

Experiment 1. In this experiment it is assumed that only two MFs are activated, D and H of Figure 10, by the inference stage . The values for their firing strenghs are assumed to be D_{up}=0.7 and D_{low}=0.4; H_{up}=0.6 and H_{low}=0.2. The output was

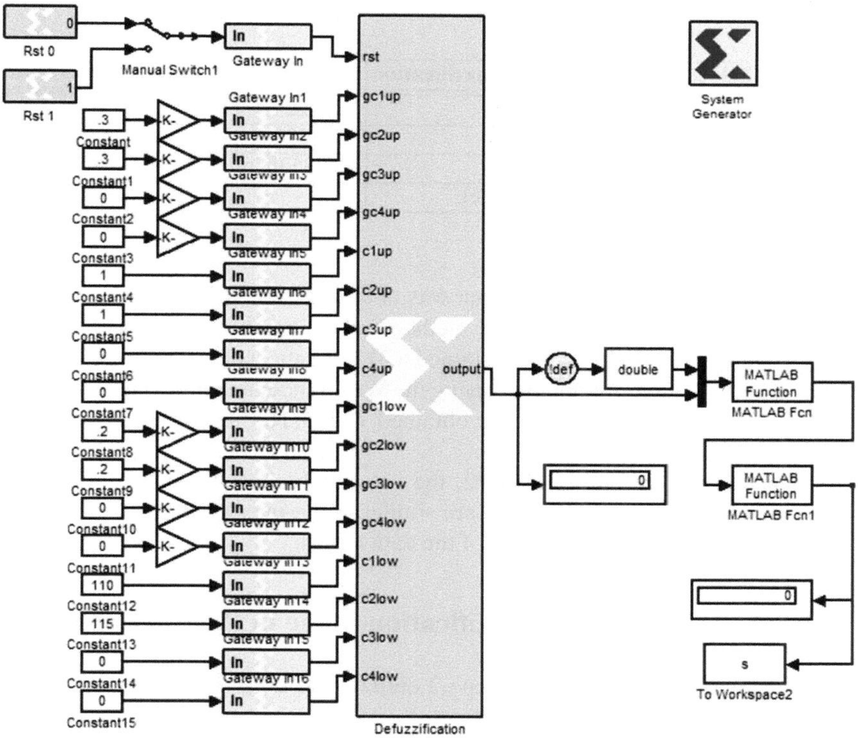

Fig. 11. Experimental design for the type-2 Defuzzification stage.

Table 1. Results of experiment 1 for the type-2 defuzzification stage.

Type-2 MF	Resolution of Defuzzification	Bits value	Output numeric value
D_{up}=0.7 D_{low}=0.4	8 bits	119	-0.06667
	16 bits	30793	-0.06026
H_{up}=0.6 H_{low}=0.2	32 bits	2018083991	-0.06028
	Wu-Mendel	—	-0.05889

obtained for 8, 16 and 32 bits of resolution using for each case the average method and the height as a defuzzifier. For example in the case of 8 bits of resolution, it was obtained a 119 value equivalent to -0.06667 in the real domain. The result obtained using the Wu-Mendel method coded in Matlab, was -0.05889. These results are shown in Table 1.

Experiment 2. In this case three MFs are activated, D, H and I of Figure 10, by the inference stage. The values for their firing strenghs are assumed to be D_{up}=0.5 and D_{low}=0.4; H_{up}=0.3 and H_{low}=0.2; finally I_{up}=0.2 and I_{low}=0.1. After the conversion

Table 2. Results of experiment 2 for the type-2 defuzzification stage.

Type-2 MF	Resolution of Defuzzification	Bits value	Output numeric value
D_{up}=0.5 D_{low}=0.4	8 bits	122	-0.04314
	16 bits	31573	-0.03645
H_{up}=0.3 H_{low}=0.2	32 bits	2069253886	-0.03643
I_{up}=0.2 I_{low}=0.1	Wu-Mendel	—	-0.0366

for the different resolutions The output was obtained for 8, 16 and 32 bits of resolution using for each case the average method and the height as a defuzzifier. In the case of 16 bits of resolution it was obtained a 31573 value equivalent to -0.03645 in the real domain. The result obtained using the Wu-Mendel method coded in Matlab, was -0.0366, almost the same as the obtained with a 16 bits of resolution. These results are shown in Table 2.

As can be seen in Tables 1 and 2, the results obtained with the Wu-Wendel method developed with Matlab code are similar to the ones obtained with the proposed method and 16 bits resolution of the data inputs.

6 Testing the Type-2 Defuzzification Stage in the Fuzzy System

Once the simulation of the VHDL type-2 defuzzification stage has been done, the next step is to test it in a type-2 FIS with the average of two fuzzification and inference stages, programmed using the appropriated Matlab function from the Fuzzy Matlab Toolbox. In the same way as in the type-1 defuzzification stage, having two parallel type-1 fuzzy systems the output is obtained as the average of the crisp values of each system. There were realized three experiments, the tests were done to verify the operation of the whole type-2 system in Simulink/Matlab using the Xilinx System Generator library. Comparisons were made between the results obtained with 8, 16 and 32 bits for the data inputs for the defuzzification stage, and those obtained with the Wu-Mendel method in Matlab code, recalling that in this one was used floating point in the input and output data. For the validation of the defuzzification stage in the compound type-2 FIS, the experiments undergo the same input conditions, with the purpose of making comparisons between the obtained results. The MFs proposed for the two input variables and for the output, are shown in Figure 12 and in Figure 10, respectively.

In Figure 13 is presented the whole Simulink model of the type-2 FIS, in which the two first stages, fuzzification and inference, were designed in Simulink/Matlab blocks, while the third one, defuzzification, using VHDL codification and imported to Simulink through the Xilinx System Generator.

Experiment 3. The type-2 FIS developed has two inputs, error and change of error, and one output. Assuming the following values for the inputs: error=-20, change of error=3, after the simulation of both type-2 FIS, the results for the 8, 16, and 32 bits

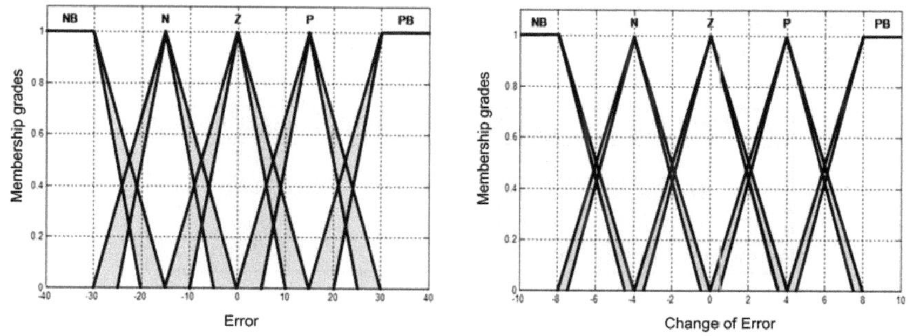

Fig. 12. Type-2 MF for the inputs of the Type-2 FIS.

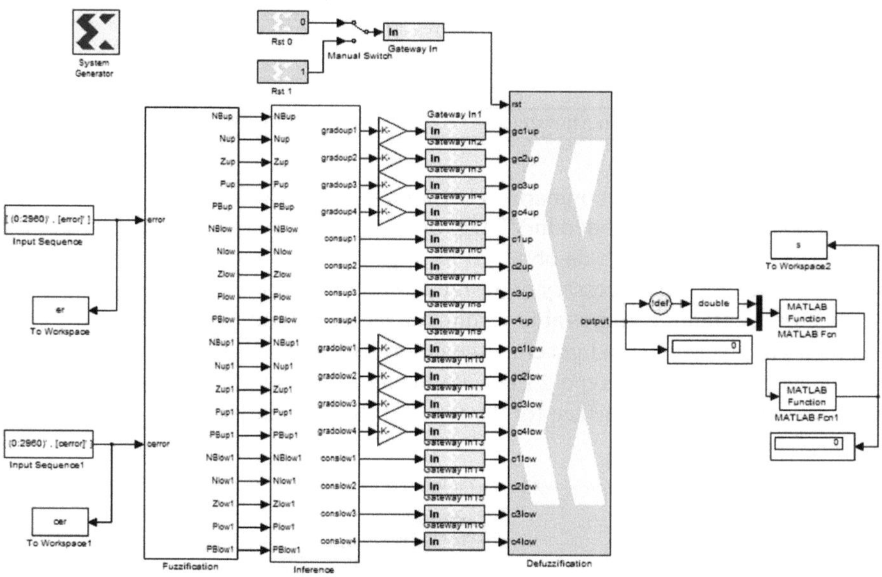

Fig. 13. Test of the type-2 defuzzification stage in a Type-2 FIS.

input data for the test of the type-2 defuzzification stage in the type-2 FIS, as well for the Wu-Mendel type-2 FIS in Matlab Model are shown in Table 3.

Experiment 4. When the values of the two inputs of the type-2 FIS were assumed to have the following values: error=10, change of error=-7, after the simulation of both type-2 FIS, the results for the 8, 16, and 32 bits input data for the test of the type-2 defuzzification stage in the type-2 FIS, as well for the Wu-Mendel type-2 FIS in Matlab Model are shown in Table 4.

The results obtained for the type-2 FIS using 16 bits for the input data are very similar to the results obtained with the type-2 FIS with the Wu-Mendel method.

Table 3. Results of experiment 3 for the type-2 defuzzification stage in the type-2 FIS, for error=-20 and cerror=3

Data Type	Bits value	Output numeric value
8 bits	139	-0.0902
16 bits	35716	-0.08998
32 bits	23640757176	-0.09
Wu-Mendel in Matlab	—	-0.09732

Table 4. Results of experiment 4 for the type-2 defuzzification stage in the type-2 FIS, for error=-20 and cerror=3

Data Type	Bits value	Output numeric value
8 bits	119	-0.06667
16 bits	30692	-0.06334
32 bits	2011476350	-0.06333
Wu-Mendel in Matlab	—	-0.06945

Experiment 5. To make a comparison between the behavior of the type-2 FIS with the defuzzification stage coded in VHDL, and the type-2 FIS using the Wu-Mendel method, a control surface was obtained for both of the fuzzy systems. To perform this experiment, it was necessary to design a Matlab function capable of generating two vectors containing all possible combinations for the inputs of each type-2 FIS, and then get the answer and generate the control surface. In Figure 14 is presented the control surface for the type-2 FIS with the defuzzification coded in VHDL, and in Figure 15 the control surface for the type-2 FIS with the Wu-Mendel method. It

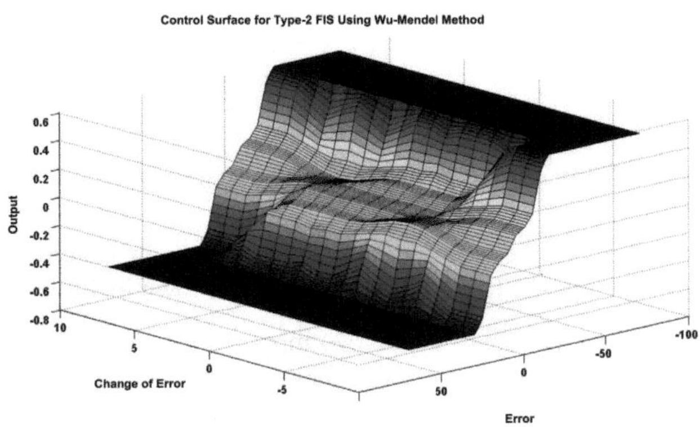

Fig. 14. Control surface using the Wu-Mendel method in the type-2 FIS.

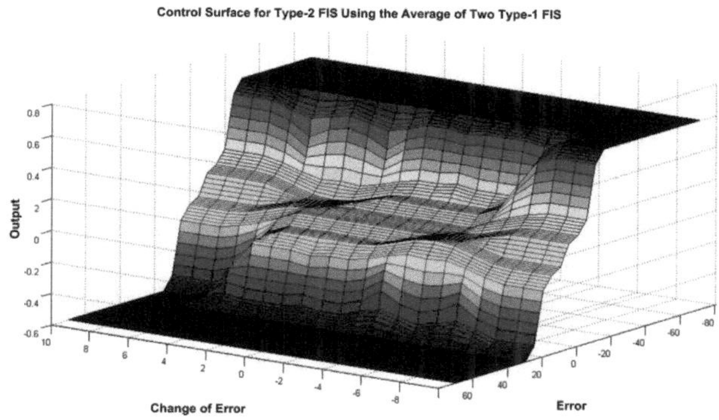

Fig. 15. Control surface using the Average of two type-1 FIS method

can be seen that the difference in both surfaces is minimum, although the input data for the latter are of floating point type, which allows to obtain more exact values.

7 Conclusions

The software implementations of fuzzy systems have been the most widely adopted, the highly flexibility is one of their characteristics, however the response time is limited by the inherent sequential execution of the programs, which is not convenient in real time applications, where the inference speed is an important factor. For this reason the number of applications using fuzzy systems into an FPGA is increasing, because one of its main advantages is the parallel processing, the low cost , low power and area consumption, etc. The methodology proposed in this paper in the design of the type-2 inference engine using a hardware description language, give a more practical and flexible model since the process of updating any change in the rule base can be made easily. So this is an alternative to quickly achieve both goals, since the developed code to describe the hardware, i.e. the system model, can be used with any modification to simulate the system and make the system implementation in the final target, for example in an FPGA.

Acknowledgements. The authors thank Comisión de Operación y Fomento de Actividades Académicas del I.P.N., Instituto Tecnológico de Tijuana for supporting our research activities, and CONACYT.

References

1. Lago, E., Hinojosa, M.A., Jimñez, C.J., Barriga, A., Sánchez-Solano, S.: FPGA Implementation of Fuzzy Controllers. In: Proc. XII Conf. on Design of Circuits and Integrated Systems (DCIS 1997), Sevilla, November 1997, pp. 715–720 (1997)

2. Lizárraga, G., Sepúlveda, R., Montiel, O., Castillo, O.: Modeling and Simulation of the Defuzzification stage using Xilinx System Generator and Simulink. Soft Computing for Hybrid Intelligent Systems, vol. 154, pp. 333–343. Springer, Heidelberg (2008)
3. Wu, H., Mendel, J.M.: Uncertanty Bounds and Their use in the Design of Interval type-2 Fuzzy Logic Systems. IEEE Transactions on Fuzzy Systems 10, 1–16 (2002)
4. Mendel, J.M.: Type-2 Fuzzy Sets and Systems: an Overview. IEEE Computational Intelligence Magazine 2, 20–29 (2007)
5. Mendel, J.M.: Type-2 Fuzzy Sets: Some Questions and Answers. IEEE Computer Society Press, Los Alamitos (2003)
6. Mendel, J.M., Bob John, R.I.: Type-2 Fuzzy Sets Made Simple. IEEE Transactions on Fuzzy Systems 10, 117–127 (2002)
7. Mendel, J.M.: Uncertain Rule-Based Fuzzy Logic Systems: Introduction and New Directions. Prentice-Hall, Upper-Saddle River (2001)
8. Karnik, N.N., Mendel, J.M., Liang, Q.: Type-2 Fuzzy Logic Systems. IEEE Transactions on Fuzzy Systems 7, 643–658 (1999)
9. Karnik, N., Mendel, J.M.: Centroid of a Type-2 Fuzzy Set. Information Sciences 132, 195–220 (2001)
10. Mendel, J.M., Hagras, H., John, R.I.: Standard Background Material About Interval Type-2 Fuzzy Logic Systems That Can Be Used By All Authors
11. Zadeh, L.A.: The concept of a linguistic variable and its application to approximate reasoning. Information Sciences 8, 199–249 (1975)
12. Zadeh, L.A.: Fuzzy sets. Information and Control 8, 338–353 (1965)
13. L.E., Jiménez, C.J., López, D.R., Sánchez-Solano, S., Barriga, A.: XFVHDL: A Tool for the Synthesis of Fuzzy Logic Controllers. Design Automation and Test in Europe, 102–107 (1998)
14. Liang, Q., Mendel, J.M.: Interval Type-2 Fuzzy Logic Systems: Theory and Design. IEEE Trans. on Fuzzy Systems 8, 535–550 (2000)
15. Manual of Xilinx System Generator, `http://www.xilinx.com`
16. Patyra, M.J., Janos, L., Koster, K.: Digital Fuzzy Logic Controller:Design and Implementation. IEEE Transactions on Fuzzy Systems, vol. 4(4) (November 1996)
17. Melgarejo, M.: Desarrollo de un Sistema de Inferencia Difusa sobre FPGA, Universidad Distrital Francisco José de Caldas, Centro de Investigación y Desarrollo Cientifico, Esfera Editores, Bogotá, Colombia (2003)
18. Melgarejo, M., Peña-Reyes, C.A.: Implementing Interval type-2 Fuzzy Processors. Computational Intelligence Magazine, IEEE 2, 63–71 (2007)
19. Melgarejo, M.A., Carlos, R., Peña-Reyes, A.: Hardware architecture and FPGA implementation of a type-2 fuzzy system. In: Proceedings of the 14th ACM Great Lakes symposium on VLSI, pp. 458–461 (2004)
20. Melgarejo, M.A., Garcia, R.A., Peña-Reyes, C.A.: Pro-Two: a hardware based platform for real time type-2 fuzzy inference. In: Proceedings of 2004 IEEE International Conference on Publication Fuzzy Systems, vol. 2, pp. 977–982 (2004)
21. Cirstea, M.N., Dinu, A., Khor, J.G., McCormick, M.: Neural and Fuzzy Logic Control of Drives and Power System, Newnes (2002)
22. Montiel, O., Maldonado, Y., Sepúlveda, R., Castillo, O.: Simple tuned fuzzy controller embedded into an FPGA. In: 2008 NAFIPS Conference Proceedings, pp. 1–6 (2008)
23. Montiel, O., Olivas, J., Sepúlveda, R., Castillo, O.: Development of an Embedded Simple Tuned Fuzzy Controller. In: Zurada, J.M., Yen, G.G., Wang, J. (eds.) WCCI 2008. LNCS, vol. 5050, pp. 555–561. Springer, Heidelberg (2008)
24. Vuong, P.T., Madni, A.M., Voung, J.B.: VHDL Implementation for a Fuzzy Logic Controller. In: World Automation Congress, WAC apos; 2006, pp. 1–8 (2006)

25. Sepúlveda, R., Castillo, O., Melin, P., Díaz, A.R., Montiel, O.: Experimental Study of Intelligent Controllers under Uncertainty using type-1 and type-2 Fuzzy Logic. Information Sciences 177, 2023–2048 (2007)
26. Poorani, S., Urmila Priya, T.V.S.: FPGA Based Fuzzy Logic Controller for Electric Vehicle. Journal of the Institution of Engineers, Singapore 45(5) (2005)
27. Iregui, S., Linares, D., Melgarejo, M.: Performance Evaluation of Fuzzy Operators for FPGA Technology. In: NAFIPS 2008, New York (2008)
28. Sánchez-Solano, S., Cabrera, A., Jiménez, C.J., Brox, P., Baturone, I., Barriga, A.: Implementación sobre FPGA de Sistemas Difusos Programables. In: IBERCHIP, Workshop IBERCHIP La Habana, Cuba (2003)
29. Sánchez-Solano, S., Senhadji, R., Cabrera, A., Baturone, I., Jiménez, C.J., Barriga, A.: Prototyping of Fuzzy Logic Based Controllers Using Standard FPGA Development Boards. In: 13th IEEE International Workshop on Rapid System Prototyping on Volume, pp. 25–32 (2002)
30. Sánchez Solano, S., Barriga, A., Jiménez, C.J., Huertas, J.L.: Design and Application of Digital Fuzzy Controllers. In: Sixth IEEE International Conference on Fuzzy Systems (FUZZ-IEED 1997), Barcelona, Spain, vol. 2, pp. 869–874 (1997)

Author Index

MIX
Papier aus verantwortungsvollen Quellen
Paper from responsible sources
FSC® C105338

If you have any concerns about our products,
you can contact us on
ProductSafety@springernature.com

In case Publisher is established outside the EU,
the EU authorized representative is:
Springer Nature Customer Service Center GmbH
Europaplatz 3, 69115 Heidelberg, Germany

Printed by Libri Plureos GmbH
in Hamburg, Germany